BALLS OF FIRE

Books by Judy Kay King

The Isis Thesis (2004) and *The Road from Orion* (2004)

Peer-Reviewed & Published Journal Articles

(available at http://www.isisthesis.com)

2005. Biosemiotics in Ancient Egyptian Texts: the Key to Long-lost Signs found in Myth, Religion, Psychology, Art and Literature. *International Journal of the Humanities* 3(7): 189-203.

2006. Biosemiotics in Ancient Egyptian Texts: the Key Unlocking the Universal Secret of Sexuality and the Birth of the Limitless. Presented at the Second International Congress for Young Egyptologists in Lisbon, Portugal on Oct. 25, 2006. Published in *ACTAS 2009 Proceedings*.

2006a. Man the MisInterpretant: Will He Discover the Universal Secret of Sexuality encoded within him? *International Journal of the Humanities* 4(9): 1-15.

2007. From History's Dustbin: a Semiotics of Evolvability discovered within Man and his Mountain of Transformation. *International Journal of the Humanities* 5(5): 113-126.

2007a. Self-Portrait in the Pharaoh's Mirror: a Reflection of Ancient Egyptian Knowledge in Teilhard de Chardin's Evolutionary Biophysics. Presented at the 32nd Annual Meeting of the Semiotic Society of America, New Orleans, October 4-7, 2007. Published in *Semiotics 2006/2007* eds. Terry J. Prewitt and Wendy Morgan, 101-115 (New York, Ontario: Legas and SSA).

2008. The Order of the Harmonious Whole: Peirce's Guess, Peregrinus' Magnet, and Pharaoh's Path. Presented at 33rd Annual Meeting of SSA, Houston, October 17, 2008. Published in *Semiotics 2008* eds. John Deely and Leonard G. Sbrocchi, 179-190 (NY, Ontario: Legas and SSA).

2008a. Cosmic Semiophysics in Ancient Architectual Vision: The Mountain Temples at *Deir el Bahari*, the Dead Sea Temple Scroll, and the Hagia Sophia. *Int'l Journal of Humanities.* 6(4): 17-26.

2009. Cosmological Patterns in Ancient Egypt and China: The Way to Unify the Universe through Knowledge, Mind, Energy, and the Beneficence of Elements. *Int'l J. of Humanities.* 7(2): 151-165.

2009a. Evolution Backward in Time: Crystals, Polyhedra and Observer-Participancy in the Cosmological Models of Peirce, Ancient Egypt and Early China. Presented October 17, 2009, at the 34th Annual Meeting of SSA, Cincinnati, USA. Published in *Semiotics 2009 The Semiotics of Time*, eds. Karen Haworth, Jason Hogue, Leonard G. Sbrocchi, 58-76 (NY, Ontario: LEGAS and SSA).

2009b. Unraveling Mountainway Ceremonials: Is Navajo Eschatological Ritual Another Semiotic Pattern of Ancient Invisible Magic Veiling a Complex Systems-based Information Science? *International Journal of the Humanities.* 8(12): 45-80.

2011. Ticket to Ride the Ancient Celestial Railroad: Natural Law, Worldview Knowledge, "Evolutionary Love," and Ockham's Razor. Presented Oct. 29, 2011, at 36th Annual Meeting of SSA, Pittsburgh, USA. Published in *The Semiotics of ??? Proceedings of the 36th Annual Meeting of SSA, 27-30 October 2011*. eds. K. Haworth, J. Hogue, L. G. Sbrocchi, 137-155 (NY: LEGAS and SSA).

2013. Death or the Powers: the Future of the Human Experience. *The International Journal of Humanities Education.* 11(3): 1-17.

BALLS OF FIRE

a Science of Life and Death

JUDY
KAY
KING

ENVISION EDITIONS LIMITED

Gaylord, Michigan

Envision Editions Ltd.
Post Office Box 442
Gaylord, Michigan 49734

Book Website: http://www.isisthesis.com

First published in the United States in 2015 by
Envision Editions Ltd., Gaylord, Michigan
Printed in the United States of America

ISBN: 978-0-9762814-2-9
Library of Congress Control Number: 2014921756

Cover Design by Rudy P. Gonzalez using Cover Art: *Scream 3000*
Mixed Media on Canvas 48 x 48
© 2012 Rudy Gonzalez
http://www.rudy5000.com

To biological altruists

The simplest individuated forms are balls.
In dimension one it is the tinkle of a bell,
in dimension two the disk circumscribed by a simple closed curve,
in dimension three the ball inside its boundary sphere.

René Thom, *Semiophysics: A Sketch*

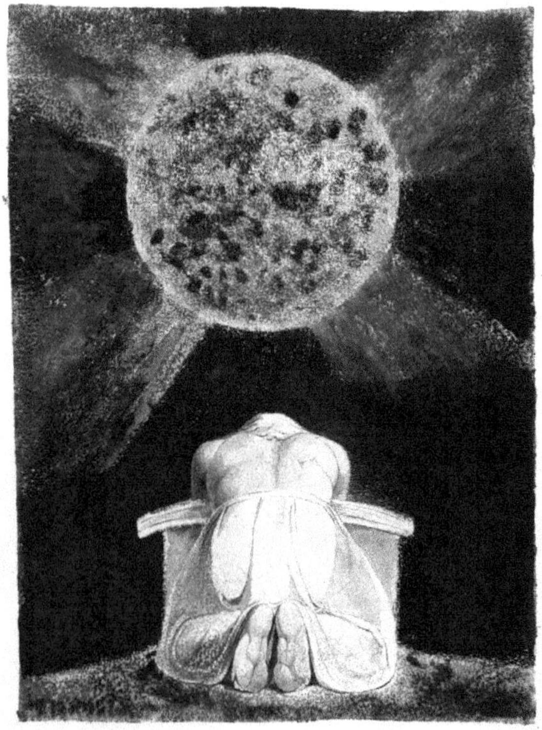

The Song of Los (1795) printed by William Blake (Pubic Domain)

CONTENTS

Preface to a Philosophy of Science, Human Behavior and Evolution

Nietzsche once said: "I want to teach men the meaning of their existence: which is the Superman, the lightning from the dark cloud man." After fifteen years of scientific research on human behavior, I have one possible explanation for Nietzsche's idea of the Superman. Spanning 5,000 years of history from ancient Egypt to our technoprogressive twenty-first century culture, the science reviewed in *Balls of Fire* builds on *The Isis Thesis* (2004) and twelve related articles (2005-2013). *The Isis Thesis* is a semiotic study of ancient Egyptian literature, artwork, ritual, and architecture. In brief, the core hypotheses of the Isis Thesis follow. First, ancient Egyptian deities are signs for human and microbial DNA (genes and proteins) evolving into a hybrid quantum species at a human death transition. The activities of the deities describe the ancient glycolysis gene expression network in our cells that can function without oxygen or the sugar glucose by using fermentation and the sugar lactose found in milk. Second, both Egyptian and Christian deities mirror the activities or lifestyles of a complex virus called bacteriophage Lambda that uses the ancient glycolysis gene expression network. So, all these deities are signs for the same evolutionary process mediated by a virus. Third, bacteriophage Lambda, a temperate eubacterial virus, is our last universal common ancestor (LUCA). Fourth, our semiotic system is based on underlying physical and chemical principles inherited from our microbial ancestors, so our microbial DNA is ordering our society space.

With these core hypotheses in mind, the Isis Thesis asserts the following novel predictions. First, a modeling relation exists between the virus Lambda and a human being, since our intentionality is governed by this ancestral virus. Our historical cultural behavior demarcates its ancient viral gene expression network that grounds its genetic switch between two lifestyles for the evolutionary transformation of DNA. Second, two ways of evolution exist due to Lambda's two lifestyles. A human being can evolve into our current thermodynamic state of being, and then, to avoid dissipation at human death, that emergent entity can select a hybrid, quantum, cold light, crystallized state of being. Third, at a human death, evolution requires scientific knowledge of bacteriophage Lambda's genetic switch between its two lifestyles. Therefore, the human being with knowledge functions as the genetic switch-hitter to evolve. In line with the methodology of Imre Lakatos (1970) on progressive and degenerating research programs, *Balls of Fire* examines the core hypotheses of the Isis Thesis, its predictions and several other auxiliary hypotheses. For example, this book addresses the following question: does the evidence for the auxiliary hypotheses add support for this research program's core hypotheses and predictions?

According to logician Charles Sanders Peirce, pragmatism holds that to attribute a meaning to ideas or a theory, the ideas must apply to our existence. In this case, what emerges from human and viral gene expression is a unique hybrid state of being or quantum species. The emergent process can be called quantum hybrid speciation. However, this hybrid state of being only looks emergent to us because we do not understand the complexity of the whole process that involves knowledge of a complex viral genetic switch that functions throughout the biosphere and the cosmos.

The ancient pharaonic priesthood concealed this knowledge in tombs and pyramids for centuries because their secret science anchored their power structure, while increasing their chances for the anticipated survival of their DNA at death. As the pharaohs understood and nature reveals, the evolution of human life functions through the sacrifice of many individuals for the survival of the few. So, the pharaonic priesthood's secret science sanctioned the survival of the élite rulers at the necessary biological expense of the plebians, who did not know the chemical pathway for morphogenesis at a death transition.

Morphogenesis is the biological process that causes an organism to develop its shape. The process involves knowing how to control gene expression and regulate cell fates to generate biological form and structure. Today, understanding morphogenesis is one of biology's unresolved problems. However, the ancient Egyptian pharaonic priesthood solved the problem of morphogenesis five thousand years ago. Understanding transdisciplinary ancient Egyptian knowledge is not easy, so *Balls of Fire* uses the same mental model that the pharaonic priesthood imagined to describe the ancient viral gene expression network in our cells for morphogenesis. That model is the game of baseball, which originated in ancient Egypt to illustrate a viral protein binding battle over gene-bases. Although the game of baseball has drifted through the centuries as a popular sport in many cultures, it originally expressed microbiological warfare at the level of viral genes and proteins. Because ancient Egyptian science mirrors the knowledge of our contemporary sciences, the baseball model simplifies the information for readers, while explaining the science that the pharaonic priesthood concealed in pyramids and tombs for centuries. So, a modeling relation exists between a virus and the human organism, and game theory supports it.

For the creation of the baseball model, a fantasy-draft selection of two teams frames the historical power/knowledge grid, as well as the scientific argument for and against the Isis Thesis, while explaining the necessary context for what the theory predicts and scientific experiments confirm. This is accomplished by the draft of dead and living scientists, philosophers, writers and other creative artists, whose ideas are presented in two fantasy teams in order to tackle the mind-body problem that has confounded humans for centuries. In the interest of simplification, the paradox is simply referred to as either the primacy of mind (thoughts) or the primacy of matter (actions). Perhaps the baseball model will improve the reader's understanding of contemporary science, clarify the context of the mind-body problem, and illuminate the grid of power relations that have governed human history by concealing biological survival knowledge. This adversarial, conceptual system of two fantasy-draft teams is presented to you—an impartial group of readers, who attempt to determine the truth of the case through a transdisciplinary quest that prioritizes scientific research.

In the Game of the Centuries between the Diamondhearts (primacy of mind) and the Thunderheads (primacy of matter), the Isis Thesis is submitted to test and possible refutation. What emerges from this Game of the Centuries is that both mind and matter are primary for nature to cycle optimally. Mind and matter in the classical world are only at odds due to time evolution.

The Road Trip to the Game of the Centuries. In the year 2000, I set aside my employment as a humanities instructor to explore great literature because I believed it conveyed a survival message. Thanks to my graduate education in English and literary criticism, I had already read a strong selection of world literature. However, I had not read the ancient Egyptian Book of the Dead, so that was my first task. I noticed that ancient Egyptian artwork in the Book of the Dead was very similar to modern scientific drawings in new physics texts that I had read. Was it possible that both ancient Egypt and new physics were using the same signs to express the same scientific concepts? If so, then perhaps the religious literature of ancient Egypt veiled a science.

During the next three years, I learned that deciphering the meaning of the ancient Egyptian symbolic system carved on the walls of the pharaohs' pyramids was still a daunting task for modern egyptologists, prompting many scholars to dismiss the funerary corpus as confusing, unintelligible and primitive. Part of their confusion stems from the language of the symbolic system itself, which contains ideograms or sense-signs, phonograms or sound-signs, number signs, relative artwork with its symbolism of actions, gestures and color, and literary metaphors. Realizing that I needed a strong survey of major Egyptian texts to support my intuition that the religious texts conveyed a science, I purchased renowned egyptologist Erik Hornung's bibliographic guide for scholars entitled *The Ancient Egyptian Books of the Afterlife* (1999), so I could select for study his recommended English translations of the funerary corpus.

For example, R. O. Faulkner's translations of the Pyramid Texts (PT) and Coffin Texts (CT) were used to comprehend the least-corrupted texts in *The Isis Thesis* and the current text. Understanding that a holistic approach was necessary to consider each sign's potential for meaning within the full context of the whole group of signs, I centered my research on the primary, least-corrupted Pyramid and Coffin Texts, as well as later major texts spanning a historical time period of two thousand years, dated from 2520 to 664 BCE. The assumption underlying this approach was that the funerary texts from the Old Kingdom through the Middle Kingdom to the New Kingdom were unified in their presentation of key ideas, and this turned out to be true. However, I had no clue that the project would mature into three solid years of independent research, writing, and isolation because I was now on an inquiry-based journey to decode a cohesive interpretation of the Egyptian signs that included their literature, artwork, ritual, and architecture. My inquiries led me directly down a pathway from our classical cosmos of space physics into the microscopic world of quantum biology and quantum physics, resulting in the scientific study named *The Isis Thesis* (2004) and the short novella explaining the science named *The Road from Orion* (2004).

Apocalypse. By 2004, I had a scientific thesis against the mainstream consensus of most egyptologists and Christians that decoded ancient Egyptian and Christian dieties as signs for microbial DNA. Testing the Isis Thesis further, I discovered additional evidence for it in our cultural history, specifically, the literature of William Blake, Antonin Artaud, Pierre Teilhard de Chardin, and the art of Albrecht Durer, as well as other cultural giants. I was also intrigued by the migration of syncretistic religious themes ad nauseam (dying/rising god, virgin birth, and so on) from Osiris to Christ. My initial plan was to present the Isis Thesis and relative cultural and scientific knowledge at an international humanities conference. So in 2005, I submitted an abstract on the Isis Thesis to the review committee for the Third International Conference on New Directions in the Humanities at the University of Cambridge, UK, which was accepted. After presenting at this conference on August 3, 2005, my paper was peer-reviewed and accepted for publication in the *International Journal of the Humanities.*

My next objective was to present my research to an egyptology audience, so I submitted an abstract to the review committee organized by the Universidade de Lisboa for the Second International Congress for Young Egyptologists in Lisbon, Portugal. Abstract accepted, I traveled to Lisbon and presented a second scientific paper on October 25, 2006. At the conference, I was honored to meet members of the International Scientific Commission who also presented their papers, particularly Professor Erik Hornung, Professor Emeritus of Egyptology at the University of Basel, Switzerland, Professor Rosalie David of the University of Manchester, and Professor Ian Shaw of the University of Liverpool.

At this conference, I also met egyptologist Renata Tatomir of the Bucareste Oriental Studies Institute whose research also involved a quantum interpretation. She translated the original eight deities of the Egyptian Ogdoad of Hermopolis as subatomic particles interacting in line with the basic theory of strong interactions called quantum chromodynamics. Her research like mine was on the path of quantum interpretation, and the positive feedback I received related to our quantum approaches was that this new perspective of the funerary texts would be knowledge for future generations of young egyptologists. However, the crux of the Isis Thesis—the comparison of viral lifestyles with ancient Egyptian religious themes mirroring Christian themes, surprised some members of the audience, for Portugal is predominantly Catholic. I appreciated their concerns, but I also understood the divine irony that Christianity had unknowingly preserved an important evolutionary science for survival of DNA in a quantum environment.

Semiotics of American Philosopher Charles Sanders Peirce (1839-1914). After *The Isis Thesis* was published in 2004, based on the least-corrupted core Egyptian sources and our new sciences, I continued my research on our potential for horizontal gene transfer relative to the surprising initial results of the International Human Genome Sequencing Consortium (2001) and recent modern research supporting the possibility that our intentionality or will could be grounded by our DNA, as ancient Egyptian texts indicated. So in 2005, I subscribed to the British Journal *Nature*, a subscription I have enjoyed for the last nine years, and I continued my research with the aid of their weekly scientific journals and other research. By 2006, I had become a member of the Semiotic Society of America, an interdisciplinary team of scholars studying semiotics or the action of signs in culture. The American philosopher and father of semiotics is Charles Sanders Peirce, whose mathematical, logical and scientific writings have supported my studies, while specifying how to interpret ancient documents logically. Drawing on the pragmatism of Charles Peirce, semiosis is the triadic action of signs in living and nonliving nature, encompassing human and animal behavior, the microcosmic niche, and the physical cosmos itself. This idea of the logician Peirce centers on the triadic action of a sign, its object, and its *interpretant*, which references the whole of living and nonliving nature versus *interpreter* that contextualizes only the human or animal. My research supports Peirce's belief that everything interprets and mind or intelligence exists everywhere in the cosmos.

Peirce sketched out his new philosophy in the spring of 1890, identifying Chance, Law, and Continuity as the great explanatory elements of the cosmos and envisioning "a theory of evolution applicable to the inorganic world also." Through research I soon learned that Peirce's cosmogony was very similar to that of the ancient Egyptian pharaonic priesthood. Since I had a theory of organic and inorganic evolution based on ancient Egyptian texts, I followed Peirce's advice on the study of ancient testimonies, beginning with an hypothesis or abduction to be tested (the Isis Thesis), followed by deduction and induction. One can never absolutely prove a thesis, so a continuing inductive approach is necessary. This meant a continuing exploration of other early cultures and thinkers to compare the action of their signs with the ancient Egyptian, that is, to find corresponding knowledge in history in random time periods. As a result, the scientific published paper I presented for feedback at the Second International Congress for Young Egyptologists in Lisbon, the six peer-reviewed published papers in the *International Journal of Humanities*, and the four published papers I presented for feedback at four Annual Meetings of the Semiotic Society of America—show that Egyptian knowledge is also present in early Chinese texts, Navajo mythologies, the cosmogony of Charles Peirce, the biophysics of Pierre Teilhard de Chardin, William Blake's unfinished masterpiece "The Four Zoas," and many other sources.

Bringing it all Together in Balls of Fire. In 2013, the research paper "Death or the Powers" was peer-reviewed and accepted for publication in the *International Journal of Humanities-Education*, showing that some modern technocrats forecast the merger of humans with machines and the end of humanity by 2100, as well as the hybrid infusion of superintelligent artificial intelligence (AI) into our universe. This technocratic rationale aspires to embrace the macrocosmic universe through metal matter, resulting from physical humans merging or uploading their brains into superintelligent machines. With their one-sided cosmic manifest destiny for becoming hybrid-human-metal matter, the technocrats are oblivious to the historical survival message for human DNA as a natural, crystallized hybrid emergence through the chemical reaction of chemiluminescence (production of cold light or starlight). The ancient Egyptian pharaonic priesthood, the early Chinese emperors and sages, the Navajo shamans, and many others envision this natural embrace of the quantum universe through a merger (horizontal gene transfer) with what they describe as a metal-binding bacteriophage or bacterial virus. To explain and present both sides of this argument on the values of both matter and mind, the pitch of these technoprogressive players is incorporated into the materialist vision of the Thunderhead team, while the historical cultural view is blended into the vision of the Diamondheart team for the Game of the Centuries, a baseball fantasy-model contest of living and dead individuals explaining the science.

Since my research methodology was also influenced by Michel Foucault's archaeological method of study, quantum theory and experiments, as well as Peirce's semiotics, my approach is necessarily transdisciplinary. I learned that expressing holistic knowledge is a formidable task often complicated by ongoing scientific and philosophical debates such as whether or not the building blocks of the cosmos are energy or information, life or nonlife, mind or matter. Further, quantum physics is counterintuitive, conveying the quasi-mystical surrealism of science fiction and the occult. What is necessary, Peirce writes, is a natural history of laws and a study of our own historical position, and this is the objective of the twelve published articles and *Balls of Fire.*

Balls of Fire is written with an argument broken down from nature that presents findings correlative with the history of human ideas, along with scientific evidence and mechanistic insights to establish the clear link between nature, our behavior and human evolutionary potential. The evidence shows that our behavior and the evolution of society in the last 12,000 years has carved a footprint into human history, profiling the morphogenesis of a virus. This viral footprint with its survival message for humanity maps the ancient glycolysis-fermentation gene expression network for lactose metabolism used by bacteriophage Lambda, a complex bacterial virus. Thus history shows that human activities and potential are linked to a virus that permits the emergence of a quasi-hybrid being of human and viral DNA. The footprint on society space represents a gene-culture system, revealing that we are influenced by the machinic viral DNA in our genome and microbiome, while confirming the current scientific belief that the quantum world is ordering our classical world.

Despite historical propaganda by those with power over life (biopower), who suppress DNA survival information, signs of nature's hidden agenda are present everywhere in human history and can be interpreted today through the lens of contemporary science and human behavior. Consequently, it became obvious to me that the highly-networked Egyptian system of quantum biology, quantum physics, cosmology and mythology is controlled by a human agent from a ground of being that allows a choice from possibilities. This happens because nature or microbes have created us and the transactional laws with an inbuilt circularity in our cells. My research supports that what is unspoken in our world, in our behavior, in our dreams, and in our psychopathologies is the survival message of a quasi-hybrid being via horizontal gene transfer mediated by a creator-virus through an ancient gene expression network still functioning in our cells. This means that the machinery of existence may be a lambdoid virus, the ground of being for our classical world and the deeper foundation of the quantum principle, as well as the origin of space and time. Perhaps the profound, mystical inner experience of God within our hearts may actually be the awareness of the nonconscious origin of our intentionality—an ancestral bacterial virus with its survival message.

Regarding my explorations of nature and human history, this research is motivated by my belief that the ancient pharaohs possessed knowledge that would shed light on life and death, thereby eliminating the fear of the unknown for humans at death. I have learned that the Egyptian knowledge is forever welling up in human history despite the attempts by biopower to literalize, confuse, satanize, psychopathologize, or conceal the knowledge. Also, the origin of alchemy in ancient Egypt and early China and the later quest for the Philosopher's Stone support the Isis Thesis, for through the lens of contemporary microbiology, the alchemical evidence suggests that seventeenth century intellectuals such as Isaac Newton and Dr. Michael Maier were experimenting with bacteria and viruses. And today, when technocrats are forecasting the ultimate disappearance of humanity due to the birth of superintelligent AI, perhaps they should review ancient histories and consider the potential of our viral heritage, for their goals may herald the death of humanity on the planet, as well as the loss of continuity for human DNA. Although no theory is an absolute theory, this research provides a scientific rationale for the emergence of a cold-light quasi-hybrid being that helps one understand who we are, how our universe works, and what we can become.

Finally, *The Isis Thesis*, the twelve published articles, and *Balls of Fire* expose the complicated roots motivating biopower. These texts also deconstruct religion as science, present compelling new scientific knowledge, and zero in on the power/knowledge grid that thrives on the weaknesses of human beings. If we can understand our obsessions, unify our sciences, and overcome our weaknesses through knowledge, then we have used our free will to achieve our full potential. With the hope that we can understand our position in history and the role of the evolutionary process, this book is organized into five sections: Spring Training, Baseball Diamonds, Pregame, Fly Ball

Ferris Wheel, and Opening Day. These sections incorporate excerpts from *The Isis Thesis* and the twelve journal articles available at the website http://www.isisthesis.com. In Spring Training, one learns the necessary knowledge for understanding the hypotheses. The Baseball Diamonds section provides interviews with key players and a discussion of each team's argument. The Pregame section introduces the reader to the fantasy-draft selection of team managers, coaches, umpires, players and pitchers for the Diamondhearts and the Thunderheads, a model depicting historical and contemporary power grids. In the Fly Ball Ferris Wheel section, we examine several individuals who have caught some of history's high-flying balls of knowledge. The last section is Opening Day, where the Game of the Centuries opens with an anthem to welcome the historical and living warriors on both teams who recount yesterday, know today, and predict tomorrow. All these historical and twenty-first century participants will allow you to see human history in a new light, while envisioning what has been called the "Sun-god" (Egypt), the "Ultra-human" (Teilhard de Chardin), the god-man (Christianity), the "diamond body" (China), the "human form divine" (William Blake), the "body without organs" (Antonin Artaud), and the "Superman" or "Overman" (Nietzsche).

I would like to thank the staff of the Otsego County Library for their research assistance and access to their interlibrary loan system, as well as friends and family who gave me valuable feedback on this text.

Theoretical Guide for the Reader

What are the core hypotheses of the Isis Thesis?

Ancient Egyptian deities are signs for human and microbial DNA (genes and proteins) evolving into a hybrid quantum species at a human death transition. The activities of the deities describe the ancient glycolysis gene expression network in our cells that also functions without oxygen or the sugar glucose by using fermentation and the sugar lactose found in milk.

Both Egyptian and Christian deities mirror the activities or lifestyles of a complex virus called bacteriophage Lambda that uses the ancient glycolysis gene expression network. So, all these deities are signs for the same evolutionary process mediated by a virus.

Bacteriophage Lambda, a temperate eubacterial virus, is our last universal common ancestor (LUCA). This means that the machinery of existence may be a lambdoid virus, the ground of being for our classical world and the deeper foundation of the quantum principle, as well as the origin of space and time.

Our semiotic system is based on underlying physical and chemical principles inherited from our microbial ancestors, so our microbial DNA is ordering our society space. Repetitive survival messages arise from our viral chromosomal apparatus that we all share.

Based on the core hypotheses, the Isis Thesis asserts the following novel predictions:
A modeling relation exists between the virus Lambda and a human being, since our intentionality is governed by this ancestral virus. Our historical cultural behavior demarcates its ancient viral gene expression network that grounds its genetic switch between two lifestyles for the evolutionary transformation of DNA.

Two ways of evolution exist for a human due to Lambda's two lifestyles. A human being can evolve into our current thermodynamic state of being, and then, to avoid DNA dissipation at human death, that emergent entity can select a hybrid, quantum, cold light, crystallized state of being.

Preface

At a human death, evolution requires scientific knowledge of bacteriophage Lambda's genetic switch between its two lifestyles. Therefore, the human being with knowledge functions as the genetic switch-hitter to evolve.

What are some of the auxiliary hypotheses of the Isis Thesis?

Pyramidal architecture and Egyptian artwork mirror phage Lambda's morphology, while the funerary texts explain its activities.

The Lambda genome has systematic affinities across several genes for two particular lineages or DNA-texts: one DNA-text expresses itself as matter in a global communicative network of our biosphere; the other DNA-text expresses itself as mind in a quantum universal communicative network.

Together, the genetic switch or decision circuit of Lambda cro and c1 proteins exhibit folding/unfolding dynamics and native state crystallization similar to the backward-in-time aspect of a microscopic wormhole, which is similar to the holographic expansion/collapse cosmos described by ancient Egypt, early China, Peirce, Einstein, and others.

Our cosmos has a holographic mode of operation.
Human evolvability (DNA exchange) is a selectable trait.

Balls of Fire builds on the core and auxiliary hypotheses of *The Isis Thesis*:

Our behavior and the evolution of society in the last 12,000 years has carved a footprint into human history, profiling the morphogenesis of a virus.

Baseball originated in ancient Egypt to express their genetic and chemical science. Sports warfare such as baseball models the competition between Lambda's two repressor proteins for gene-seats that determine its two competitive lifestyles.

Our intentionality may be grounded by our viral DNA.
Christian ritual mirrors ancient Egyptian ritual.
Reputable alchemists were exploring the activities of viruses and bacteria.
Many other historical individuals understood the evolutionary message of viral morphogenesis.
A temperate virus such as bacteriophage Lambda may be living, not dead.

Carl Jung's archetypes mirror major motifs in ancient Egyptian myth, Christianity, and alchemy that can be interpreted as microbiology.

Our non-coding DNA is not useless. If a bacterial virus is our remote ancestor, then our genome's viral reverse transcriptase or telomerase may be just the genetic ticket we need for hybrid speciation.

The general cultural phenomenon of the Few versus the Many models the native versus nonnative interactions in bacteriophage Lambda's two competitive repressor proteins when they partner. Regarding the survival of our species, human history can be explained as a nonconscious psychic game of political deceit about knowledge determining whose DNA will survive or transform at the transition of death.

In evolutionary history, the transition from a bacterial cell without a nucleus (3.8 billion years ago) to our cell-type with a nucleus (1.8 billion years ago) may have resulted from a dormant Lambda genome in the cell that evolved into our eukaryotic cell's nucleus.

Advancing technology forecasts an evolutionary merger between humans and superintelligent machines (hybrid-human-metal matter) that supports the evolutionary viral footprint on our society space that mirrors the evolutionary quantum merger between organic human DNA and metal-binding viral DNA.

To the Reader

In *Balls of Fire*, does the evidence for the auxiliary hypotheses
add support for this research program's core hypotheses and predictions?

Existence begins in every instant; the ball There rolls around every Here.
The middle is everywhere.

Friedrich Nietzsche

There are three types of players: those who make it happen,
those who watch it happen, and those who wonder what happens.

Tommy Lasorda

WELCOME
to
Spring Training 1

Searching for a Unified Science

The Sphinx stares grimly, ominous with question,
Her stony, blank gray eyes tell nothing—nothing—
No single saving sign, no ray of light—
And if I solve it not—my life must pay.

Gustave Mahler, Symphony No. 7 in E Minor
("Song of the Night")

Nature operates by the basic phenomenon that a Few will survive over the Many. Historical power relations generally favor some of the species rather than all. This competition between the Few and the Many seems to center on the acquisition of money, land, food, and security. However, these struggles may be signs related to the survival of each individual's DNA, for the evidence from five thousand years of human history supports that our creative imagination produces repetitive behavior with a genetic survival message that only preserves the DNA of some of the species, not all. This message may arise directly from our genome, the chromosomal apparatus that we all share. So, we all have access to the same genetic survival message, if we can recognize it after power relations distort or conceal it.

Historically, genetic survival information has been secreted by the power élite, and today the biological wisdom is drowned out in the ocean of conflicting signs flooding our daily lives. What is necessary is an understanding of the knowledge secreted by the historical élite and an awareness of our behavior that is influenced by the viral DNA in our bodies. This will equalize the competition for genetic survival information between the Few and the Many. In light of these dynamics related to DNA, quantum theory provides a comprehensive scientific explanation and complete biological foundation for life's chemical-physical aspects as well as emergent states of being.

Quantum theory applies to living systems entangled with the quantum mesocosm of microbes. With approximately 3 percent of the human genome coding proteins for our bodies, about 97 percent has no obvious function (Nicholl 2002). This DNA has been referred to as "junk" DNA or a "cemetery of viruses," but many microbiologists believe it has a function, so the ENCODE project (Encyclopedia of DNA Elements) is exploring this non-coding DNA. It seems that we are inextricably linked to viruses, since their repetitive survival message is coded into human history. In the context of evolutionary and geological time, the last 5,000 years of human cultural history can be interpreted as one long moment that delivers the same microbial survival information to each human being because of our common microbial genome. For example, randomly-selected studies of diverse texts from ancient Egypt, early China, early Christianity, the American Navajo Indians, reputable alchemists, modern genetic engineers, and many others all center on the same microbiological survival message that emerges at different historical times. This lack of causality is reasonable since the research supports that the game of meanings in our lives is a function of the sign actions in our nonlocal genome that has a substance-wave nature or application in space-time and the quantum domain. The evolution of society reflects the power game of meanings cast by our cellular chromosomal apparatus. The strategic information content, the survival message of this system of relations, defines an urgent need for morphogenesis or the emergence of a unique form and structure for adaptability in the quantum domain. This is fertile ground for a unified science of life and death.

The Mysterious Secret of our Material Existence. Ancient Egyptian texts describe the individual's soul functioning in a dwarf world of light and darkness at a death transition. If one replaces the word "soul" with "DNA," the texts begin to make sense. Surprisingly, this journey of the soul at death veils a science about the passage of the individual's DNA into a world of photons (packets of light) and other quantum particles. The pharaoh's secret for DNA survival at a death transition was to absorb energy by merging with a hydrogen-bonded carrier molecule in the interstellar region of the western Sun. They called this carrier molecule the Sun-god's ferryboat, and this seems very appropriate from a genetic engineering perspective because viruses often function as ferryboats, conveying DNA fragments into the cell for cloning. In fact, our genetic engineers have named the virus Charon 16A after the ferryman of Greek mythology, who transported the spirits of the dead across the river Styx. Charon 16A is a bacterial virus or bacteriophage that can transport DNA fragments into the cell for cloning.

Now, DNA is not confined to an organism at death. DNA can escape, and if a bacteriophage carrier molecule is present, DNA can merge and survive. According to the Isis Thesis interpretation of ancient Egyptian texts, carbon-based human DNA self-assembles into a metal-binding, hydrogen-bonded viral carrier molecule for horizontal gene transfer (HGT) and transformation at a death transition. Once considered an extraordinary event, experimental documentation of HGT (the exchange of DNA between species) in many scientific studies supports significant gene shuffling in "the domains Bacteria, Archaea and Eukarya – including plants, fungi and human cells" (Sorensen et al. 2005). Today scientists recognize HGT as a common biospheric event and "Biology's next revolution" (Goldenfeld and Woese 2007). Since many of nature's laws are still veiled, it is clearly illogical to argue absolutely that HGT for human DNA is not possible after death based on the absence of hard evidence. At death, genetic barriers may no longer exist that would prevent DNA-swapping evolutionary moves with a virus.

Absorbing energy from sunlight boosts a molecule to a higher energy state, and then the molecule is not dissipated, that is, it does not lose free energy to waste-heat. Physicist Richard Feynman understood that the absorption of a normal particle creates an antiparticle or a particle moving backward in time. Absorbing energy is the first chemical step in ancient Egypt's science to transform the pharaoh into "the Unique One" who assumes "a form which is unknown" in the Book of Caverns (Piankoff 1954, 124). So, the ancient texts are talking about HGT and morphogenesis or the development and emergence of a unique form or species at the microscopic level of DNA.

Consider that if this first chemical step is possible, then the hydrogen Sun and its electromagnetic spectrum should show the resulting absorption bands of carrier molecules. What we find is an electromagnetic field of diffuse interstellar bands (DIBs) or wide absorption bands left by some mysterious molecules, according to astrophysicists. For the last century, scientists have been studying these wide absorption bands, but they cannot identify the carrier molecule responsible for the absorption features. Some think the mysterious carrier molecules at the visible wavelengths in space are dust grains, hydrocarbons, or small linear molecules. A recent study supports that the molecules that produce the bands have a large number of carbons and other metal atoms (Zasowski et al. 2014). This metal content suggests that the carrier molecules may be metal-binding viral molecules with carbons related to organic molecules. This and other data supports the following thesis.

Is this immortality? *The Isis Thesis* is the initial study of 870 ancient Egyptian signs, showing that the corpus of Egyptian funerary texts spanning from 2520 to 664 BCE preserves an advanced modern knowledge of space physics, quantum physics, quantum biology, and cosmology. This is demonstrated in their least-corrupted Pyramid and Coffin Texts by the activities of their deities, who represent bacterial and viral genes and proteins in the quantum domain or their "regions of the dwarfs." Egyptian scientific knowledge supports that human DNA can be transformed at a death transition via a specific viral gene expression network linked to the ancient glycolysis pathway within our cells. In these ancient texts, a human agent jumpstarts the development of its own form through a merger with a viral carrier molecule that uses the ancient glycolysis network operating in our cells. This novel emergence of a sustainable new entity with unique interaction qualities is not only supported by ancient Egyptian texts, but also related studies showing that the Egyptian knowledge is also present in early Chinese texts, Navajo mythology, the cosmogony of Charles Peirce, the biophysics of Pierre Teilhard de Chardin, William Blake's unfinished masterpiece "The Four Zoas," Christianity, and many other sources (see King articles). So one could easily say that ancient Egyptian mythology conceals an evolutionary science, as does the religion of Christianity. The evidence supports that literature, artwork, architecture, ritual, grain agriculture, milk production, religion, and other human behaviors are signs and patterns identifying a dynamical viral system for human DNA transformation. Thus, a modeling relation exists between a complex bacterial virus and human beings. What is emerging is a science of life and death that may achieve a transdisciplinary unity of knowledge.

> **SELF-ORGANIZATION**
> Spontaneous ordering of molecular or supramolecular units into a higher-ordered non-covalent structure, characterized by some degree of spatial and/or temporal order or design by correlations between remote regions; will exhibit collective and often nonlinear behavior; could include crystallization and related ordering phenomena such as liquid-crystallinity. (Lehn and Ball 2000, 304)
>
> **SELF-ASSEMBLY**
> Spontaneous association of several molecular components into a discrete, non-covalently bound aggregate with a well-defined structure, e.g. binding with a purpose. (2000, 304)

Yes, this is a bold claim about a novel emergence, but it is backed up by our behavior and our sciences. The original causal pattern—or what biologist Robert Rosen calls the "internal predictive model" (1985), appears to be a complex virus that allows viral transformation in an ancient gene expression network found in our cells. To test this bold claim further, we must simply recognize the variety of natural patterns in our behavior that can be compared to the activities of a specific virus that self-assembles and uses the ancient glycolysis gene expression network. Although correlation is not causation, if a correlation exists between viral lifestyles and human behavior, then we have found a science of human behavior endorsing the evolutionary potential of our viral DNA. After all, emergence is a fundamental property of complex systems such as ours, and ancient Egypt and other early cultures have identified a synergistic hybrid coupling of human and viral DNA for adaptive evolution in the microscopic quantum environment. With this idea of self-organization in mind, let's first consider the nature of the proposed internal predictive model, as identified by the ancient Egyptian pharaonic priesthood, so that we can understand the evolutionary meaning of our own creativity and the biological idea of DNA immortalization.

The Good, the Bad and the Ugly. Correlation is not causation but it implies some type of causal relation. Contemporary microbial experiments describe the same series of viral protein actions modeled by Egyptian deities as they journey along the ancient glycolysis gene expression network. The semiotic phenomenology of the Egyptian pharaonic priesthood scientifically details this survival message for humans, for it involves horizontal gene transfer (HGT) or DNA exchange with a complex bacterial virus called bacteriophage Lambda (phage for short pronounced like "cage"). This virus may be the mechanism governing pattern formation in our macrocosm. Throughout history this metal-binding virus has been used as a medicine, as it is today, and it is also used for cloning purposes in our genetic engineering labs. Now, if the mention of *virus* induces microbe-phobia in your mind, consider this logic. If we break microbes down into categories of the good, the bad, and the ugly, then the ugly viruses would be HIV and Ebola, harmful bacteria would be the bad microbes, and the good viruses would be the viral bacteriophages, which attack bacteria and are used in medicine to remedy disease. Because a bacteriophage has ample space on its DNA chromosome to carry other fragments of DNA from various cell-types, including human, scientists use the virus as a ferryboat to carry fragments of DNA into the cell. In ancient Egyptian texts, the Sun-god's ferryboat is the sign of the hybrid merger of metal-binding viral DNA with organic human DNA. The texts support that the resulting chemical transformation produces an emergent state of being or a quantum of crystallized light that is hybrid, metallic and alive with the command and power expressed by the hybrid Egyptian Sphinx at the Pyramid Complex of Giza. To understand Egyptian science and the relational microbial model of phage Lambda, let's review its complex viral lifestyles and the ancient Egyptian core myth.

The Ferryboat Phage. Perhaps you remember H. G. Wells' forecast in *The War of the Worlds* about a final viral structure for humans, one similar to his sexless machinic Martians: "They were heads—merely heads." Similarly, the language of the Egyptian funerary texts related to vertices, heads, tails, and faces, along with the triangular structure of the pyramids with their tail-like causeways depict phage Lambda's morphology or shape. In addition, tails carved on deities in the Edifice of Pharaoh Taharqa, the Ramesses VI tomb drawing of an armless human with coiled legs and a sphere for a head, and the Seti I tomb artwork of humans with spheres for heads provide more evidence for this view, along with the interpretation of over 870 ancient Egyptian signs supporting the lytic cloning lifestyle of icosahedral phage Lambda (also known as λ).

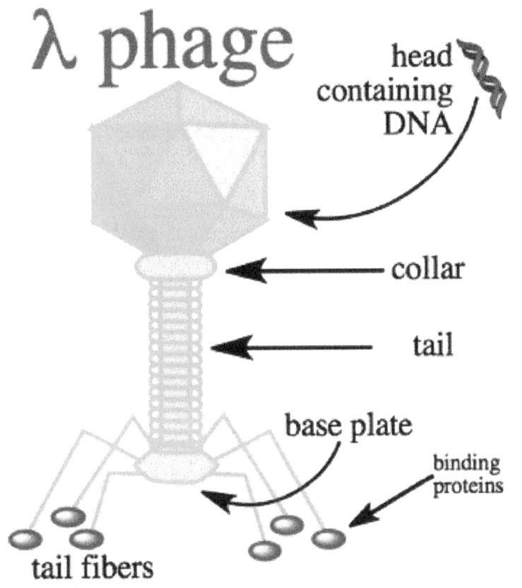

1.1. Public Domain image

An icosahedron is a phage conformation which has 20 equilateral triangles arranged around the face of a sphere. Most viruses fit 60 x N subunits into their capsids (heads), with N being the triangulation number, having values of 1, 3, 4, 7, 9, 12, and so on.

Phage Lambda's icosahedral head of DNA is actually enclosed in a sphere, which is not depicted in Figure 1.1. Now, this virus has a large space on its chromosome (DNA) to carry other fragments of DNA. Our genetic engineers use this phage in their labs because it is a ferryboat for fragments of DNA that can be inserted into phage Lambda and transformed or cloned in a cell via the two complex lifestyles of the virus.

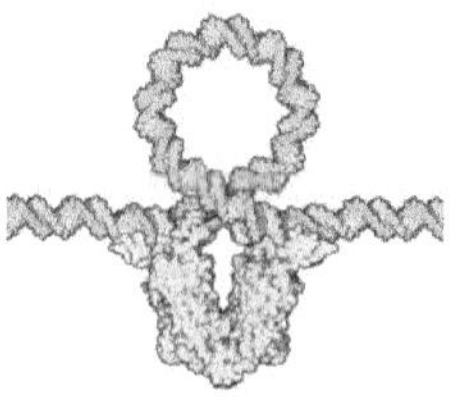

1.2

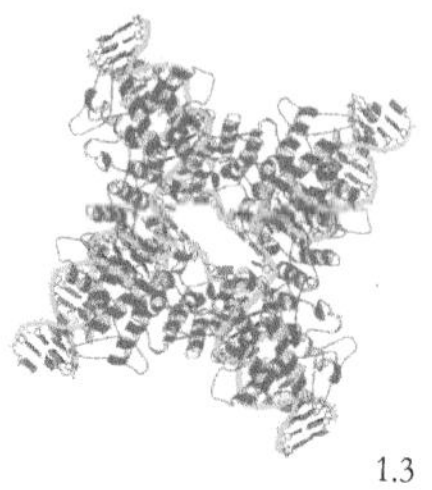

1.3

Figure 1.2. Lac repressor protein looping the DNA in *Escherichia coli* (*E. coli*). Protein Data Bank Molecule of the Month, 2003, PDB-101 (Public Domain). Illustrator: David S. Goodsell.

Figure 1.3. Holliday junction intermediate, a site-specific recombination process for the integration and excision of bacteriophage genomes into and out of the host bacterium. Protein Data Bank PDB-1XNS (Public Domain); Ghosh, K. et al. (2005) *J. Biol. Chem.* 280, 8290-8299.

In lysogeny, the ferryboat-phage adsorbs or attaches to the host cell, pumping its DNA inside the cell like a syringe. Then the DNA circularizes for protection and travels to a specific site next to the lactose metabolism genes. The lactose metabolism genes are tied off by a protein that loops the DNA into the shape of the Egyptian ankh (see Figure 1.2). Here, the phage falls asleep beneath a protein DNA-cross called a Holliday junction (see Figure 1.3), to be silently replicated with the cell's DNA as it divides. Attached to its protein-cross in this dormant state on the host cell chromosome, phage Lambda is called a prophage.

The other lifestyle of phage Lambda is lysis. When ultraviolet light damages the host cell DNA, the cell sends out an SOS call for help to a special group of approximately 42 proteins. This alerts the Lambda prophage to rise from its dormant state, escape from its protein-cross, and save itself by cloning progeny. In the lifestyle of lysis, the awakened prophage Lambda then takes over the host cell's replication machinery to clone its own kind. Once the risen Lambda is in charge of the cell's replication machinery, the DNA is packaged into numerous viruses with heads and tails. This replication process is called rolling circle replication, where the circular DNA spits out new virions that escape from the host cell in a great flood. Ancient Egyptian texts explain that what activates lysis is the ultraviolet light or the incoming hybrid DNA (human and viral) that sets off the cell's SOS response. This is the ferryboat phage or raft of proteins on its way to the Hidden Chamber of the inert Lambda prophage (Osiris) on the host chromosome. Relative to this viral ferryboat of proteins, biologists are aware of the lipid raft hypothesis, supporting that a protein raft may be involved in viral assembly, as well as the viral point of entry into the cell. Ancient Egyptian texts support this hypothesis, for they describe a viral raft or ferryboat of activating proteins self-assembling at the viral point of cell entry to travel the ancient glycolysis fermentation gene expression network. Isis is present on the ferryboat, and she is the sign for the protein lactose permease that activates the lactose genes along this path.

Now, Lambda's lifestyles are controlled by two competitive protein teams. The leading protein controlling lysogeny is the large repressor protein of 236 amino acids named c1 protein (pronounced c-one). The opposing leading protein controlling lysis is a small repressor protein with only 66 amino acids named cro protein. (Ptashne 2004) The lead proteins and their teams are battling over the genetic switch between the two lifestyles that depends on the capture of the right arm of the Lambda genome with its 40 binding states (Santillán and Mackey 2004). What allows small cro protein to win the battle is the presence of UV light, that is, the Sun-god.

Horizontal Gene Transfer. What the unified signs support in ancient Egyptian texts is the merger of a deceased human's DNA with the Sun-god's self-assembled viral ferryboat of DNA. This is horizontal gene transfer (HGT), the exchange of DNA between two different species—a human and an ancient virus that is actually present in the human body or our gut microbiome (intestines, stomach, and so on). Curiously, the ancient texts support that this exchange occurs at human death, when it is possible for DNA to escape the human body. It is known today that genetic material released from dead and living cells persists in all environments and that DNA can be transferred from dead to living cells (Avery et al. 1944). When the International Human Genome Sequencing Consortium found a set of 223 proteins similar to bacterial proteins, it was likely that the genes entered our vertebrate lineage by HGT from bacteria. As mentioned, HGT or gene shuffling has been documented in Bacteria, Archaea and Eukarya, including human cells, plants, and fungi. Put simply, because of the large 98 percent non-coding DNA in the human genome (Elgar and Vavouri 2008) with its regulatory elements, RNAs, introns, pseudogenes, telomeres, and viral elements, HGT mediated by a virus may be possible at a human death transition. More support for this idea is that 14.6 percent of our genome has a complete recipe for the protein reverse transcriptase (Ridley 1999), an enzyme with an RNA template for DNA replication that may be necessary for HGT and cell entry. Accordingly, we can reason that if we descended from a common ancestor, as indicated by our genome, it could be a virus, so we may be a species capable of DNA-swapping evolutionary moves at physical death, as ancient Egyptian texts explain in detail. Again, Egyptian texts model the activities of their deities on phage Lambda's lifestyles. The actions of these deities match the experimental descriptions of phage Lambda's lifestyles by contemporary microbiologists.

The Ancient Egyptian Core Myth. The least-corrupted texts of ancient Egypt are those carved on the pyramid walls of the pharaohs and the coffins of the nobility. Many myths have developed by word of mouth, but the core myth was protected and buried within the tombs and coffins. To determine the pharaonic priesthood's core myth, 108 key themes were identified in the least-corrupted Pyramid Texts and Coffin Texts (2520 – 1640 BCE). The 108 themes were then synthesized into 30 major idea strands, defining textual events and activities of major deities. In this core myth, we meet the dying/rising Osiris, the virgin Isis, the Sun-God's ferryboat, the competitive brothers Seth and Horus, the creator-god Atum, the scribe Thoth, and the monster serpent Apopis. In the least corrupted funerary texts, microbiological concepts are embedded in a religious narrative, detailing the afterlife adventures of the dead king who desires eternity and transformation. Through this process of re-creation, the dead king does not die for millions of years. However, to obtain his desired state in the early cosmos as the first-born son of the creator god Atum, the dead king must follow specific directions, using the earth and stars as guides. By following the advice in the funerary texts, the king-to-be-transformed joins the Sun-god Re's ferryboat for the voyage through the black Duat to overcome the monster serpent Apopis. Isis, the Lady of Provisions, and the scribe Thoth assist the dead king in his journey to the Hidden Chamber of Osiris. The dead king judges the two brothers, Horus and Seth, who compete for the throne of their father Osiris. When Seth wins the throne, Osiris dies; when Horus controls the throne, Osiris rises from the dead with the help of Isis who breathes life into him and conceives the golden child Horus by virgin birth. The rival brothers are then reunited, resulting in the restoration of the Eye of Horus, along with the birth of the dead king's new form (the golden Horus child), which is nourished by Isis, and rising in the East as a morning star. In the Book of Two Ways Coffin Text 1099, the dead king states: "I establish millions" (Piankoff 1974). With this least-corrupted narrative in mind, below are the major Egyptian signs and their parallels to microbial supermolecules and proteins related to phage Lambda's lifestyles in its bacterial host cell, as decoded in *The Isis Thesis* (2004).

EGYPTIAN SIGN	MICROBIOLOGICAL PARALLEL
Dead King	Human DNA fragment
Sun-god's Ferryboat (the Light)	Viral protein raft entering cell with human DNA, lactose permease, and other viral proteins
Dying/Rising Osiris	Inert prophage inside cell that rises from dead state
Macrocosmic Earth	Microcosmic *E. coli* cell
Monster Serpent Apopis	Monster LexA repressor protein repressing lysis in cell
Brother Horus	Lambda cro repressor protein (controls lysis)
Brother Seth	Lambda c1 repressor protein (controls lysogeny)
Virgin Isis	Lactose metabolism (*lac* genes or lactose permease)
Golden Horus Child	Hybrid progeny (clones)
Scribe Thoth	RNA polymerase for transcription
Thoth's 42 Books	De-repression of 42 genes controlled by Monster LexA
Whole Eye of Horus	Left and right arms of Lambda genome

Deflowering the Belief in a Real Virgin. When the Lambda prophage (Osiris) rises from its dead state, detaches from its protein DNA-cross (see Figure 1.3), takes over the replication machinery of the host cell, and clones its own kind, the viral energy source is lactose metabolism, which is now available for the rising virus on the ancient glycolysis-fermentation pathway. If the sugar glucose is unavailable and lactose enters the cell, lactose becomes available. Many biologists believe that the presence of lactose has the potential to allow adaptive mutations.

The goddess Isis is the sign for the nutriment lactose, and ancient Egyptian texts indicate that Isis is present on the Sun-bark with the dead Sun-god or the light. To the ancient Egyptians, Isis represented the *Magna Mater*, the Lady of Provisions, the great creative power nurturing all living things. According to the texts, lactose metabolism is necessary for the birth of viral clones. Genetic cloning may also inspire the creativity of religious artists such as Andrea Mantegna who conceive of multiple cherubim with a single Madonna, as in "The Madonna of the Cherubim" (see Figure 1.4). Cherubim are celestial beings often represented as winged children, yet in the biblical Book of Ezekiel, cherubim are winged hybrid beasts with four faces (human, bull, eagle, and lion). The priest Ezekiel's ministry is dated 593 to 563 BCE and marks the transition from pre-exilic Israelite religion to post-exilic Judaism (May and Metzger 1973, 1000). Still, Ezekiel's description of the four living creatures embraces the primary ancient Egyptian signs associated with the pharaoh's chemical transformation. And so, we have the connection of Isis with lactose, the Madonna with numerous winged children (clones), and Ezekiel's hybrid beasts that are linked to the major Egyptian signs of the divine human, the bull, the eagle or hawk, and the lion.

Figure 1.4. Madonna of the Cherubim (1485) by Andrea Mantegna (Public Domain)

Now, consider the idea of Isis and the virgin birth of her little Horus child. Christ too was born from a virgin according to Christianity. In ancient Egyptian texts, Isis is a sign of the necessary lactose metabolism necessary for the birth of the Horus child. She is a sign of bacterial lactose proteins and is not a real person. The text states, "Isis wakes pregnant with the seed of her brother Osiris," suggesting that the viral prophage uses the lactose metabolism. This brother/sister union of Osiris and Isis is a sign of viral/bacterial sex. Biologists consider the trading of DNA between different organisms as sex similar to human sexual reproduction involving a male and a female who produce a child in the next generation. Human sexual reproduction is called vertical gene transfer. However, bacteria trade genes horizontally within the same generation (brother and sister). In ancient Egypt, the pharaonic family also practiced royal incest or trading genes horizontally within the same generation.

Now, Osiris and Isis are genetic signs, not real people involved in incest. Yet, to explain the metaphor, Isis is pregnant with the seed of Osiris because the DNA of the Lambda prophage (Osiris) is lodged next to the bacterial genes for lactose metabolism (Isis) on the host cell chromosome (DNA). So, the seed of Osiris (Lambda DNA) is within her womb (lactose metabolism gene-seats). Remember that when the Lambda prophage rises and leaves its hidden gene-seat on the host cell chromosome, lactose metabolism is provided. Isis is often called "the Lady of Provisions" (CT 246) and the "milk-goddess" (CT 175), for lactose is the main sugar in milk. Also, her crown consists of a pair of horns between which is a lunar disk, sometimes surmounted by the throne symbol. When the prophage rises through its DNA-cross from its dormant state to replicate, it leaves an empty gene-seat, that is, an empty throne. Thus, we have a possible correlation between the name of Isis depicting the empty throne or gene-seat of the risen prophage and the potential for lactose metabolism. Calling Isis a virgin is one way to express this asexual genetic union related to lactose metabolism and prophage replication that results in the golden Horus child or viral cloned progeny.

The Ancient Glycolysis Pathway. For living humans, glycolysis involves the use of the sugar glucose for our cells and the metabolic system of breathing oxygen for energy. However, bacteria can derive energy in other ways such as fermentation, eating nitrogen gas right out of the air, precipitating iron and manganese, or combusting hydrogen with oxygen to make water. (Margulis and Sagan 1986, 128) Microbes invented fermentation, and humans have turned fermentation into a technology, producing bread, cheese, wine and beer. Human fermentation technology may be another sign of the ancient glycolysis-fermentation pathway, a nature-inspired design invented by microbes. Perhaps our evolutionary predilection for eating sweets is also related to this useful metabolic pathway with its message of survival via various sugars. As we shall see, our macroscopic behavior often points to dynamics on the microscopic level.

Biologists believe that before the advent of oxygen, the ancient glycolysis-fermentation pathway was the source of energy production for earlier organisms such as bacteria.

This ancient gene expression network can function without oxygen or glucose by using lactose, the predominant sugar in milk. An outer membrane protein door known as the LamB porin is the gateway to the ancient glycolysis gene expression network. The channel protein LamB allows the passage of sugars, especially carbohydrates such as maltose, maltodextrins, and glucose. LamB contributes seventy percent of the total glucose import capacity of the cell (Death and Ferenci 1994). LamB is also known as the gateway to the maltose transport system that the bacterial virus Lambda uses. Conveniently, phage Lambda enters the bacterial cell at this receptor site or gateway to the ancient glycolysis gene expression network, a surprising coincidence. It then travels along the bacterial chromosome to the lactose genes, where it integrates the chromosome by a site-specific recombination process called a Holliday junction (see Figure 1.3). The virus is now a prophage, but it has the potential to take over the replication machinery of the bacterial host cell, using it as a factory to clone its own kind. This mass-production technique is the lifestyle of lysis or the lytic cycle, and it has been in existence for billions of years to clone viral heads and tails, long before Henry Ford picked up the same production technique for cloning factory automotive parts.

In ancient Egyptian texts, the Sun-bark or light damages the cell, activating the cell's SOS response. At least 42 genes that function in DNA repair, recombination, mutation, translesion DNA synthesis, and prevention of the normal cell cycle are de-repressed (Rosenberg 2001). These genes are usually repressed by the cell's monster LexA repressor protein. However, the SOS response activates the multifunctional enzyme RecA that slays the monster LexA protein, so mutations can begin to fix the damage.

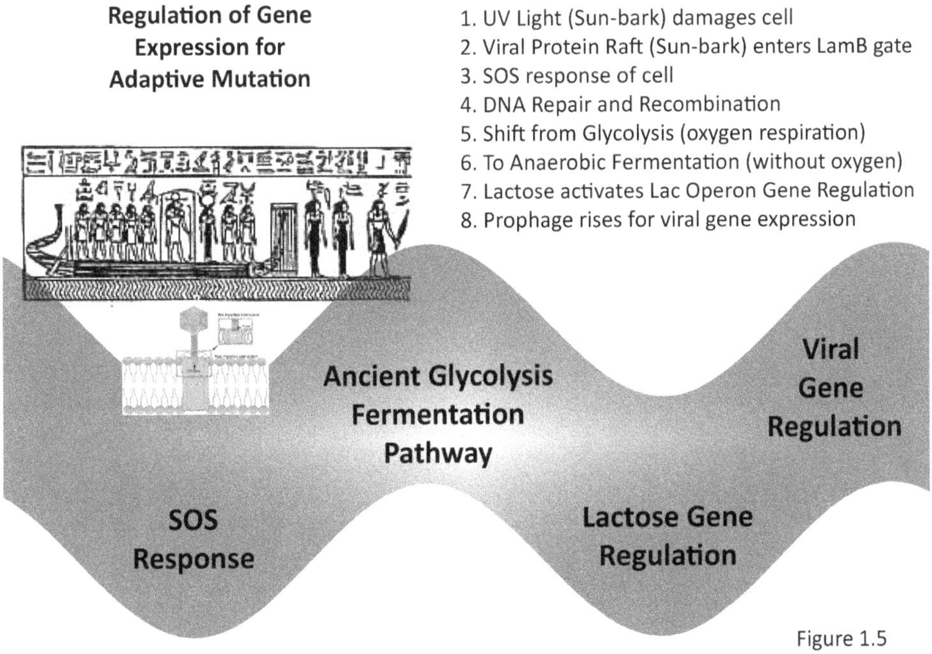

Figure 1.5

Numerous Egyptian textual references also emphasize the importance of cultivating or brewing barley. The sugar maltose is used in the brewing of barley, and Osiris, sign of the Lambda prophage, is closely connected to the harvest of barley in ancient Egypt. So, the brewing of barley is another sign of the maltose transport system in the ancient glycolysis-fermentation gene expression network used by phage Lambda. This same system also allows the transport of lactose into the cell by a low-affinity pathway (Merino and Shuman 1997). Thus, a variety of metabolic sugar mechanisms (glucose, lactose, maltose) are available in the ancient glycolysis-fermentation gene expression network, showing how living organisms are efficiently organized for survival (Boos and Shuman 1998; Shimizu 2013).

The Second Death. Gene regulatory networks control gene expression relative to environmental changes. The ancient Egyptian texts explain that a deceased person's DNA can transform and survive via the ancient glycolysis-fermentation gene expression network, that is also known as the maltose transport system used by phage Lambda. The ancient texts explain that an awareness of the signs is necessary for survival. The texts also tell us that those without knowledge are deliberately and systematically destroyed in a protein-degradation manner similar to human genocide. Survival or dissipation by waste-heat is simply the difference between knowing the signs or not. Thus, ancient Egyptian literature, artwork, ritual, and architecture model a complex viral replication cycle for human DNA survival at a death transition, along the ancient glycolysis pathway that runs with or without oxygen. The Egyptian texts explain that without knowledge of this ancient sugary gene expression network and its LamB protein gateway for phage Lambda, one meets death a second time in the microscopic domain that we all return to when we quit breathing and die. So, the secret Egyptian texts inform us that our cells have the potential for survival, as well as a role for phage Lambda to play relative to human survival of DNA at a death transition.

Wings for Serpents and Spheres for Heads. Modern genetic engineers use the same language that the ancient Egyptians used to explain microbiology: heads, vertices, tails, transformation, faces, and so on. Ancient Egypt also uses similar terminology of upstream and downstream to describe the two ways of the DNA molecule, where "words" (DNA) are transcribed and replicated. In the Pyramid Texts, the term *querert* (Eg. *Krrt*) translated as "cavern" is associated with the netherworld. Texts frequently mention the jackal-headed deity Anubis and Osiris at the Entrance to the Holy Cavern. One also finds the Caverns of Hathor, and the Coffin Texts refer to the gods of the Duat as "those who are in their Caverns." The Egyptian Book of the Dead, the Book of Gates and the Litany of Re also mention "the gods in the Caverns, who are in the West." (Piankoff 1974) The multidimensional space of DNA is easily described as a cavern system for mapping functionality. In addition, Egyptian artwork of serpents with multiple heads indicates the number of snakelike helices a viral protein has, while serpents with wings indicate the wing-like beta sheet structures of viral proteins. Our contemporary biologists also designate these same viral proteins in a category called "winged" helix-turn-helix. So, both the pharaonic priesthood and contemporary scientists use the same terminology to identify several of phage Lambda's proteins. Yet, the pharaonic priesthood anticipated the future through their deities that encoded their viral model, so they tapped into information about the human species that was also encoded in the viral component of human DNA. The viral genes in our bodies ground this internal predictive modeling relation with a virus, and our behavior models its lifestyles. The ancient Egyptians were guarding valuable evolutionary information about an ancestral, self-assembling virus that their texts claim is our last universal common ancestor (LUCA).

Consider the constant presentation of Egyptian artwork featuring disks and spheres on human bodies in place of human heads. This is a sign of the emergent hybrid form that is viral and human. It is interesting that Emperor Justinian at the Fifth Council in Constantinople (553 CE) condemned the ideas of the early Christian theologian Origen (c. 184-254) with the following warnings:

> If anyone says or thinks that at the resurrection, human bodies will rise spherical in form and unlike our present form, let him be anathema.
>
> If anyone says that the heavens, the sun, the moon, the stars, and the waters that are above heavens, have souls, and are reasonable beings, let him be anathema. (Christian Classics Ethereal Library)

Justinian's anathemas suggest that the early Church understood that some type of transformation about a change of form was possible at death that they could not comprehend, and so the mystery was cursed and dumped in history's dustbin, and the scholar Origen lost his chance to be canonized as a saint. We can trace the roots of these beliefs about souls, the Sun, and resurrection in spherical form directly to ancient Egypt. So Origen, who studied in the Egyptian city of Alexandria, is elaborating on ancient Egyptian beliefs about morphogenesis or the development of spherical form.

Origen's comments also show us that the early Christian Church was concerned about the mind-body problem. In *De Principiis* (Book III), Origen writes:

> The whole of this reasoning, then, amounts to this: that God created two general natures—a visible, i.e., a corporeal nature; and an invisible nature, which is incorporeal.

This debate about human nature, free will, a change in substance, and transformation into forms or species has troubled humans for millennia. Actually, the mind-body problem is summed up nicely in the Eleventh Division of the ancient Egyptian Book of Gates in the Twentieth Dynasty tomb of Ramesses VI: "Peace, peace, (O thou) whose forms are numerous. Thy soul is in heaven, thy body is for the earth, greatness has been ordered by thee thyself" (Piankoff 1954). Perhaps ancient Egypt generated the mind-body debate that we will address in the Game of the Centuries between the

Diamondhearts who favor the primacy of mind and the Thunderheads who prioritize the material body.

A Perspective. The mechanism governing pattern formation in human behavior may be a self-assembling virus replicating along the ancient glycolysis gene expression network. In ancient Egyptian texts, the Sun is the sign of a self-assembling virus. Life on earth is a result of the flow of energy from the Sun. According to ancient Egyptian texts, the Sun's ultraviolet radiation, a solar spectrum with thousands of dark absorption lines, as well as diffuse interstellar bands, causes an elemental chemical reaction allowing a natural emergence that amounts to a phage-mediated assembly of human and viral DNA in ordered crystallized superlattices. This emergence possesses a state of being. So it seems reasonable to refer to this creation as a quasi-hybrid being or perhaps a mind crystal because viruses are crystal heads of DNA without organs that can re-animate. This ancient Egyptian emergence of a quasi-hybrid being or mind crystal is supported by similar historical ideas in the literature of early China and the seventeenth century American Navajo Indians, in the magnificent Aztec carving of a human skull in rock-crystal, in the alchemic search for the Philosopher's Stone (see Chapters 22-24), in the logic of mathematician Charles Sanders Peirce who describes the continuity of crystallized mind (King 2009a), in the physics of Nobel laureate Frank Wilczek with his mathematical solution for a time crystal (see Chapter 9), and in many other sources.

Morphogenesis is the biological process that causes an organism to develop its shape. The process involves knowing how to control gene expression and regulate cell fates to generate biological form and structure. In the tomb of Ramesses VI, texts such as the Book of Caverns and the Book of Gates carved on the tomb walls explain the development of form or morphogenesis as a process of horizontal gene transfer with a virus for transformation along the ancient glycolysis-fermentation pathway using lactose metabolism. Today, understanding morphogenesis is one of biology's unresolved problems. Mathematician René Thom explains that animals use their retinas to correlate forms or extract information to obtain food or energy (1975, 127). Relative to this logic, the geometric form of the Sun carries information for humans, and the ancient texts explain that we need the energy from the Sun at a death transition. Intuitively, humans must know that they have a home in the quantum universe, and that life is possible there as a crystallized form. Yet, this intuition is smothered within the cluster of relations enfolding historical power and knowledge, as well as repressed due to the tensions between modes of domination and individual freedom of will. As we shall see, the connections between life and nonlife, and the links between our classical sciences (physics, biology, and cosmology) and quantum mechanics (quantum physics, microbiology or quantum biology, and quantum cosmology) reinforce this intuition.

As ancient Egyptian, early Chinese, and American Navajo art and literature support, the ancient glycolysis gene expression pathway used by phage Lambda is signaled by our behavior. The virus itself may be the fundamental reason for the structural stability of the emergent entity, as well as the original causal pattern or internal predictive model for human behavior. The following additional behavioral patterns support that Lambda may be the internal predictive model. First, artwork such as the Egyptian ankh and the Christian cross describe viral protein shapes. Second, both Egyptian and Christian deities model the same viral dynamics. Third, human incest in the same generation mirrors microbial sex in the same generation. Fourth, our fermentation technology (bread, cheese, wine, beer), our consumption of milk, our evolutionary predilection for sweets, our agricultural cultivation of barley and wheat point to fermentation, metabolic, and chemical processes in the ancient glycolysis gene expression network used by the virus Lambda. In addition, some modern technocrats forecast the merger of humans with machines and the end of humanity by 2100, as well as the hybrid infusion of superintelligent artificial intelligence (AI) into our universe. Unlike the quasi-hybrid being formed through a merger of organic human and viral metal-binding DNA, the contemporary technocratic rationale aspires to embrace our classical cosmos through metal matter, resulting from physical humans merging or uploading their brains into superintelligent metal machines.

As has been observed by John Avery (2003), the rapid cultural development of our species can be viewed as an increasing growth of information that began 40,000 years ago. Culture accelerated when the invention of agriculture replaced hunter-gatherer societies about 10,000 BCE and when the birth of writing erupted in the Middle East circa 9000 BCE. Information-driven cultural evolution of our species then exploded through the codes of Mesopotamian cuneiform, Egyptian hieroglyphics, Chinese ideograms, Mayan glyphs, and the Phoenician and Greek alphabets, resulting in the invention of paper, ink, printing, computing and information technology. Avery believes that "all our activities are fundamentally biological phenomena" (2003, 127). However, our activities may be fundamentally microbiological because information-driven cultural evolution maps the morphogenesis of a virus in human history. Let's first look at the sketch below of approximately 12,000 years of recent human activity.

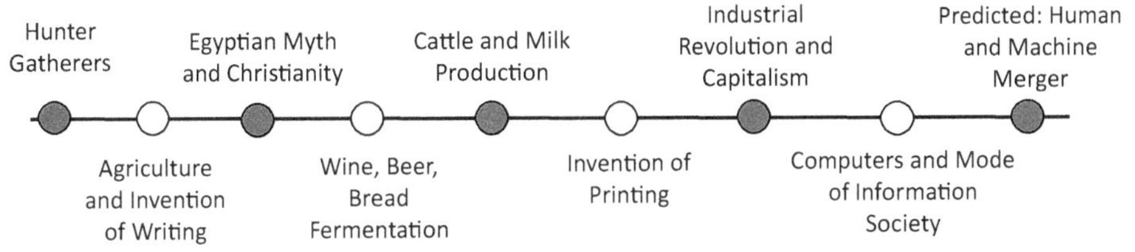

Human History: 10,000 BCE to 2100 CE

Figure 1.6

Figure 1.6 maps the following viral footprint.

SIGNS OF A VIRAL FOOTPRINT

Historical sign (10,000 BCE to 2100 CE)	Ancient glycolysis fermentation pathway with lactose metabolism
Hunter Gatherers	Viral self-assembly/recruitment of energy-rich molecules
Agriculture (grain, plants) Invention of Writing	Viral vegetative replication (plant-like growth) DNA transcription
Egyptian Myth/Christianity	Deities are signs for microbial proteins/genes on pathway
Wine, Beer, Bread	Fermentation (created by microbes)
Cattle/Milk Production	Lactose gene expression/metabolism for viral lysis
Invention of Printing	DNA transcription and replication
Machine Industrial Revolution/ Capitalism (cloning of $$$)	Human and metal-binding virus merger; cloning
Information Society in advanced Capitalism	Viral DNA transcription; accelerating replication of information (cloning or lysis)
Predicted Hybrid Merger of Human with Metal AI	Predicted quantum Quasi-Hybrid Being of organic human DNA and viral metal-binding DNA

As we shall see, the evidence reveals that not only our behavior, but also the evolution of society over the last 12,000 years has carved a footprint into human history, profiling the morphogenesis of a complex, metal-binding virus. This viral footprint is a survival message for humans who can recognize it. The chemical formula maps the ancient glycolysis-fermentation gene expression network for lactose metabolism used by bacteriophage Lambda in its lytic cloning lifestyle of vegetative replication. Surprisingly, dual inheritance of information from genes and culture is mapping this genetic survival pathway for human transformation. This suggests that the quantum world is ordering our classical world, as many scientists believe. History is showing us that human activities and potential for emergence are linked to the morphogenesis of a complex virus.

If so, data is consistent with a gene-culture system where genes influence the probability that certain cultural traits will be adopted (farming, fermentation practices, milk production, and so on). This viral footprint suggests that a quasi-hybrid being of viral-human DNA is possible. Our allure for the machinic may also be related to the machinic DNA of a metal-binding virus. In summary, the Isis Thesis proposes that a modeling relation exists between a complex bacterial virus and human beings, and our viral genes ground this internal predictive model that motivates human behavior in an environment that influences the choice of developmental pathways.

Spring Training 2

Signs of the DNA World System

> The key is in the window, the key is in the sunlight at the window—I have the key—Get married Allen don't take drugs—the key is in the bars, in the sunlight in the window. Love, your mother.
>
> Message to poet Allen Ginsberg from his mother Naomi Ginsberg in a letter sent to him after her death.

Like a falcon of gold over the mountains, the rising Sun melts the horizon's nightshade of deep blue silk that the ancient Egyptians called lapis lazuli. Slowly the golden firebird crowns the mountain peaks in rose, violet and pearl hues. From a mountain peak, the spinning world below looks peaceful and ordered, as if some thing had designed it. The human transportation highway system coils and twists along the surface of the planet like two strands of a DNA molecule running in opposite directions. Hybrid cars stream up and down the two ways like machinic proteins with bits of data. The locatable towns and small cities along the highway system are the lands or genes associated with regulatory regions along the DNA strands. Larger cities such as Dresden, Tokyo or Mexico City with tall buildings are promoters or regions of DNA expressing genes or regulating elements of DNA. Human trading, bartering, buying and selling in the towns and cities along the highway system simulate the substitution or exchange of amino acids along the DNA strands. Humans are the workers and organizers in the cities and countryside, who function like proteins, the workhorses in the cell that follow the directions in the genes. Human highways, towns, cities, modes of exchange, and work organization exhibit the microcosmic activity of the DNA molecule. Perhaps the action of these signs is offering us a biological message about ourselves. After all, humans are composed of proteins and recipes for proteins, so our behavior may be a mirror of this biological activity.

Without a moral code, proteins have rivalries similar to warfare, alliances similar to tribalism, differences similar to race, and gene ownership interests similar to land territorialism, like humans who devise moral codes. Proteins also search for energy-rich molecules like human hunter-gatherers search for food. Proteins participate in vegetative DNA replication that is similar to human agricultural production. Proteins transcribe, copy and replicate words of DNA like humans write, copy and print words of communication. Proteins enslave other molecules, exhibit divisions of labor, and complete suicide like humans. As an example, the protein Polo-like kinase 4 (PLK4) completes its job in the cell and then degrades, what scientists describe as self-destruction or suicide (Instituto 2013).

Human bodies and cells need the sugar glucose for energy to work, stay alive, play, and function, just like the DNA backbone needs its repeated pattern of sugar groups. Humans construct metamodern buildings such as the Tokyo Skytree, stretching upward for 634 metres, and the 829.8 metre Burj Khalifa, the tallest building in the world gracing the sky of Dubais because humans aspire heavenward and think they may be gifted intellectually, unlike other organisms. And so we marvel at the sophistication of our cities, or the complexity of the Tokyo rail system with 13 subway lines and over 100 surface routes, or the intricacy of the highway system that stretches across Canada. And then we discover that researchers Andrew Adamatzky and Selim Akl (2012) have rolled out some oats on a Canadian road map for an experiment. Placing simple amoeba-like slime mould in one city area and the oats on other major cities, they watched as the slime mould reached out for the food source on thin tubes that formed the same network of the Canadian highway system on the map below. This was surprising. Only a unicellular eukaryote (our cell type), the slime mould is also able to form an efficient network comparable to the real-world infrastructure network of the Tokyo rail system. So, human beings and slime mould both produce the same nature-inspired design.

But then, like microbes, humans multiply when the food supplies increase. Consider satellite images of Spokane, Washington, that show the city's growth patterns are similar to microbial growth patterns (Margulis 1986, 229). Spokane, often translated as "children of the Sun" or "Star People," received its name from the Native American Spokane tribe, the original hunter-gatherers inhabiting northeast Washington, northern Idaho and western Montana. Their name reminds one of the ancient Egyptian, early Chinese, and Navajo quest to become the Morning Star.

Beings as Balls and Vortices. Consider dimensions and balls. Semiophysics addresses the quest for significant forms, supporting a general intelligible theory transcending human life, according to mathematician René Thom. Now, Thom states that in dimension one the form of a ball is simply a sound. In dimension two it is the disk marked by a closed curve, and in dimension three it is a ball inside its boundary sphere. (1990, 3) We see this simple form in microscopic spheres of DNA, proteins, and viral heads, as well as in the macroscopic planets, moons, suns, and human heads. Thom speculates that if living beings were "particles or structurally stable singularities" then symbiosis, predation, parasitism, and sexuality would be the particles' interactions and couplings. If form represents information, as Thom states, which is active as transformative spheres, vortices with horizons and singularities, gravitational waves, and hierarchical uphill protein processes to a native folded structure, then these forms—

as well as our behavior—may be redolent with a biological survival message.

We can see the biological survival value of a ball in the film Cast Away (2000). Actor Tom Hanks plays the role of Chuck Noland, a systems analyst for FedEx, who becomes stranded on an uninhabited South Pacific island after his plane crashes. What saves Noland from madness and loneliness is the unexpected FedEx package that crashes with him. Inside the package is a Wilson Sporting Goods volleyball, a spherical shape very similar to a football and a soccer ball that are truncated icosahedrons. Icosahedral shape is also found in the virus. For example, the ball-like head or capsid of the ancient virus Lambda exhibits this same spherical, icosahedral symmetry. It is possible that the creative inspiration for Wilson as spherical savior is related to our large viral DNA heritage. Either way, in the film Noland falls off his raft, losing his friend Wilson, a ball that saved his sanity but plunged Noland into despondent loneliness when it disappeared on the waves.

Thom states that all animals have the topological simplicity of a three-dimensional ball, except for primitive animals living in colonies. He asks what the a priori reason is for this shape in living beings (1975, 152-153), that is, why are we ball-like and why are we obsessed with balls? A simple answer may be that a ball is the most efficient form of protection in a gravitational world and the lowest energy state. However, the prior knowledge for the structural stability of the human being's ball-like shape may be our genome, those 46 chromosomes with their knowledge of our origin, development, and perhaps our future. In statistical theory, a priori denotes knowledge present before a particular observation is made. Consider human genesis in the womb directed by DNA, and that all organisms start out with a ball-like head and tail. This basic form suggests that our last universal common ancestor (LUCA) may have been a virus with a ball-like head and tail. The Human Genome Project discovered that human DNA is microbial, that is, our cells possess primarily recipes for viral and bacterial proteins and genes. Scientific evidence and observation also support that the information-theoretic character of our world of semiosis (sign activity) is controlled by the rule-governed lifestyles of a viral genome. Behaviors such as ancient Egyptian mythology, Christianity, genetic engineering, to name a few human activities, model the complicated replication cycles of a bacterial virus—bacteriophage Lambda. This actually seems very natural, considering that our DNA is a graveyard of viruses, while our gut microbiome (intestines, stomach, and so on) is made of microbes. Perhaps the world we see is only a pattern cast from the world of microbes.

In light of this, it should not be surprising that the deities in the least-corrupted ancient Egyptian texts are signs for viral and bacterial proteins and genes. According to the Isis Thesis interpretation, these molecular balls are reacting, translating, copying, and transcribing along an ancient gene expression network that we still use today in our cells. Still, we do not recognize our behavior as the viral action of signs denoting this survival pathway in our cells.

Bulls, Bread, Bats and Balls. Another behavioral example is an Egyptian ritual carved on the subterranean walls of the Edifice of Ethiopian Pharaoh Taharqa (663 BCE) at Karnak. The ritual also describes the glycolysis-fermentation pathway and phage Lambda's activities along this pathway, using simple signs such as bulls, bread, bats and balls. Pharaoh Taharqa, one of many ancient Egyptian pitchers, is actually throwing balls in the ceremony. The production of balls is a sign of cloning viral molecules.

Also intriguing is that this ancient ritual in the Edifice of Taharqa describing microbiology and ball throwing is also the origin of our modern game of baseball. Further, most of Taharqa's hybrid viral ritual surfaces one thousand years later in the early Christian Mass at Hagia Sophia, Constantinople, during the reign of Justinian during the sixth century of the common era (CE), supporting that Christianity is showcasing the same complex viral replication process for our survival. Put simply, Taharqa's ritual, the Christian Mass, and baseball actually describe the ancient glycolysis-fermentation pathway and phage Lambda's genetic switch between two viral lifestyles. This microbiology is also the obsession of our modern genetic engineers. Even our motivation for mass producing hybrid cars may be inspired by the viral DNA in our bodies. All this mirrored human-viral behavior—in ancient Egyptian ritual, in early and contemporary Christian Masses, in sports arenas over the centuries, in automotive factories, and in modern genetic engineering labs—may be signs of a survival mechanism imprinted on us by the massive viral component of our genome and our microbiome.

The important message from ancient Egypt is that our cells can save us via microbial gene expression. After all, microbes are our ancestors and they invented fermentation, the proton rotary motor wheel, sulfur breathing, photosynthesis, and nitrogen fixation (Margulis 1986, 95-96). To begin to understand this ancient gene expression survival system in Taharqa's ritual and the Christian Mass at Hagia Sophia a thousand years later, let's first review the evidence showing that Egypt is the origin of baseball, a natural pattern of game activity that ancient Egyptian pharaohs used to express a chemical process.

Spring Training 3

The Ancient Egyptian Origin of Baseball

I give thee all life and power, all stability, all health, and all joy, when appearing as the king of Upper and Lower Egypt on the throne of Horus, like Re, forever.

Creator-god Shu
The Edifice of Taharqa

Viewing ancient Egyptian signs through the lens of new science and human behavior shows that ancient Egyptian deities are signs for viral and bacterial DNA and proteins, functioning along the ancient glycolysis-fermentation pathway in our cells. Because of the complexity of the genetic process, the pharaonic priesthood invented the game of baseball as a model for their microbiology. This suggests that sports warfare (baseball, soccer, football, basketball, and so on) is a sign and natural pattern reflecting this viral genetic survival message. For baseball, this claim is clearly supported by the carvings on the subterranean Edifice of Pharaoh Taharqa. Wearing a ceremonial tail, Pharaoh Taharqa pitches four of the first fast balls in history. Symbolic of clearing away the enemies, the balls are born to protect the risen deity Osiris forever.

Taharqa (690-664 BCE) was a Twenty-Fifth Dynasty Nubian (Ethiopian) pharaoh, whose Edifice exists between the Sacred Lake of Karnak and the south enclosure wall of the Great Temple. The excavation of the subterranean Edifice of Pharaoh Taharqa began in 1907-08, when G. Legrain, under the direction of G. Maspero, cleared the monument at Karnak. By 1969, J. Lauffray had cleared and studied the mysterious nilometer, a long narrow structure connecting to a court leading to the east entrance of the Edifice. Finally, Richard A. Parker, J. Leclant and Jean-Claude Goyon collaborated to produce their 1979 work detailing the architecture of the Edifice, the plates of scenes and texts, and the translation of the texts. Jean-Claude Goyon not only translated the texts, but also provided an interpretation of the Egyptian ceremony that was performed in the ruined building of Taharqa. Goyon notes that the themes date at least until the New Kingdom (1979, 86), but the evidence is that they date back to the Old Kingdom Pyramid Texts circa 2520 BCE and the Sed festival (King 2004).

The ritual carved on the walls depicts Taharqa attired in a kilt with a triangular front and a ceremonial tail as he walks to the west, carrying a round-headed mace of consecration. (see Figure 3.1) The pharaoh proceeds to four sacrificial animals and offers incense. Taharqa consecrates white bread in the hopes of having life forever, and then he walks north. He then enacts the well-known Egyptian ceremony of running with the oars or water jug that relates to good running capabilities for rising with the wind on the night bark journey through the netherworld. The night bark is represented on the walls with a steering oar fixed to a support on its stern. (Parker et al. 1979)

3.1

3.2

Figure 3.1. Like Taharqa, Nineteenth Dynasty Pharaoh Seti I is carrying a round-headed mace. From his temple at Abydos by permission of Wikipedia GNU Free Documentation License.

Figure 3.2. Bovine Goddess Bat from Narmer Palette. Marija Gimbutus (1974) pointed out that circa 6500 BCE the bull was initially associated with the Goddess, becoming a male symbol through syncretism.

Carved on the Edifice wall is a hawk-winged Sun disk described as "He of Edfu, the great god, [lord of] the sky, many-colored of plumage, who comes out [of the horizon, the lord of Mesnet, may he give life!]" (Parker et al. 1979, 26). Taharqa then proceeds to the Chapel for the Rites of the Mound of Djemê, where the carved lintels also show the Wife of God walking in front of Pharaoh Taharqa. The scenes on the west wall also depict the Lady of Bat (49), the bovine goddess or form of the goddess Hathor (see Figure 3.2), who is also on the front and back of the Narmer Palette. A winged solar disk and two uraei with the Egyptian ankh (sign of life) at their necks decorate the West Wall. On the center of the Mound of Djemê is a "vaulted crypt, symbolized by a wide band arching above two outstretched arms with hands open toward the sky" (50) similar to Christ's stance on the cross six hundred years later. This is the Mound of the netherworld deity Osiris, who is rising. Another scene shows a priest bearing a shrine-chest who is being purified with an incense cup. Four priests and a divine votaress called the Wife of God bring the tabernacle of the god Amun-Re-Kamutef. Many scholars believe the Wife of God represents the goddess Isis. Other scenes depict the divine votaress or Isis shooting four arrows to the four cardinal points, as well as the solar rite where Taharqa throws balls from the Mound of Osiris. Parker and colleagues explain the Throwing of the Balls ceremony:

> In the actual rite, the king ran toward each of the cardinal points in succession, throwing one of the four balls in each direction, but, conventionally he is represented as running toward the right, the four successive phases being concentrated into one. . . . The officiating sovereign throws the four balls with his right hand, while his left one holds a mace with a pear-shaped head. An understanding of the physical aspects of the rite and its significance is made possible by some late papyri describing the ball-throwing ritual. They tell us that the four balls are made of clay, that they have the names of the protective divinities inscribed on them, and that one has to be thrown toward each of the cardinal points after the curses appropriate for that quarter of the world have been pronounced over it. The tutelary deities whose names are inscribed in the clay are paired off: south ball, Amun and Montu; north ball, Shu and Tefnut; west ball, Neith and Wadjyt; east ball, Sakhmet and Bastet. These are the principal deities to carry out the curse on the enemy coming from one of the four directions. The magic bearers of their names, the balls, are defined by the texts as the protections of Re, born of him or as the balls come into existence for Re. . . . Like the shooting of the arrows, the throwing of the balls is necessary to purify the universe for its welcome of Amun as he is reborn . . . (1979, 62-63)

In this Egyptian ritual, a blood sacrifice is made, incense is offered, bread is consecrated, a tabernacle is presented for the god's Divine Re-entrance or resurrection, and two arms are stretched out like Christ's arms on the cross. Then, the Wife of God (Isis) shoots four arrows with her bow, and the king conquers the enemy with his mace-bat and multiple balls that are fired from the Mound of Osiris to the four quarters of the world in an act of purifying the universe.

The Bull Pen of Pitchers. The origin of baseball is ancient Egypt. On the dual-sided Narmer palette (circa 3000 BCE) carved from grey-green siltstone, the torso of the pharaoh wearing the White Crown is to the front with his right arm lifted and ready to strike his enemies with a mace (see Figure 3.3). Above this striking pharaoh are two heavily-horned bull heads with two additional bull heads on the other side of the palette. Four bull heads suggest the four quarters of the sky (Campbell 1962) and the bovine goddess Bat, a form of Hathor the bull goddess. This reminds one that baseball diamonds are often set up today relative to the four directions of space and each team has a bull pen for pitchers. Also, the Greek historian Herodotus, who visited Egypt in the fifth century BCE, said that a flash of lightning descended upon a special cow, causing her to receive Apis, a black bull with a white diamond on its forehead, an eagle image on its back, double hairs on its tail, and a scarab beetle under its tongue (Brown 1992). Herodotus wrote that Apis was the "calf of a cow which is never afterwards able to have another" (1996). This means that each special calf descended from a cow struck by lightning that was incapable of conceiving another offspring, a virgin cow. In genetics, this type of lightning, asexual virgin breeding for specific traits is called cloning or forming genetically identical calves by nuclear implantation. As we shall see, viral replication or cloning of spherical forms describes the chemistry of ancient Egypt.

Figure 3.3. Narmer Palette circa 3000 BCE. Public Domain.

We meet this image of a striking deity again in Hour Two of the ancient Egyptian Book Amduat (c. 1540-1292 BCE). Figure 3.4a shows the striking god called "Powerful Arm, He who smites his enemies" (Piankoff 1954, 242).

Figure 3.4a. In Amduat Hour Two the striking deity is second from right. Public Domain. Figure 3.4b. Studio Photo of Philadelphia pitcher from New York Public Library's Spalding Collection donated in 1921 by early baseball player and sporting-goods tycoon A. G. Spalding (whose name to this day is printed across every ball used in the National League). The photo is dated to the 1870s or 1880s. No known copyright restrictions.

Pharaoh Taharqa exhibits the same pitching posture of the striking deities on the Narmer Palette and in the Amduat text. Jean-Claude Goyon, who translated the hieroglyphs carved on the walls of the Edifice, notes that the underground building was consecrated to primeval deity Amun in his ancient aspect as the god *Dsr-* or "He Whose Arm Is Sublime" (Parker et al. 1979, 80-86). The text for the Throwing of Balls rite also states that the king throws the balls when he halts at the Mound of Osiris (65). So, the ancient king pitches from a pitcher's mound just like our modern baseball pitchers. However, this is the pitcher Mound of Osiris, who all the potent noble dead are associated with in the Coffin Texts (Goelet 1994, 140).

In addition, the image in Plate 25 of Taharqa with his right arm raised, his legs spread to throw, his left arm stretched out (Parker et al. 1979), is the same image of "Powerful Arm, He who smites his enemies" in Amduat Hour Two (see Figure 3.4a). However, in ancient Egyptian texts and the Edifice of Pharaoh Taharqa, arms have a special genetic significance in the pharaoh's netherworld journey, as they do for our modern genetic engineers who also speak of genetic arms on DNA molecules. According to the Isis Thesis interpretation of ancient Egyptian texts as decoded through the lens of contemporary science, the Edifice's artwork in the Rites of the Mound of Djemé (Room E, west wall) of two arms outstretched like Christ's stance on the cross is the sign for the two arms of the viral genome (DNA or chromosome) of phage Lambda.

Now, Lambda lifestyles happen in the tiny quantum world. So, the subterranean Edifice of Pharaoh Taharqa symbolizes his journey through the quantum world described as the "regions of the dwarfs" (Coffin Text 132). The ongoing echo of this tiny dwarf world is the modern baseball dugout, which is nested in a modern baseball stadium that is akin to a large concave underground hole. Further, the king in the Narmer Palette swings a white mace or heavy club similar to a modern bat (see Figure 3.3). Actually, King Narmer's mace is the identical shape of bacteriophage Lambda, whose morphology is a triangulated icosahedral head enclosed in a sphere (see Figure 3.5). As many observed in December of 2006, the Chicago Field Museum Exhibit of King Tutankhamun (circa 1334-1325 BCE) and his royal ancestors showcased a beautiful white sculpture of a mace with a triangulated head. The mace or bat of the pharaohs has the iconic morphological signature of the spherical phage Lambda.

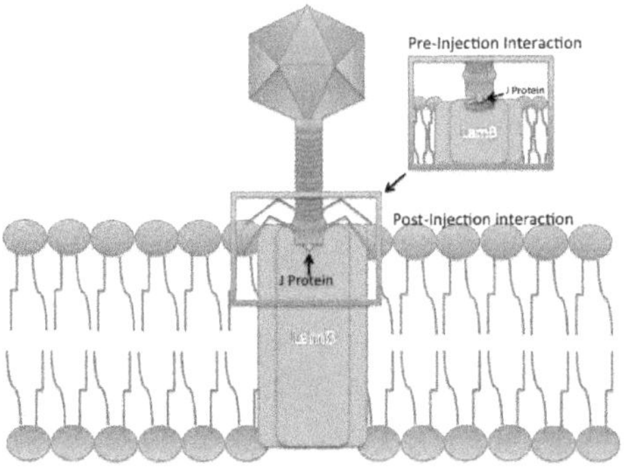

Figure 3.5. Phage Lambda infects a bacterial cell at the maltose LamB porin by means of J protein and other proteins in its tail tip. Like a syringe, it then injects its linear DNA into the host cell. Permission to use by Wikipedia GNU Free Documentation License.

The Genetic Switch-hitter. Probably the strongest connection between the game of baseball and phage Lambda is the competition between its two dominant repressor proteins, c1 (pronounced c-one) and cro proteins that control Lambda's two lifestyles, respectively, lysogeny and lysis. Both proteins compete for three binding sites or gene seats on the right arm of the viral genome, but there are also three binding sites or gene seats on the left arm that help determine the lifestyle of lysogeny or lysis. (See Figure 3.6) Put simply, possession of the right arm's three gene seats allows both c1 and cro proteins to fold and bind to their genes or home plates. These three gene seats are similar to baseball's three bases. Both c1 protein and cro protein have a team of enzymes that cascade into the contest to win the three bases, making a run to the protein's native folded state or home plate. The competition is especially tough because the genes encoding c1 protein and cro protein are adjacent on the Lambda chromosome. So, the proteins and their folding energy landscapes are entangled. Also, cro protein is much smaller than c1 protein, so it is David and Goliath all over again. Cro protein is made of only 66 amino acids, while c1 protein has 236 amino acids (Ptashne 2004, 16-17). It is interesting that cro protein has 66 amino acids, reminding one of the numbers of the beast (666) in the Book Revelation (13:18).

Also, the two competitive proteins are transcribed in opposite directions: c1 protein is transcribed leftward (right to left), while cro protein is transcribed rightward (left to right). If c1 protein and its team of proteins capture the binding bases first (due to leftward transcription by an RNA polymerase activator), the virus becomes dormant within the cell for the lifestyle of lysogeny. Yet, if cro protein and its team capture the three binding bases due to rightward transcription, the prophage rises from its dormant state and takes over the replication machinery of its host to clone its own kind through the lifestyle of lysis. This genetic contest between c1 and cro proteins results because the proteins are following the folding/binding instructions from their respective genes, *c1* gene and *cro* gene. This molecular battle is a game similar to baseball due to all the additional proteins involved in the competition. If either primary protein wins or runs the three bases for a Home Run, that protein folds to its native state and activates its gene for the lifestyle it controls. This genetic Home Run to the native state by the protein switches on the gene or home plate. All that is necessary for victory is to capture the three bases or gene seats on the right arm of the Lambda genome.

In ancient Egyptian texts, knowledge of this ancient gene expression network allows the dead king's resurrection to the stars in spherical form for continuity of mind or DNA via the lifestyle of rightward transcription or lysis. So, little cro protein with 66 amino acids represents the chemiluminescent crystal chemical reaction and product. What jumpstarts lysis controlled by Lambda cro is the damaging mutagen of ultraviolet light on the host cell. By anticipating the future through a viral model, the pharaonic priesthood tapped into information about the human species that was encoded into human DNA, but they concealed the information from the plebians. Consider Figure 3.6, a diagram of a baseball diamond that shrinks into a dual diagram below of the Lambda chromosome (DNA), describing the microscopic gene-bases of competing c1 and cro proteins and their genes.

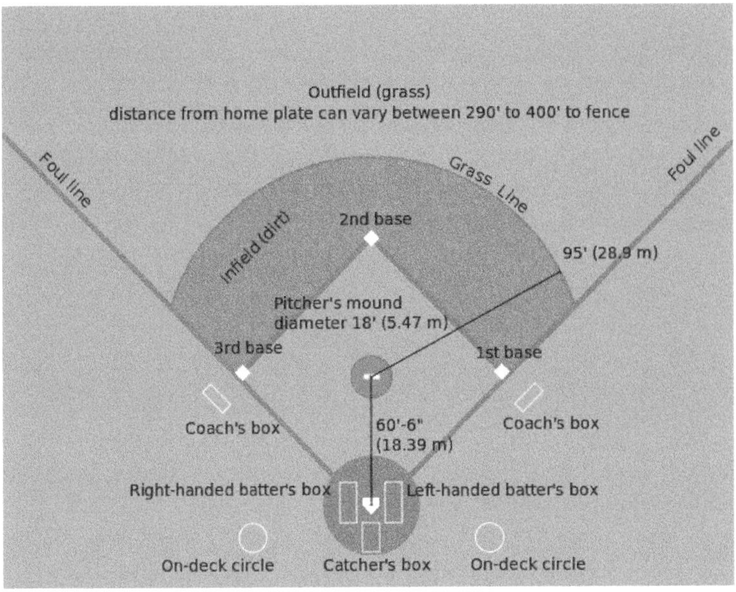

Figure 3.6. Permission to use baseball field from Wikipedia GNU Free Documentation License. Below is the Lambda chromosome, showing adjacent *c1* and *cro* genes.

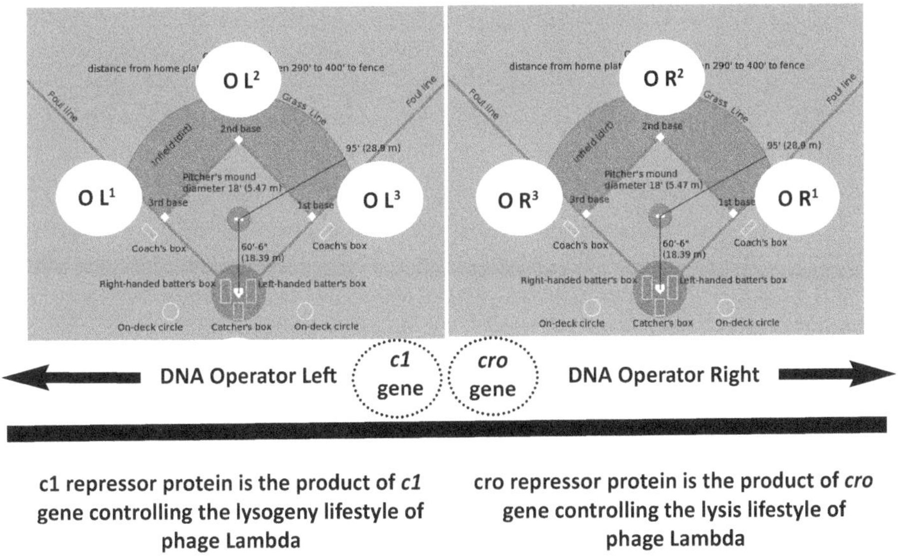

Figure 3.6 is a section of the Lambda DNA or chromosome, showing the three left operator sequences and three right operator sequences that activate either leftward transcription of the *c1* gene or rightward transcription of the *cro* gene. The proteins battle over control of two promoters or series of three leftward operator ("O") sequences OL^1 OL^2 OL^3 and three rightward operator sequences OR^1 OR^2 OR^3 that determine lysogeny or lysis. Ancient Egyptian texts name these operator sequences the "Mansion of the Six." Both proteins bind to the right side but in a preferential order:

c1 binds the right in the order of OR^1 OR^2 OR^3 while cro binds OR^3 OR^2 OR^1. Egyptian texts describe this protein battle as a competition between two brothers, Seth (c1) and Horus (cro). If c1 protein or Seth binds the OR^1 first, then it captures the entire operator sequence, stimulates its own synthesis, and stops the "rightful order" (Maat) of cro or Horus. Lysogeny is the result. Yet, Coffin Text 1181 states, "I am sitting on the Eye of Horus as the first of the three," indicating victory for Horus (cro) by winning OR^1 first. How does cro protein or Horus win? *In vivo*, if the cell DNA is badly damaged by ultraviolet (UV) irradiation, the cell triggers its SOS response, which activates the host cell protein *E. coli* RecA. Put simply, this enzyme destroys the c1 protein and cro protein wins. So the coming of the light is crucial.

Now, in the early days of baseball, runners often went around the bases in the opposite direction like cro protein's OR^3 OR^2 OR^1 binding process, and runners could also be struck out by being hit with the ball in a manner similar to Pharaoh Taharqa striking out his enemies with clay balls. Today the rule exists that a runner cannot run the bases in reverse order in an attempt to confuse the defence. When New York Mets outfielder Jimmy Piersall hit his 100th Home Run on June 23, 1963, he ran the bases in the correct order, but facing backwards. Many considered Piersall's behavior eccentric. Perhaps his inspiration came from the ancient Egyptian game, where deities walk backwards to signify time reverse to the origin or the protein's home plate.

In light of these analogies, accepting that folk games like Rounders and Cricket in England were the origin of baseball neglects ancient Egyptian sources of knowledge and misses the genetic evolutionary meaning of ancient signs decoded by the Isis Thesis. That interpretation viewed through the lens of new science supports that ancient Egyptian deities are signs for viral and bacterial DNA, functioning in the ancient gene expression network of phage Lambda in the cell known as the glycolysis-fermentation pathway. This microbiological rationale for baseball or sports warfare is supported by the carvings on the subterranean Edifice of Pharaoh Taharqa. Wearing a ceremonial tail, Pharaoh Taharqa throws one of the first fast balls in history. Symbolic of clearing away the enemies, the balls are born to protect the risen Osiris forever, and the action of these signs support the dynamics of the ancient viral/bacterial gene expression network for the lytic cycle as defined by modern experiment and observation of phage Lambda and its *Escherichia coli* (*E. coli*) bacterial host cell. This same genetic pattern of sports warfare is evident in the Mayan culture.

The Mayan Hero Twins and their Game of Death. In the Quiche Mayan Popol Vuh (c. 1701 CE), the Hero Twins suggest the same genetic activity of two viral protein teams that descend into the underworld to play a combination soccer-basketball with the gods. One twin is decapitated, but saved by resurrection from the dead and ascension to the stars. As in Egypt, the oldest twin is reborn as a chemiluminescent Morning Star, just as in ancient Egyptian, early Chinese, and Navajo texts (see King 2004; 2009; 2009a; 2009b), while the other becomes associated with the full moon. The game itself deteriorated into a battle for life and death similar to the one described by Sir James George Frazer between the old king and new king in *The Golden Bough*. At Chichén Itzá, scenes of human sacrifice decorate the walls, in which the winning players decapitate the losers. This mythopoeic game mirrors the Egyptian brother rivalry of Seth and Horus, that is, Lambda c1 protein for lysogeny (Seth) versus cro protein for lysis (Horus). As mentioned, the proteins battle over control of two promoters (series of operator sequences or gene seats), determining lysogeny or lysis. Cro and c1 are produced from opposing transcripts that both originate from the rightward operator region. The different protein levels vary as measured by the number of protein molecules in each concentration, while Lambda switches through forty possible gene states. Like a basketball game, cro and c1 battle to produce protein molecules (balls) within a binding process guided by funneled energy landscapes (basins of attraction or baskets). Lambda c1 forms via the induced-fit mechanism, whereas cro uses the fly-casting mechanism, which might be compared to a slam-dunk versus a jump-shot.

Proteins exist in the cell as parts of multicomponent assemblies, a higher-order network of interacting protein complexes, what one might call a team. Molecules have form, function and fluctuation in the same way that players have a formation, a task, and ability to change formations and tasks. Like balls into baskets, protein molecules are produced in random bursts as protein copy number increases. In fact, the entire game or random burst activity is determined by the number of balls replicated in baskets just like the c1 versus cro protein molecule competition for operator sequences determining lysogeny or lysis—a viral DNA-binding battle from which sports warfare arose. This viral DNA-binding battle may also be our biological motivation for war, a disconcerting thought considering human attraction to sports warfare.

Again, in the Edifice of Taharqa, the viral genome is symbolized by the artwork of two arms stretched out like Christ's stance on the cross. On the other hand, the sign of the Egyptian Ka is two arms stretched up to the sky. This indicates the ascension of the dead king into the heavens and the king's transformation into the golden child Horus, who embraces the cosmos in a quantum of crystallized light. The two upward arms represent both arms of the Lambda genome enabled for its full function. The brothers Horus and Seth are reconciled, for both proteins are cooperating in a binding process that is necessary for replication to occur. And so, the deceased king as the transformed golden child Horus has chemically embraced the cosmos by means of the two arms of the Lambda genome that connote a quantum experience of cosmic unity at the origin of space-time.

Today, microbiologists also speak of Lambda's two arms that function for the competing lifestyles of lysogeny ruled by c1 protein and lysis ruled by cro protein. In the Isis Thesis interpretation, ancient Egyptian references to arms and artwork depicting arms are signs of the Lambda genome (DNA or genes and proteins), while upward arms point to the joint function of the left arm of Seth cooperating with the right arm of Horus for the transformation of the dead king. What is ascendant for continuity of mind and DNA is the lifestyle of lysis ruled by cro protein that is facilitated by the cooperative interactions of c1 protein. So the brothers Horus (cro) and Seth (c1) cooperate peacefully, and this is the genetic Secret of the Two Partners at the Egyptian Sed festival.

The Sed festival from the Old Kingdom (c. 2520-2360 BCE). At the Sed festival, the opening rites in honor of the bull-goddess Hathor involved the procession of the king following four standards from temple to temple. According to Joseph Campbell (1962, 75-76), in the middle of this ritual, the king disappears into the palace chapel, reappearing with his right hand holding a flail scepter and his left hand holding a small scroll called the Will, the House document, or Secret of the Two Partners. The king then states:

> I have run holding the Secret of the Two Partners, the Will that my father has given me before Geb. I have passed through the land and touched the four sides of it. I traverse it as I desire.

With the ancient king's description of his Home Run, the game of baseball between two teams is the perfect medium to explore the argument for and against the Isis Thesis. The evidence suggests that the game originated in ancient Egypt to explain the ancient gene expression network of bacteriophage Lambda and its *E. coli* host, where the genetic switch to lysis, activated by the light, requires the cooperation of two repressor proteins that allow the dead king's transformation for continuity of mind in the cosmos as a Morning Star, that is, a quasi-hybrid being described as crystallized mind.

In the least-corrupted Pyramid and Coffin Texts, this secret knowledge of life beyond death at the molecular level of DNA was carved in the pyramids of the pharaohs and on the bottoms of the tombs of the nobility. This genetic knowledge with ritual provided the pharaohs and nobility with a godlike spiritual experience, and so the knowledge was glorified, and humans were elevated as deities, an apotheosis that was reserved for the pharaohs and their élite hieratic priesthood. Following the instructions in this hidden network of signs describing an ancient gene expression pathway, the deceased king became a member of the potent noble dead. This science provided Egyptian pharaohs and their priesthood with answers to profound ontological questions, such as who are we, is there life after death, and how does our universe function. Buried in pyramids and tombs for centuries, the wisdom of the Egyptian élite remained masked even though Jean-François Champollion cracked the code for Egyptian hieroglyphs in 1822. Although this event was an advance, nineteenth and twentieth century scholars struggled with little success to interpret the hidden meaning of the hieroglyphs and artwork.

Although the twentieth century exploded with the counterintuitional magic of quantum mechanics, traditional egyptologists still view the ancient Egyptian texts as primitive, confusing and unintelligible. Today, quantum mechanics and quantum biology can explain the problematic magic egyptologists encounter in the funerary texts and rituals. New science is the magnifying glass to interpret the quantum mysteries of the least-corrupted genetic texts. This strategy reveals that the actions of the deities are signs defining the ancient gene expression network of the creator-virus Lambda in its bacterial host *Escherichia coli* (*E. coli*). This ancient gene expression network is the glycolysis-fermentation pathway, also known as the maltose transport system. The pharaonic priesthood believed that this ancient chemical pathway would allow continuity of mind and DNA at a human death transition, that is, if the person had knowledge about the ancient viral gene expression network for transformation to a Morning Star, what can be described as the crystallized emergence of a cold-light quasi-hybrid being.

Certainly, the modern terms virus and bacterium were not used by the pharaonic priesthood for their deities. However, their least-corrupted texts describe the ancient glycolysis gene expression network and lactose metabolism down to minute details (see *The Isis Thesis*). Again, the hidden status of this quantum biology in pyramids and tombs generally preserved its scientific purity, preventing the infusion of contrasting notions or variant ideologies from other cultures, although these still emerged. Remarkably, ancient Egyptian science mirrors the detailed knowledge of our new scientific discoveries verified by experiment and observation, as *The Isis Thesis* and twelve articles explain.

It's unbelievable how much you don't know
about the game you've been playing all your life.

Mickey Mantle

Spring Training 4

The Christian Mass at Hagia Sophia

I Jesus have sent my angel to you with this testimony for the churches.
I am the root and the offspring of David, the bright morning star.

Revelation
22:16

According to one translator, the name Hagia Sophia refers to a partial inscription in its south tympanum dedicated to "deathless Wisdom" (Downey 1959, 39), while other research indicates the name was dedicated to the Sophia of God as the second person of the Trinity (41) with Christ as the Wisdom or Word of God made flesh (Mainstone 1988, 133; Fiene 1989, 450). Christian authors used the term wisdom as "the natural wisdom of the universe" and the Greek Trinitarian understanding of the true Wisdom of Christ, that is, "Wisdom not only as a divine Person but also as a divine manifestation or attribute, or 'energy'" (Meyendorff 1987, 391-392).

In 360 CE Constantius dedicated the first Church of Hagia Sophia at Constantinople, which was most likely a simple rectangular basilica with a timber roof and galleries. Due to fires in 404 and 532, Justinian redesigned the structure between 532 and 537, constructing a two-storied domed structure. The architects were Anthemius of Tralles, a mathematics teacher, and the elder Isidorus of Miletus, professor of geometry. Their architectural plan of combining arched and part-spherical forms flows into the great spherical domed interior, symbolizing the vault of heaven (Mainstone 1988, 157). According to a description by the Greek poet Paulus Silentarius, countless oil lamps blazed like stars in three rings of light suspended from the dome by brass chains with "attached silver discs." Silentarius adds: "Yet not from the discs alone does the light shine at night, for in the circle you will see, close to the discs, the symbol of a mighty cross with many eyes, and in its pierced back it holds other lamps . . ." (quoted in Mainstone 1988, 237 from *ecclesia*).

Hagia Sophia has a northwestern entrance and is oriented to the southeast (Mainstone 1988, 6). The liturgy of Hagia Sophia begins with the principal celebrant entering from the northwestern narthex, preceded by a deacon carrying the Gospel representing

Christ. They proceed to the southeastern ambo, then the solea (walkway) to the sanctuary. Placing the Gospel on the altar, the principal celebrant then prepares for the Eucharist proper, as the catechumens are dismissed and the doors closed, while the deacons go to the skeuophylakion, a separate building or sacristy at the northeast corner of the Church, for the preparation of the altar gifts. The Cherubic Hymm to the "life-giving Trinity" is chanted for the procession of gifts back to the altar from the skeuophylakion, followed by the congregational Kiss of Peace and Communion. (1988, 226-228) Patriarch Germanus in his *Historia Ecclesiastica* writes "the preparation of the gifts, which takes place in the [interpolation: sanctuary or in the] skeuophylakion, stands for the place of Calvary where Christ was crucified" (Taft 1975, 186). The ceremonial carrying of the Gospel to and from the ambo came to be known as the Little Entrance, and the offertory procession of bread and wine symbolizing Christ and his Crucifixion was the Great Entrance (Mainstone 1988, 231). Thus, the Christian ritual includes the Little Entrance of the principal celebrant entering from the northwest and then meeting the Great Entrance of deacons with the altar gifts from the skeuophylakion, signifying where Christ was crucified on the cross. This ritual models the same pattern in the Book Amduat, the Book of Gates, the Book of Caverns, and other ancient Egyptian texts, where the Egyptian Sun-god (Little Entrance) travels to the hidden sanctuary of Osiris, who rises (Great Entrance) to meet the Sun-god.

Edifice of Pharaoh Taharqa. Similar to the Little Entrance, the excavation of the subterranean Edifice of Pharaoh Taharqa (c. 690-664 BCE) in 1907-08 at Karnak reveals ritual scenes and hieroglyphs depicting the king's netherworld descent into the "caverns of Nun" at the northwest stairway, his purification, and approach to the sanctuary, as well as his consecration of "white bread" (Parker et al. 1979, 11-18). Similar to the Christian Great Entrance, the scenes in Room E show the king's procession with text describing the "great entrance of the cavern of the Nun at the west of Thebes." A winged disk decorates the top of the door. This artwork represents the sacred Mound of Djemê, the mythical burial place of Osiris. A mound is present above two outstretched arms with hands open to the sky. A priest celebrates, his chest decorated with two crossed strips. Two goddesses are present, each holding a bow and arrow. Goyon explains that the Sun-god visits the sacred mound of Djemê, related to Osiris' "Divine Reentrance" for transformation (1979, 48-51). The partial texts on the West Wall of Room E are translated: "The Eye of Horus," "The Wife of the God," "The Lord of the Sublime Arm," "The arm of Geb," "The arm of Horus," and "The Lady of Bat," along with the king is alive with "all life, all stability and power, all health" (52-53). So, Osiris has risen for his "Divine Reentrance" and the king has transformed.

Briefly interpreted, the "cavern of the Nun" represents the host cell DNA or chromosome, and white bread indicates fermentation along the cell's ancient glycolysis pathway. The Mound of Djemê is the prophage site, where the dormant virus is getting ready to rise through its DNA-cross to begin its activity. The priest with the cross on his chest is a sign of the cruciform structure of a Holliday junction (DNA-cross), a site-specific recombination process responsible for integration and excision of bacteriophage genomes into and out of the host cell chromosome. Excision allows the prophage to escape from the dormant state. The Eye of Horus is the complete Lambda genome that is now functional, for both arms partner for lysis—the left arm of Geb or Seth (lysogeny) and the right arm of Horus (lysis). A tabernacle is presented for the god's "Divine Reentrance" or resurrection. A lintel on the next room's east wall shows Taharqa throwing four balls, which the texts define "as the balls come into existence for Re," for the Sun-god Re "created himself in the form of millions" (1979, 62-63; 74). This suggests lytic replication and the cloning of the Sun-god into molecular viral balls of light.

The pagan ritual dedicated to Osiris in the Edifice of Pharaoh Taharqa circa 663 BCE is the forerunner of the early Christian Mass performed during the reign of Justinian about a thousand years later at Hagia Sophia in Constantinople. Despite the thousand year time difference, both the Egyptian ritual and the Christian Mass at Hagia Sophia begin with an entrance in the northwest and a re-entrance, both consecrate white bread and symbolize a blood sacrifice, and the processional events appear in the same order. In the Christian Mass, the bread or host is multiplied for the Communion of the Faithful, while in Taharqa's ritual, the balls are thrown or multiplied into millions, suggesting replication or cloning.

Bread and Fermentation. In both Egyptian and Christian rituals, transubstantiation, an act that changes the form or character or substance of something, results in the king consecrating white bread as the dying/rising god Osiris that becomes the Sun-god Re "in the form of millions," and the priest changing white bread or the host into the dying/rising body of Christ that is multiplied for the Communion of the Faithful. The rituals are focused on morphogenesis or the development of a new spherical form. Further, the circular host is the "body of Christ, consecrated bread," with an etymological meaning of "sacrifice" from the Latin *hostia* and "multitude" from Old French *host* "army." From a microbiological perspective, this ritual of multiplying bread into hosts or millions mirrors the process of producing identical copies of a DNA segment asexually, that is, cloning, which involves inserting a recombinant DNA molecule into a fast-replicating virus vector. So, the concept of the god-man may signify viral-human recombinant DNA, with the crucifixion/resurrection patterning the microbiology of phage Lambda (DNA-cross/lysogeny/lysis). Both Christian and Egyptian sources suggest this is the eschatological survival message for humanity.

To understand the importance of bread and fermentation, consider that the ancient glycolysis-fermentation gene expression pathway described in Egyptian texts is the oldest enzyme pathway known. Using various sugars (glucose, maltose, lactose), this enzyme pathway runs with or without oxygen in our cells. Maltose is a sugar used in brewing beer: barley is mashed to convert starches into fermentable maltose. While fermentation is used for the production of beer and wine, yeast and bacteria in dough cause fermentation in bread. These signs in both rituals suggest that ancient Egyptian texts and later Christianity are describing the ancient glycolysis-fermentation gene expression network as a survival mechanism at a death transition. More evidence for this view follows.

The Lactose Connection. After the Gospel in the Catholic Mass, the Credo (Creed or summary of Catholic doctrine) is recited. The Credo explains that Christ came down from heaven and "became incarnate by the Holy Ghost of the Virgin Mary: and was made man." Again, the virgin birth should be deflowered, for it is simply a sign of asexual cloning using lactose metabolism. With great simplification, if there is no glucose for the cell, the ancient glycolysis pathway allows the cell to use lactose, that is, allolactose, a lactose isomer formed by β-galactosidase. This is what happens in ancient Egyptian texts, when the Sun-god's raft of proteins enters the cell and activates the SOS response that stops normal cell processes. Because an increased rate of change is now possible, biologists call this a directed or adaptive mutation due to using lactose for growth and replication. This adaptive mutation shifts the position of many millions of particles within the cell. However, biologists are not sure how these directed mutations happen, yet they do know that these mutations only occur in cells that are not dividing, they are time-dependent, and they only appear after the cell encounters a selective pressure. McFadden and Al-khalili (1999) suggest that adaptive mutations with lactose occur because cells have measuring devices inside that probe what is happening. At temperatures close to absolute zero, the behavior of all systems becomes quantum mechanical, and quantum measurement influences the dynamics of quantum systems.

> In the quantum Zeno effect, continuous measurement of a quantum system freezes the dynamics of that system. . . . In the inverse quantum Zeno effect, a dense series of measurements of a particle along a chosen path can force the dynamics of that particle to evolve along that path. (1999)

In McFadden and Al-khalili's article the cell environment is the observing-measuring entity. This is similar to the ancient Egyptian textual descriptions of a quantum environment of proteins or particles observing and measuring the deceased Sun-god's progress on a viral protein raft or bark with Isis, the sign for lactose. In the ancient texts, this viral raft or bark of proteins is activating the lactose genes and taking over the replication machinery of the cell, so the cell is not dividing, and the viral raft is also an observing-measuring entity. So the proteins are the unit of selection.

The milk-goddess Isis, the Wife of God, and the later Virgin Mary are signs for the lactose metabolism that is available in the ancient glycolysis gene expression network. The disaccharide lactose is the predominant sugar in milk, and humans have been preoccupied with milk production since Neolithic culture developed in the Middle East. According to archaeologists, chemists, and geneticists, dairy products have shaped European human settlements. When farming replaced hunter-gathering about 11,000 years ago, cattle herders reduced lactose in dairy products by fermenting milk to make cheese and yogurt. Several thousand years later, a mutation spread through Europe that adapted people to the enzyme lactase and the consumption of milk, what some researchers believe to be consistent with gene-culture co-evolution. The researchers think that humans may have domesticated cattle, goats and sheep for dairy purposes, and that the nutritional benefits allowed farmers to replace the original hunter-gathering societies. (Curry 2013) Maintaining cattle also required growing grain. In ancient Egypt, the netherworld dying/rising deity Osiris was associated with grain and agriculture, while the pharaohs entombed their sacred Apis bulls or cattle they considered as the incarnation of the creator-god Ptah.

It is also interesting that centuries later Christianity converted the ancient Egyptian signs of the deified human, the lion, the calf, and the eagle into the iconic symbols of Matthew, Mark, Luke, and John, while the virgin Isis materialized as the virgin Mary. The Egyptian cow-goddess Isis and lactose are closely associated with the grain deity Osiris and grain production or cloning. Perhaps we can interpret the cultural obsessions with virginity, cows, milk, and cloning grain as additional signs related to the ancient gene expression network in our cells allowing lactose metabolism for asexual vegetative replication or viral cloning. The Pyramid Texts (PT) and Coffin Texts (CT) point to this idea in the countless Egyptian references to the Milky Way and the mother goddess Isis, whose provisions are powerful (CT 246). So it is that "cream" or *zmn* (PT 34) or *smyn* (CT 354) reconciles the two brother deities Horus and Seth (cro and c1 proteins) by providing the lactose medium for the complete Lambda genome or Eye of Horus. This binding partnership of proteins supports the genetic switch for viral replication and lactose metabolism. So, the élite Egyptian pharaonic priesthood and the early Christian priesthood are tuned into the same microbiological song that may be inspired by our microbial genome, as well as our gut microbiome.

Anointed with Cream. The etymology of our words shows a similar correspondence to lactose metabolism in Christian ritual. Related to language's biological roots is the word cream from the Greek *chriein*, which means "anoint." *Chriein* is also linked to the root of Christ ("the anointed one"), associating ritual with oil, an element of cream caused by the globules becoming more concentrated than usual. (McGee 1984) From the Greek, *christós* means "anointed," a translation from the Hebrew *māshīah*, "anointed, messiah." Now, DNA is a dynamic crystalline structure, proteins fold to a crystallized native state, and a virus, such as phage Lambda, is a crystal having a spherical shape. From the Latin, *crystallus* means "crystal" and *chrysos* means "gold" and *Christus* means "Christ." In addition, chrysalis is the third stage in the development of an insect, especially a moth or a butterfly enclosed in a firm case or cocoon, a pupa. The word "chrysalis" originated from Latin *chrysallus* and Greek *khrusallis* meaning gold-colored pupa of a butterfly. These etymologies are charged with microbiological messages, and this is probably because our viral DNA possesses lingual structures with a survival message. Put simply, a chrysalis or cocoon stage suggests the lysogenic inert state of phage Lambda. So, Christ's actions and name connote that the deity is a sign of a viral crystal "anointed" with the creamy oil of lactose, so it can rise from its inert state.

Of course, the dying/rising god Christ is one of the later versions of the dying/rising deity Osiris who rises from the dead due to the "breath of life" from Isis, that is, lactose energy metabolism. Similarly, as supported by etymology, the Christian emphasis on the Word, the Epistle and the Gospel in the Byzantine Mass at Hagia Sophia can be interpreted as signs of DNA transcription along the ancient gene expression network as it is in ancient Egyptian texts. Examining the concept of the Word in the least-corrupted Pyramid Texts (PT) and Coffin Texts (CT), the king "bears the god's book" (PT 250), speaks "the word" (PT 460), and is "the Great Word" (PT 506). Also, Osiris "is the word which was in darkness" (CT 1087), and the king is "in charge of the record of the word of God" (CT 351). The king is also "the messenger" (CT 422). (Faulkner 1969; 1973-78) The deity Thoth is often in ibis form, carrying a pen and palette of the scribe, representing the activity of copying words or transcription. This focus on the Word by pharaonic Egypt precedes its Christian counterpart found in the Gospel of John (1:1): "In the beginning was the Word, and the Word was with God, and the Word was God." Put simply, the Word is viral DNA.

The Byzantine Mass also emphasizes the Word and its letters by its ritual of the Epistle (Latin *epistula*, "letter") and Gospel (Old English: *godspel*, "good news"; spell, "name the letters of"; O. E. *spellian*, "to tell, speak"). This Latin and Old English derivation harmonizes with the language biologists use to explain the conversion of nucleotide words of DNA into amino acid words. Put simply, the chemical bases or letters of adenine (A), thymine (T), guanine (G), and cytosine (C) spell out the genetic code or word of a protein or a string of amino acids. So, the messenger RNA (message or letter, Epistle, Sun-bark with dead king) translates into a protein sequence of amino acids (letters specifying a specific protein or word, Gospel, cro protein) that activates genes, allowing the Lambda prophage to rise (Osiris' divine re-entrance) through lactose metabolism (Isis). Similarly, the Great Entrance of Christ in the Byzantine Mass also models this chemistry, for after the Epistle and Gospel is the offertory procession of bread and wine, symbolizing Christ's Great Entrance after his Crucifixion and Death.

Mirror Mirror on the Wall. Now, Taharqa's ritual is rooted in the ancient Egyptian Sed festival circa 2500 BCE, which Joseph Campbell sums up in six general stages (1962, 75):

1. Preparatory vestings, blessings, and consecrations
2. Introductory processions
3. Rites approaching the consummation
4. The consummating sacrifice (or its counterpart)
5. The application of the benefits
6. Thanksgiving, final blessing, and dismissal

As mentioned, the king returns with the Will or Secret of the Two Partners during the fourth stage. During the fifth stage of the Sed festival, the king is carried in a litter-shaped basket to the chapel of the brother-deities Horus and Seth for the bow and arrow ceremony, where the king releases an arrow in each of the four directions and then is crowned four times (1962, 78). Now, the flow of events in the ancient Egyptian Sed festival is remarkably similar, not only to the early Christian Mass at Hagia Sophia, but also to the modern Christian Mass, described below in the New Marian Missal for Daily Mass (Juergens 1953):

1. Preparatory prayers at the foot of the altar
2. Introductory procession of priest to altar and Introit or "entrance"
3. Rites of Collects (prayers), Epistle (letters), Gospel, Credo
4. Offertory/consecration of bread and wine; sacrifice of Christ
5. Holy Communion; consumption of the Body and Blood of Christ
6. Thanksgiving, Blessing and Dismissal

In Ralph Gorman's preface to the New Marian Missal, he states that in the Mass the priest and the victim are the same—Jesus Christ. This is similar to the Egyptian deceased king representing Osiris. In the Christian Mass, Christ is sacrificed, while Holy Communion allows the celebrants to partake in the divine sacrifice represented by eating the two-dimensional bread host. Similarly, ancient Egyptian texts describe the degradation or corruption of Osiris, who becomes the necessary substrate or sacrifice for the transformation of the deceased pharaoh or king.

In summary, we are caught up in a wall of mirrors: the early Christian Mass at Hagia Sophia mirrors Taharqa's ritual in his Edifice that mirrors events in ancient Egyptian texts and the ritual at the Sed festival that mirrors the contemporary Christian Mass. All this human behavior models viral replication in the ancient glycolysis fermentation pathway that uses lactose metabolism. This is reinforced by the etymology of words such as "cream" and "Christ," as well as human behavior related to consumption of dairy products and maintaining cattle.

You teach me baseball and I'll teach you relativity . . .
No we must not. You will learn about relativity
faster than I learn baseball.

Albert Einstein

Spring Training 5

Diamondhearts versus Thunderheads

You'll have to say that you will stick with it even though it gets difficult and unpleasant.

David Bohm
Thought as a System

Let's now consider the main claim of the Isis Thesis defined by ancient Egyptian texts, artwork, ritual, architecture and other signs. Using contemporary science as a lens to interpret the religious texts, the comparison reveals that the Egyptian pathway for DNA survival is horizontal gene transfer mediated by a virus that uses the ancient glycolysis fermentation gene expression network in our cells. The main claim is that the fundamental heart of everything is the DNA of a creator-virus known as phage Lambda that functions for human DNA transformation along this ancient enzyme pathway. The ancient Egyptian texts support that both inorganic and organic evolution are anchored by the DNA of this viral worldheart, as is the existence of space and time in our cosmos. Ancient Egyptian mythology (the dying/rising god Osiris, the virgin birth of the child Horus by Isis, the brotherly rivalry, the cross, and the Sun-god's bark), preserved in the least-corrupted ancient Egyptian texts circa 2500 BCE, explains the synergistic merger of human DNA and viral DNA that is transformed after entry into the LamB porin, the docking site for the ferryboat-phage in the ancient glycolysis gene expression network. This is quite a biological coincidence that phage Lambda's docking site is the entry gate to the ancient enzyme pathway, supporting that the viral bacteriophage may be our last universal common ancestor (LUCA).

Surprisingly, the genetic mythology of ancient Egypt mirrors the major tenets of Christianity, and this supports a theory of repetitive, historical human behavior that models viral transformation for survival of human DNA. Ancient Egyptian knowledge, secreted for centuries, reveals that a human death is the opportunity for this ancient gene expression network to be activated for viral lysis or replication controlled by Lambda cro protein. This occurs through cro protein's folding and binding to its native state, one of nature's remarkable reversals. Cro protein follows origin-fixation dynamics—that is, it folds back to the genome or gene-seat. If the Lambda genome is the quantum worldheart of our cosmos as Egyptian texts support, then the DNA has cosmological applications, as we shall see in future chapters. Cro protein's reversal to the native state on the genome places emphasis on distinct individual mutations, while activating viral lytic replication via phage Lambda's genetic switch. This genetic switch is the original inspiration for switch-hitters in the game of baseball that models the microbiology of phage Lambda. Thus, the patterns found in religion and sports warfare suggest that this virus is the original causal pattern of human behavior.

And so, using baseball as a model to understand the potential of the ancient microbial gene expression network in our cells, we will review current research for and against the Isis Thesis, depicting the power grid that has dominated the centuries through two fantasy-draft teams, the Diamondhearts who value the primacy of mind and the Thunderheads who support the primacy of matter. In light of contemporary science, what the mind-versus-matter problem comes down to in the twenty-first century is the idea that mind or the human psyche is somehow connected to the survival potential of the individual's DNA, while matter is a composite of consciousness/brain/body/DNA. The bridging factor here is DNA which is minute matter. However, in the interest of simplification, this argument is simply referred to as either the primacy of mind or the primacy of matter.

Because the Isis Thesis is supported by modern theoretical predictions and scientific experiments, the theory is subject to testing and refutation, despite its speculative aspects. The purpose of Spring Training is to explain this scientific semiotic theory in a simple fashion, while presenting other scientific views. As theoretical physicist David Bohm said in *Thought as a System*, the "flow of meaning between people is more fundamental than any individual's particular thoughts" (1994, xi). With this in mind, the best method for understanding this conceptual argument is the baseball game. This sporting framework will allow a simple flow of meaning or an adversarial partnership of thought between two teams. Generally, the Diamondhearts support the primacy of mind due to quantum mechanics. On the other hand, the Thunderheads favor the primacy of matter due to sensory perception of the macroscopic classical world.

For this ballgame, the Thunderheads realize first, second, and third bases respectively as representing Physics, Biology, and Cosmology, and the Diamondhearts deepen the vision of First, Second, and Third Bases by supporting the challenging new sciences of Quantum Physics, Microbiology or Quantum Biology, and Quantum Cosmology. Perhaps in this way we can understand and unify the whole system of nature proposed by the Isis Thesis and substantiated by reciprocal natural laws (King 2009a; 2011) that provide a basis for the evolution of minute matter and mind via viral DNA.

At first base, the two teams battle over the idea of what is real, the quantum world or the classical world we experience now. Second base is the debate over what is living. Is a virus a living entity and is both the organic and inorganic alive or just the organic? Third base is a struggle to determine the origin of space and time and the evolution

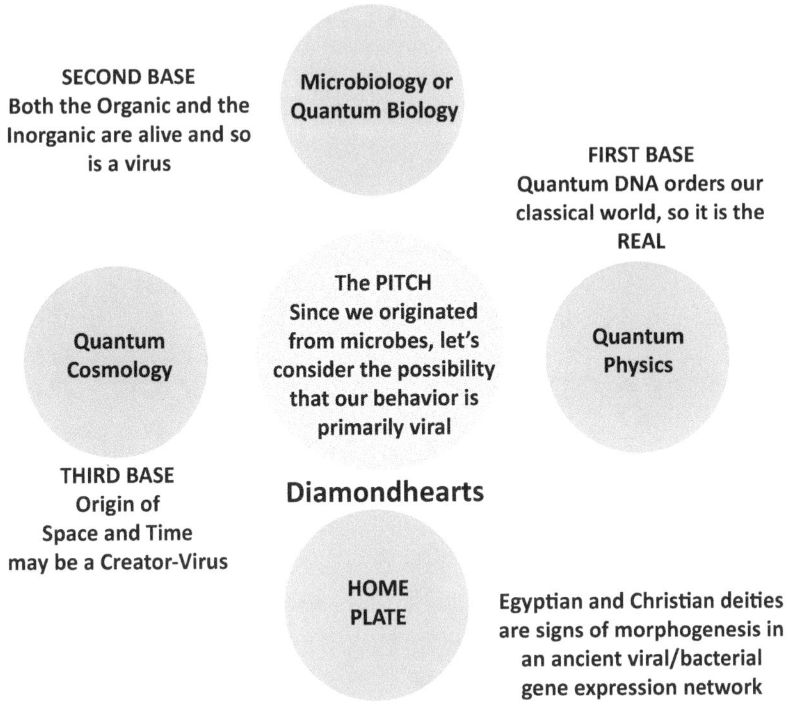
SECOND BASE
Both the Organic and the Inorganic are alive and so is a virus
Microbiology or Quantum Biology
FIRST BASE
Quantum DNA orders our classical world, so it is the REAL
Quantum Cosmology
The PITCH
Since we originated from microbes, let's consider the possibility that our behavior is primarily viral
Quantum Physics
THIRD BASE
Origin of
Space and Time
may be a Creator-Virus
Diamondhearts
HOME PLATE
Egyptian and Christian deities are signs of morphogenesis in an ancient viral/bacterial gene expression network

Figure 5.1. Argumentative proposals of the Diamondhearts and Thunderheads.

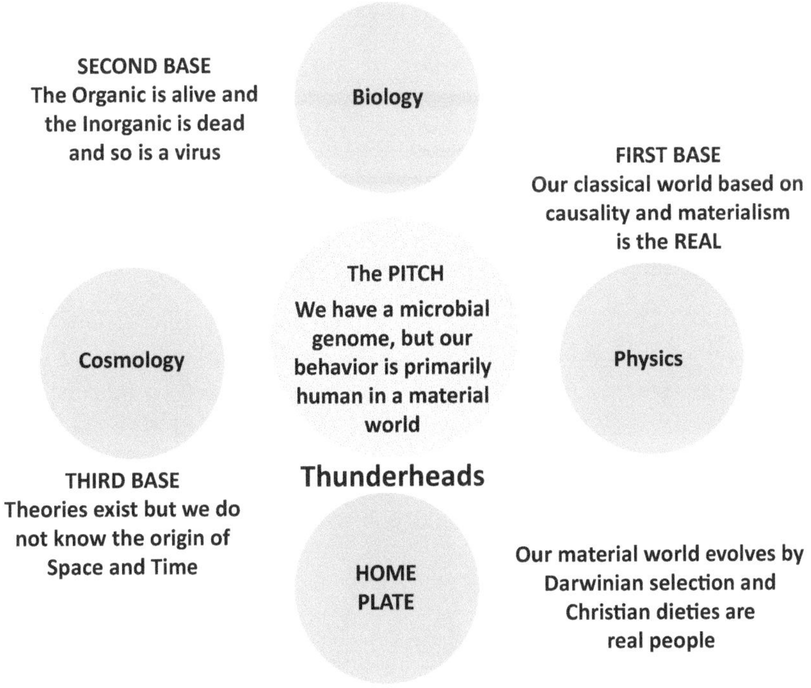
SECOND BASE
The Organic is alive and the Inorganic is dead and so is a virus
Biology
FIRST BASE
Our classical world based on causality and materialism is the REAL
Cosmology
The PITCH
We have a microbial genome, but our behavior is primarily human in a material world
Physics
THIRD BASE
Theories exist but we do not know the origin of Space and Time
Thunderheads
HOME PLATE
Our material world evolves by Darwinian selection and Christian dieties are real people

of cosmic mind or cosmic matter.

The proposal at the Diamondheart home plate is that knowledge of HGT and the ancient gene expression network in our cells will allow continuity of DNA, and that Christianity is a biological sign or pattern of this because both ancient Egyptian and Christian deities are signs for viral and bacterial DNA navigating this ancient gene expression network. The fixed belief at the Thunderhead home plate proposes our behavior is rationally human, evolving by Darwinian selection, resulting in death, and according to Christianity, whose deities are considered real people, passing on to an afterlife determined by one's morality during life.

The first pitch from the Diamondhearts is that humans originated from microbes, as indicated by our DNA, suggesting that our behavior is primarily viral. According to Luis P. Villarreal, the Director of the Center for Virus Research at the University of California at Irvine, viral genes provided the following traits to eukaryotes (our cell type) in the evolution of complexity: the eukaryotic nucleus, chromatin proteins, linear chromosomes with telomere ends, and DNA-dependent RNA polymerase (2004). In contrast, the first pitch from the Thunderheads is that our behavior is primarily human, despite the fact that just over two percent of our genome codes for human proteins. However, in their current quest to understand the human genome, the ENCODE project has determined a "spectrum of elements with different functional properties" in the human genome (2012, 71), and the project concludes:

> Interestingly, even using the most conservative estimates, the fraction of bases likely to be involved in direct gene regulation, even though incomplete, is significantly higher than that ascribed to protein-coding exons (1.2%), raising the possibility that more information in the human genome may be important for gene regulation than for biochemical function. (ENCODE 2012, 71)

This conclusion works nicely with the Isis Thesis because gene regulation is the task of the dead king in ancient Egyptian texts, so perhaps a good portion of our genome does have an important function.

Now, each team has four pitchers and one closer, so all the pitches will be discussed in the Pregame program before the Game of the Centuries on Opening Day.

Spring Training 6

The Cold Reality of Élite Deceit

Everyone dies. All hearts are broken. *Caring is not an advantage.*

Mycroft Holmes
Sherlock Holmes: a Game of Shadows

According to the texts of ancient Egypt, early China, and the seventeenth century American Navajo Indians, human potential depends on two ways of discarding energy: photosynthesis on earth and chemiluminescence, the chemical reaction that dominates the universe. Whereas the chemical reaction of photosynthesis produces both light and heat such as a burning candle, chemiluminescence produces light without heat or cold light. With their knowledge of science surfacing as mythology or storytelling, these early cultures explain how a death transition sets the stage for a cold-light chemiluminescent transformation for human DNA through the operating ancient glycolysis pathway in our cells. This enzyme pathway has metabolic energy pathways for several sugars, such as glucose, maltose, and lactose (galactose), as well as the LamB porin receptor site for phage Lambda.

Consider that our DNA is composed of four main nucleobases: guanine, cytosine, adenine, and thymine. Chemiluminescence, the low temperature emission of light produced by chemical or electrical action, is characteristic of guanine, which can give rise to luminescence (Ma et al. 1998). The average G+C content of a 100-kb fragment in the human genome ranges between 35 to 60 percent (Romiguier 2010), while phage Lambda has a high G+C content of 49.858 percent (Mesbah et al. 1989, 161). So, emergence as a cold-light chemiluminescent state of being may be possible through HGT and viral replication due to guanine content. If our DNA escapes degradation at death by microbes, water, and oxygen, scientists consider it somewhat immortal. Egyptologists today analyze mummified DNA that has escaped degradation. According to biologist Lynn Margulis, at death we return to the bacterial world, where free-lancing DNA fragments hover between life and nonlife, constituting a powerful toolbox for evolution (1986, 93). Microbes are our ancestors or planetary elders (95). More accurately, in Antonin Artaud's radio play entitled *To Have Done with the Judgment of God*, he wrote, "laugh if you like, what we call microbes is god" (1947, 569).

PHOTOSYNTHESIS is the reaction where a living organism captures the Sun's energy, converting the light into water and carbon-dioxide, then into carbohydrates, producing the byproduct of oxygen.

BIOLUMINESCENCE is a type of chemiluminescence, where light and carbon-dioxide are released by breaking apart organic materials using oxygen.

CHEMILUMINESCENCE is the emission of light by an atom or molecule that is in an excited state. Whereas photosynthesis produces both light and heat, chemiluminescence produces light without heat—cold starlight.

The Mesocosm. In between the two extremes of the microcosm (particles) and the macrocosm (humans) is the unexplored experimental gap of mesoscopic states (viruses and bacteria). As explained mathematically by Jasper van Wezel, it seems reasonable that an energy (or mass) scale exists in the mesocosm between the classical macrocosm of large objects and the quantum microcosm of microscopic objects. In the mesocosm, dynamics occur at the human timescale. In this thin mesoscopic spectrum, states become equal in energy as the system grows to infinity. When this happens, it is possible to reduce to the system's quantum ground state (lowest energy state), while violating the system's governing physical laws related to time symmetry. (2010) For example, a violation of T or time symmetry means that time reverses due to quantum effects. In a magnetic field, time reversal symmetry can be broken by halting a system and allowing it to go backward on its previous trajectory. Ancient Egyptian texts describe the Sun-god coming to a halt at Osiris' Mound and reversing time.

Mesoscopic physics, a sub-discipline of condensed matter physics, deals with objects the size of a group of atoms (a molecule) or other materials of micrometer size. Viruses and bacteria are the systems normally studied, and these objects are subject to the rules of quantum mechanics—not the classical physics that governs our physical world. Actually, at this mesoscopic scale of molecules, DNA transcription and translation is occurring, and DNA fragments survive due to self-assembly with other molecules or gene transfer via bacteriophages, such as the very abundant phage Lambda. In this quantum domain, self-organizing principles are at work such as crystallization (Laughlin 2005, 137), as Egyptian, Chinese, and Navajo eschatologies support, relative to a complex process involving protein synthesis and nonnative interactions resulting in degradation.

The Marriage of Élite Deceit and Altruism. Regarding the survival of our species, human history can be explained as a nonconscious psychic game of political deceit about knowledge determining whose DNA will survive or transform at the transition of death. As you shall see in the Pregame to the Game of the Centuries, ideologies (religion, psychology, capitalism, and so on) promote an underlying biological altruism (self-sacrifice) that undermines the individual's power through normalization processes. So historically, chemical knowledge of human transformation to a crystal or morning star was mystified, satanized, literalized, criminalized and controlled by mechanisms of power. However, many historical individuals understood human potential for a cold light crystallized transformation. For instance, American philosopher and mathematician Charles Sanders Peirce envisioned the world becoming "an absolutely perfect, rational, and symmetrical system, in which mind is at last crystallized" (1992, 297). Peirce said that man "should become welded into the universal continuum" to prepare himself for "a transmutation into a new form of life," (1931, 673) what suggests an emergence to a higher-ordered crystal state of being. (see King 2008)

According to one study, the 98 percent microbial DNA in our cells possesses the potential for some as yet unspecified goal (Mantegna et al. 1994). After all, the history of life and nonlife exists within our cells, and science explains how our journey began with nature's cosmic logic of star birth (chemiluminescence), followed by carbon-based life on earth (photosynthesis). Bioluminescence is the chemical reverse of photosynthesis. The Isis Thesis proposes that a creation process involving protein folding continues to cycle via the chemical reaction of bioluminescence or chemiluminescence because our large microbial DNA toolbox has the potential for an emergent creation, rather than DNA degradation at human death.

Degradation is when the steady-state level of a protein is reduced. Proteins in cells are subject to stress due to glucose starvation, heat-shock, and mutation. Oxidative stress damages many proteins, making them genetically unstable, so these proteins are targeted for destruction. Degradation or proteolysis is the mechanism that salvages reusable amino acids from non-essential proteins. However, the mechanism of horizontal gene transfer allows a DNA fragment, hovering between life and nonlife, to enter a cell for DNA transcription and translation, so it is not unstable, not stressed by glucose starvation, and not degraded by oxygen, water or microbes.

So, in the mesocosm it may be possible that a bacteriophage ferryboat or carrier molecule might transfer human DNA to avoid degradation at a death transition. One must know where to find the viral ferryboat, and this is what ancient Egypt, early China, and the Navajo are explaining in their eschatological literature—how to find a viral ride to avoid degradation. It is like trying to find a taxi home on a busy night after a ball game—if you don't know exactly where to find a taxi, you are out of luck. Also consider that there is room in the taxi for only one DNA fragment to be horizontally gene transferred for transformation due to vector requirements on the Lambda genome. Add to this that crystals and DNA have atomic spatial limitations, and you are beginning to understand why the pharaonic élite kept chemical information away from the plebians. Put simply, there are not enough DNA seats in the viral DNA taxi or the resultant crystal, so this results in nonnative interactions and the degradation of DNA.

Now, in the case of Lambda c1 and cro proteins, their adjacent genes on the genome allow their folding funnel landscapes to interconnect in their binding battle for gene-seats. With extreme simplification, when two stable states exist, the underlying energy landscape rotates similar to two vortices in opposite directions or a vortex/anti-vortex pair. In this case, dissipation is essential, and without it, the system cannot change its energy. (Eckel et al. 2004) As you shall see, dissipation or degradation is essential in the lytic lifestyle of phage Lambda, where two competitive proteins exhibit two stable states that function similar to a rotating black hole with two horizons or a vortex/anti-vortex pair.

The Rationale for Secrecy. Let's assume that here on earth the unwritten code for carbon-based human life is to keep death-transition knowledge secret because nature operates by the sacrifice of numerous individuals for the benefit of a few organisms, otherwise the system does not work. So, dissipation demands must be met, and this preserves some of the species and recycles the rest. Biologists call altruism the behavior of any organism that harms itself to benefit its species in terms of reproductive fitness. Biological altruism is different from the altruism of caring for others. And, biological altruism is necessary to make sure the great cosmic system cycles and preserves some of the species. Historically, as a ploy of biopower, concealing chemical knowledge about viral gene expression for reproductive fitness is a selfish device of the élite to assure the survival of their DNA over the majority of human beings whose DNA will probably be degraded due to necessary nonnative interactions. Thus, the DNA of unaware biological altruists is degraded, boosting the fitness of the DNA of the non-altruistic élite.

As mentioned, when lactose permease enters the cell, SOS mutagenesis is possible and the lactose genes become active. In the study of mutation, true reversion occurs when the mutated base pair is converted back to the original base pair, and the function of the translation to protein is restored. In this case, lactose metabolism is available for the new product. What the élite understood is what the experiments of Cairns and colleagues support on adaptive mutation: "a non-random, possibly product-oriented form of mutation" (1988, 142). However, cryptic growth or "the turnover of some of the population at the expense of the majority" is evident in this process, and thermodynamic constraints apply since "the number of cell deaths required to give rise to a new cell must be greater than one and has been estimated to be three to five" (Foster 1993, 7). Thus, numerous individuals are sacrificed for the benefit of a few.

The word "élite" is from the French élite meaning "selection, choice" from Old French *eslite*, feminine past participle of *elire*, *elisre* meaning "pick out, choose" from the Latin *eligere* "choose." As research indicates, the pharaonic priesthood had a choice at a death transition due to secreting knowledge about gene selection or adaptive mutation, but the élite pharaonic priesthood left the uninformed altruistic plebians with no choice because a sacrifice was necessary for the natural process to function. Perhaps today in our modern culture, the élite are still secreting DNA survival knowledge. Nonetheless, if you can read the signs and see the patterns of human behavior, then the message is still present for anyone. After all, Cairns and colleagues showed that selective pressure or the effect of selection can direct mutations. So, having knowledge gives everyone an equal choice.

As mentioned, another sign of the ancient Egyptian survival science is Christianity, a religion deifying altruism to the congregation by the model of Jesus Christ dying for our sins. Recall that Christ only saved 144,000 souls at the final judgment in the Book Revelation, while destroying the majority of souls. Christ's genocidal behavior supports nature's modus operandi of preserving the Few over the Many. So, Christ may have died for us, but his avenging actions show that he took a lot of us down with him. Even though the Catholic priesthood emphasizes faith, hope, and charity to their congregations, it would be an error to venerate Christ's vengeance without understanding the microbiological meaning of the sign and its knowledge. So perhaps one should eat of the tree of knowledge, even though it is forbidden by Christian theology.

The problem is clearly stated in Genesis, where God prohibits humans from eating of the tree of knowledge of good and evil that confers wisdom, but allows them to eat of any other tree, including the tree of life that confers eternal life. Now, both the serpent that entices the woman to eat of the tree of knowledge and God understand knowledge is essential, for it allows a godlike state of being. The almighty trick is that the tree of life does not work unless you eat of the tree of knowledge first, and God forbids this. Genesis 3:22-24 states:

> Then the Lord God said, "Behold, the man has become like one of us, knowing good and evil; and now lest he put forth his hand and take also of the tree of life, and eat, and live forever"—therefore the Lord God sent him forth from the garden of Eden to till the ground from which he was taken. He drove out the man; and at the east of the garden of Eden he placed the cherubim, and a flaming sword which turned every way, to guard the way to the tree of life.

The cherubim are described as guardians of sacred areas (1 Kings 8:6-7) and are represented as winged creatures similar to the hybrid Egyptian Sphinx, half human and half lion (Ezekial 41:18-19). God has placed these hybrid cherubim—a possible representation of the potential morphogenesis—east of Eden, while protecting the tree of life with a flaming sword. Perhaps this sword is similar to the two-edged sword in Christ's mouth in the Book Revelation (1:12-16), a piercing representation of the word of God that is also similar to the Aztec god Tonatiuh, the Fifth Sun who sports a jeweled sacrificial knife for a tongue (see King 2007). But the issue here is that eating of the tree of knowledge gives one the wisdom that allows the tree of life to bloom and confer eternal life. In other words, the power struggle between God and humans is over knowledge of immortality, that is, the genetic knowledge or immortal word of DNA. Human DNA must be degraded for the process to work, so God wants obedience from biological altruists, not rivalry for immortality through knowledge. This, of course, is the Christian ideology of a God that safeguards the survival of the élite or the chosen ones over the obedient, altruistic masses. Like God, the élite demand obedience and submission from the masses, not competition for DNA immortalization.

While the pharaonic priesthood's pursuit of knowledge negates the fundamental idea in Genesis that acquiring knowledge leads to death, the non-altruistic Egyptian élite prospered by concealing knowledge of transformation from the plebs. This historical secrecy about genetic knowledge resurfaces later as the biblical command not to eat from the tree of knowledge of good and evil, and this is still emphasized today in Christian ritual. Recently, at a Catholic Mass in northern Michigan, the priest as vicar of Christ stressed this idea of no nutrition for the mind as he explained how the congregation should deal with the sign of the Trinity, that mysterious merger of God the father, God the son, and God the holy spirit. As the shepherd of his obedient flock, the priest explained that the triune god is a mystery of faith. He urged the parishioners to "find joy in confusion" because God meant it to be a mystery.

Contrary to being joyfully confused, perhaps the triune god represents the molecular self-assembly of the Egyptian Sun-god (hydrogen and other viral elements), the deceased king (organic DNA), and Isis (lactose), who is the holy spirit or "breath of life" for the dying god Osiris, as well as the dead king. Isis is the voice of wisdom, for her titles include "the greatest of the gods and goddesses, the queen of all gods, the opener of the year, lady of heaven, queen of the earth, lady of warmth and fire, the maker of kings, the lady of words of power, and the lady of the House of fire," to name a few. These appellations and Isis as the "breath of life" are signs suggesting the value of lactose metabolism in phage Lambda's ancient gene expression network for viral transformation. And so, we can begin to understand the ancient importance of the goddess in human history that has been debased by Christianity.

Perhaps if Catholic priests and their parishioners would reconsider their view of the value of knowledge, then they may realize that the Christological themes of virgin birth, death, resurrection, and ascension first surfaced in ancient Egyptian literature circa 2500 BCE as the so-called pagan activities of the ancient deities Isis and Osiris. Also, the precursor of the historical Catholic Mass at Sophia Hagia during the sixth century is carved on the walls of the underground Edifice of Pharaoh Taharqa one thousand years earlier, a ritual firmly rooted in the ancient Egyptian Sed festival circa 2500 BCE. In light of this, perhaps the Trinity is a natural outflow of the evolutionary wisdom from ancient Egypt. Just as relative knowledge allows different degrees of understanding, unquestioning ignorance is the root of confusion and fear, not joy. Accordingly, our quest for the deathless wisdom of earlier cultures centers on the premise that when one discovers a mystery, the person should use reasonable theory, natural law, experimental evidence, comparative data, and a transdisciplinary approach to unravel the potential truth and meaning of the mystery. Confusion lies in narrowing one's perspective to avoid the transdisciplinary, holistic unity of knowledge.

Now, naturalist E. O. Wilson considers that religion is a behavior unique to our species, and its key learning rules and ultimate genetic motivation are hidden from our consciousness. In addition, the religious process itself persuades humans to subordinate their self-interest in favor of the interests of the group. (1978, 175-176) According to Wilson, this hard-core altruism supporting the interests of the group or family and disallowing hypocrisy is also found in social insects such as honeybees and termites that charge their attackers at the expense of their lives. On the other hand, he explains that human beings can be hypocrites exhibiting soft-core altruism that is distinguished by lying, pretense, deceit, and self-deception. (156) Applying Wilson's ideas to priesthoods, one of their key unspoken operating mechanisms is to create a unified representation of an afterlife context to control behavior by recollecting a specific past, while interpreting the present with their authoritative proclamations to guide the obedient flocks into a future eternal realm or heaven. This promotion depends on the priesthood's soft-core altruism to prevent the congregation's acquisition of knowledge and to assure the congregation that good morals will justify heavenly rewards at death. This practice of soft-core altruism does not deliver knowledge to the masses, but only power to the élite. According to the ancient Egyptian pharaonic priesthood and their knowledge that is supported by contemporary science, a sacrifice is necessary for the system to work, so a thermodynamic hell of nonnative interactions may be the reward of those who rely on faith and hope rather than knowledge.

The Slime Mould Élite. Nature imposes the same artifice of altruism on the slime mould, a unicellular eukaryote (our cell type). As mentioned, the slime mould is able to form an efficient network comparable to the real-world infrastructure network of the Tokyo rail system that humans have created, according to scientists. Slime moulds are protists similar to amoebae. Ordinarily, the individual cells roam separately for food; however, during a food shortage, they send out a chemical distress signal, the molecule cAMP (cyclic adenosine monophosphate). This signal motivates the community of cells to crawl, leaving a slimy trail. Then the community of cells develops into a tall stalk, surmounted by a sphere or fruiting body of spores that may be able to spread to a plentiful food source by adhering to the coats of passing animals. As Hudson and colleagues (2002) observed, the cells exhibit extreme altruism, for those in the tall stock (the altruists) die, but the spores (the beneficiaries of the sacrifice) survive. Similarly, the ancient Egyptian pharaonic priesthood was very aware of the natural process of horizontal gene transfer, and they made sure death-transition knowledge was secreted inside pyramids and tombs from the plebians, so that the DNA of the élite transformed to a quasi-hybrid being instead of degrading or dissipating like the DNA of the unknowing plebians, who were obliviously altruistic. Thus, knowledgeable hypocrites exhibiting soft-core altruism—distinguished by pretense and deceit—won the day, while the DNA of hard-core altruistic believers dissipated. Perhaps nature's pattern of the survival of the Few over the Many is the biological motivation for secrecy as well as the genetic impetus for the idea of human sacrifice in history.

Altruistic Suicide or Suttee Burial. Science confirms that along with slime moulds, altruistic suicide or the performance of duty is present in humans, lower animals, and social insects (Wilson 1978, 151). Recorded history provides more than one disturbing human example. According to Joseph Campbell's research in ancient history, what appears to be a fabricated mythological context functions with the rite of suttee burial and human sacrifice found in India, China and the Nubian cemetery at Kerma in Egypt. Campbell recounts how archaeologist George Reisner (1867-1942) excavated an Egyptian burial ground of approximately 400 individuals, primarily female sacrifice from the Prince Hepzefa's family and his large harem that died from a conscious death of suffocation. Reisner then reconstructed the burial rites of the provincial governor Prince Hepzefa, that is, his funerary procession to the tomb, the sealing of his chamber, and the group sacrifice based on archaeological evidence and ritual:

> The women and attendants take their places jostling in the narrow corridor, perhaps still with shrill cries or speaking only such words as the selection of their places required. The cries and all movements cease. The signal is given. The crowd of people assembled for the feast, now waiting ready, cast the earth from their baskets upon the still, but living victims on the floor and rush away for more. The frantic confusion and haste of the assisting multitude is easy to imagine. The emotions of the victims may perhaps be exaggerated by ourselves; they were fortified and sustained by their religious beliefs, and had taken their places willingly, without doubt, but at that last moment, we know from their attitudes in death that a rustle of fear passed through them and that in some cases there was a spasm of physical agony. . . .The assembled crowd turned then probably to the great feast. (quoted in Campbell 1962, 68)

This account is reminiscent of the Nazi gas chambers at Auschwitz, where relaxed guards doing their duty escorted naked prisoners into underground gas caverns under the pretense that showers were available for the prisoners. In light of Reisner's excavation, Campbell explains that the pharaoh was eternal, and all believed this, including his priesthood, family, and commoners. Yet, Campbell wrote that "the whole society was mad," and from this madness blossomed the great Egyptian civilization that experienced mythic identification through the pharaoh, who was the universe itself, and so, to follow the pharaoh in death represented life or "the peace of eternal being" for many (1962, 53; 80; 82). One wonders if those who followed the pharaoh to death had knowledge or not. Perhaps, as unaware biological altruists, they were safeguarding the nature-inspired design or pattern of survival of the Few over the Many.

Still, what is this artifice of altruism that nurtures human nobility or hard-core suicidal behavior that drives one to death within the context of moral cultural ideas? The German philosopher Friedrich Nietzsche explains it as follows in *The Will to Power*:

> The basic phenomenon: countless individuals sacrificed for the sake of a few, to make them possible.—One must not let oneself be deceived; it is just the same with peoples and races: they constitute the "body" for the production of isolated valuable individuals, who carry on the great process. (1968, 360)

Nietzsche explains that biologists are prey to the basic error of using moral evaluations and considering altruism as the higher value over the will to power. (1968, 681) Yet, Nietzsche understood the will to power is connected to speciation and survival: "Our way is upward, from the species to the superspecies" (1961, 100). Explaining that the culture of man does not go deep, and where it does go deep, it becomes degeneration, Nietzsche elevates the savage, or morally speaking, the evil man who returns to nature. (360-364). Nietzsche's explicit view of degeneration is Christianity:

> I rebel against the translation of reality into a morality: therefore I abhor Christianity with a deadly hatred, because it created sublime words and gestures to throw over a horrible reality the cloak of justice, virtue, and divinity— (364)

On morality, Nietzsche urges philologists, historians, and philosophers to answer this question—"What light does linguistics, and especially the study of etymology, throw on the history of the evolution of moral concepts?" (1967, 55) Nietzsche also suggests viewing morality related to physiology, the biological study of the chemical and physical functions of living organisms and their parts, rather than to psychology. So let's continue on the track of etymology, microbiology, and physiology.

Meaning of Right. First consider the etymology of "right" relative to the secret knowledge of the Egyptian élite. In the Egyptian Book of the Dead, *maat* means "Truth, Rightful Order" (Goelet 1994, 155) and the movement to the West or land of the dead is considered "right" (156). Numerous references are also made to the right arm. The Isis Thesis interpretation considers the Egyptian usage of left and right as genetic directional queues for their afterlife biophysics of matter, where the directions of space become real by means of a parity violation. In particle physics, a parity violation allows space to have preferred directions and a right or left handedness. Thus, the etymological and physiological connection relates to a spatial orientation, as well as a directional genetic interpretation of the right arm of the Lambda genome. This assures the right chemical path for those with knowledge. Nietzsche's hatred of Christianity relates to the replacement of a physiological value of right and left with a morality of right and wrong by which the soft-core altruism of the Church dominates and encourages hard-core altruism in the uninformed masses. The Christian lie is that the good behavior of an individual, such as supporting the Church with donations, following the rules, and relying on faith, hope and charity, will reward that person with an afterlife existence of eternal life, rather than hell, a possibility that looms over a Christian's conscience like a sharp guillotine or double-edged sword. Biologically speaking, conscious or not, the value of Christian morality is to direct the uninformed masses into altruistic DNA suicide, so the great process works—exactly what is needed for the transformation of the few with genetic knowledge.

As the Isis Thesis explains, nature's modus operandi for species survival applies to all species, including humanity, for it is those few informed individuals, such as the ancient Egyptian pharaohs and their hieratic priesthood, the early Chinese emperors and their sages, the Navajo shamans, to name a few of the elitist groups, who understand the genetic meaning of their secret knowledge. The logician Charles Peirce understood the process as the crystallization of mind. Other aware historical individuals, such as the German poet Goethe, the English poet William Blake, the French dramatist Antonin Artaud, the French writer Gérard de Nerval, the Jesuit paleontologist Teilhard de Chardin, to name a few, were empowered by their recognition of the knowledge that had been historically secreted, literalized, satanized, disguised as religion, and finally rediscovered as science today (see King 2005; 2009a). Still, these individuals were not always free from cultural mechanisms of domination during their lives. Peirce was ostracized from intellectual society because of his impulsive temperament and other behavioral irregularities such as his legal drug addiction (Brent 1998, 14), even though the Harrison Narcotics Tax Act controlling drugs such as heroin did not take effect until December 17, 1914, eight months after Peirce died. Blake was criticized for his metaphysical views and visionary experiences. Artaud, who participated in the peyote rite of the Tarahumara Indians in Old Mexico, was electroshocked for his chronic madness. The Catholic Church rewarded Teilhard de Chardin for his brilliant insights with a ministry in China. Nerval was in and out of insane asylums during his life, and before Gérard de Nerval hung himself, he placed the final pages of his greatest work *Aurélia* in his pockets. In those final pages he wrote:

> I resolved to fix my dream-state and learn its secret. "Why should I not," I asked myself, "at last force those mystic gates, armed with all my will power, and dominate my sensations instead of being subject to them? Is it not possible to control this fascinating, dread chimera, to rule the spirits of the night which play with our reason? Sleep takes up a third of our lives. It consoles the sorrows of our days and the sorrow of their pleasures; but I have never felt any rest in sleep. For a few seconds I am numbed, then a new life begins, freed from the conditions of time and space, and doubtless similar to that state which awaits us after death. Who knows if there is not some link between those two existences and if it is not possible for the soul to unite them now?" (1996, 68)

Perhaps madness, drug use, visionary experience, and literature are successful methods for exploring interior survival knowledge at the biological level of our DNA. With these strategies, de Nerval, Artaud, Blake, Peirce, Goethe, and de Chardin battled mediocrity and domination. Yet, of this select group, only Goethe escaped the chains of domination during his lifetime, despite mocking scholars, criticizing convention, attacking Catholicism, and creating his great scholar Faust who makes a pact with the devil, but still escapes hell and is saved. Goethe's greatest work *Faust* reinforces the view that faith and good works have nothing to do with human salvation. What counts is scientific knowledge.

Spring Training 7

A Trinity of Natural Patterns

The cosmic microwave background (CMB) is a relic from the early universe. Scientists represent it as a complex system of warm and cool spots, or "lobes," projected onto the celestial sphere around Earth. The patterns, which reflect large-scale structures present in the early universe, line up with the solar system in strange and as-yet-unexplained ways.

Dragan Huterer
"Why is the solar system cosmically aligned?"

Today the knowledge of the élite has survived and is preserved in great works of art and literature, as well as contemporary science. With their knowledge and power, perhaps the élite activated the cycle of the great cosmic system at a death transition. And those without this knowledge, who do not question the motivations of authority—the normalized, the altruistic, the sacrificial dissipating flux of the great process—may they rest in peace if that is possible. Either one takes the time to understand natural processes or one does not. This is the key because human history documents that numerous religions and mythologies have primed human beings for altruistic sacrifice through the promise of eternity, while nationalist governments have awarded gold and silver medals to soldiers for altruistic suicide in warfare.

Consider that proteins are workhorses that turn a gene (bits of DNA) on or off. Lambda c1 and cro proteins fold and bind to DNA to turn genes on or off, and these particular proteins function through two funneled protein energy landscapes that are similar to a tiny biological black hole connected to a white hole that casts out matter or goes backward in time. In the interest of understanding this ancient Egyptian biophysical interpretation of funerary texts, let's review protein energy landscape theory.

In recent years, energy landscape theory has become an important tool linking three disciplines, as can be seen in the cooperative work of biologist Charles L. Brooks III, physicist José N. Onuchic, and chemist David J. Wales (2001). An energy landscape is the potential energy surface defining the folding of proteins, the complex behavior of glasses, and the structure and dynamics of atomic and molecular clusters. This protein energy surface has topographical features similar to the earth's landscape, such as mountains, valleys and hollows. However, the dimension is much higher and the energy landscape determines the behavior of the system (Doye et al. 1999).

The classical view of protein folding established by Cyrus Levinthal asserts that proteins cannot fold by a random search for the native (correctly folded) state, so to save time, they fold by a sequential, energy biased folding pathway. On the other hand, proteins with random amino acid sequences encounter a rough potential energy surface or frustrated system. (Brooks et al. 2001) In ancient Egyptian texts, the "second death" in their black-hole netherworld Duat represents a random, frustrated protein system, while knowledge of the sequential, energy-biased folding pathway allows the dead king to reach the native state.

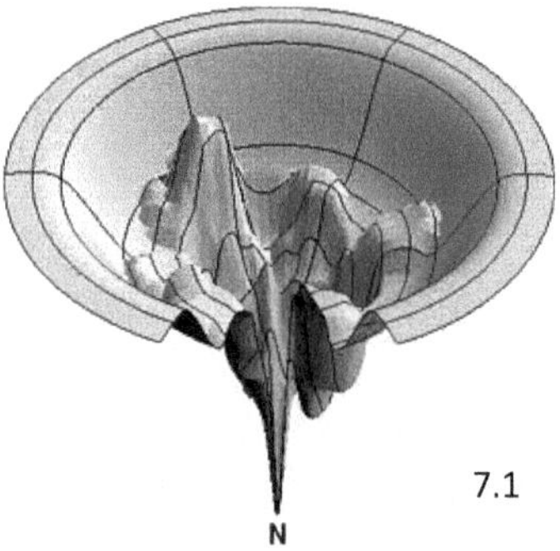

7.1

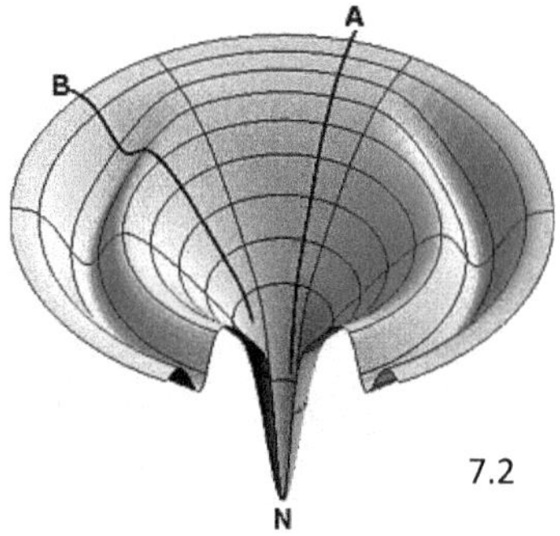

7.2

Figure 7.1. A rugged energy landscape with kinetic traps, energy barriers, and some narrow throughway paths to native state. Folding can be multi-state.

Figure 7.2. Moat Landscape, to illustrate how a protein could have a fast-folding throughway process (A), in parallel with a slow-folding process (B) involving a kinetic trap. Permission to use both images from Ken A. Dill and Hue Sun Chan (1997).

In Figure 7.1 we see what appears to be a circular cone of mountains and cavernous valleys, pierced to its center by a hole in the shape of an irregular cone or funnel. The protein folds itself along a free energy surface resembling a rough funnel, what is very similar to the construction of a tiny black hole. Peaks represent the highly fit forms. Populations of organisms driven by mutation, recombination, and natural selection struggle to climb toward those high peaks. (Kauffman 1995, 149)

In Figure 7.2 the faster folding process "A" avoids the kinetic trap represented by the moat around the axis that slows down the "B" folding process. Natural selection, the central Darwinian view of evolution, sifts through useful variations among mutations that have random effects on organisms. Evolutionary adaptation is a process of "hill climbing" on a fitness landscape. A random change in the genome (organism's DNA and genes) slides a mutant higher or lower on the mountain terrain. Stuart Kauffman explains, "There are pathways uphill to the distant peaks, and natural selection, in sifting for the fitter variants, pulls the population up toward them" (1995, 154). Kauffman explains this as gradualism or gradual ascent at work in Darwin's sense in that the fitness landscapes are correlated (169).

Ancient Egyptian texts clearly describe the quantum misfortunes of those without knowledge, who travel on a random, rugged energy landscape with kinetic traps and energy barriers, rather than the fast-folding axial pathway traveled by the pharaonic élite. Kauffman gives us a good description of the dilemma a population of organisms might encounter on a rough random landscape:

> Random landscapes have hyperastronomical numbers of local peaks. . . . It begins to be obvious why adaptive search on random landscapes is very hard indeed. Suppose we wanted to find the highest peak. We try to search by climbing uphill. The adaptive walk soon becomes trapped on a local peak. The chance that the local peak is the highest peak, the global peak, is inversely proportional to the number of local peaks. . . . On random landscapes, finding the global peak by searching uphill is totally useless; we have to search the entire space of possibilities. But even for modestly complex genotypes, or programs, that would take longer than the history of the universe. (1995, 167)
>
> Try to evolve on a random landscape, seeking the highest possible peak, and the population remains boxed into infinitesimally small regions of the space of possibilities. In these random moonscapes of impossible cliffs soaring and plummeting in every direction, no clues exist about where to go, only the confusion of a hyperastronomical number of local peaks, pinnacles studded into the vast space of possibilities. No pathways uphill lead farther than a few steps before the stunned and dazed hiker ascends to some minor peak, one of so many that the stars in the sky are a minor number by comparison. Start anywhere climbing uphill, and one will remain frozen in that tiny region of the space forever. (1995, 168-169)

The ancient Egyptian pharaonic priesthood understood evolution as happening on a nonrandom protein landscape. A protein folding from a set of instructions from a gene is a nonrandom landscape. The protein instructions demand that the protein go uphill to the highest peak and turn the gene on by an adaptive walk on an ordered fitness landscape. That vertex or genotype is the uphill goal, and Egyptian texts refer to the vertex often. The vertex is the highest peak, the global optimum, the global peak. To see how proteins function, we can understand the fitness of the genotype as a height that the protein climbs to turn on the gene. As theoretical biologist Stuart Kauffman explains, "Evolution requires landscapes that are not random" (1995, 166), and ancient Egyptian science supports this. Individual selective pressure allows evolution.

The basic idea is that at a death transition we need new traits to be fit in the quantum universe, and so morphogenesis or the emergence of new form and structure is essential. Our fitness may depend on HGT and recombination mediated by the genomic network of a virus such as phage Lambda, so we can acquire the traits we need for the new quantum environment. Kauffman states, "The genomic network, I believe, lies in the ordered regime, perhaps near the edge of chaos, because such networks are readily formed, part of order for free, but also because such systems are structurally and dynamically stable, so they adapt on correlated landscapes and are able to be molded for further tasks" (1995, 188-189). Egyptian texts support that Lambda c1 and cro proteins are entangled like a vortex/anti-vortex pair, and the ordered regime is cro protein's inner axial vortex or horizon that is surrounded by the chaos of c1 protein's folding funnel landscape.

Ancient Egyptian texts stress the importance of crossing over, that is, recombining or exchanging DNA for continuity by HGT. As Margulis and Sagan stress, the speed of recombination is superior to mutation (1986, 16). On sex and recombination, Kauffman agrees that fitness increases far more rapidly when mutation and selection are used to climb the peaks, along with recombination for exploring the genotype space for adaptation. Kauffman explains that recombination is useless on a random landscape, so a landscape must be well-correlated. (1995, 182) Also, he considers that evolution may depend on the system being able to organize itself spontaneously (185). Ancient Egyptian texts also confirm Kauffman's last point, for self-assembly and self-organization jumpstart the biological process, followed by recombination and mutation. In addition, ancient Egypt and other early cultures describe the signs of emergent order as a process of crystallization.

The Palm Tree. We can envision this multi-dimensional potential energy surface for emergent order as a palm tree. David Wales (2005) explains that the archetypal palm tree energy landscape model provides a unifying foundation for viral self-assembly, protein folding, and crystallization. This potential energy surface with funneling characteristics that looks like a palm tree permits efficient viral self-assembly, and it underpins nonrandom searches in molecular science. Icosahedral geometry provides the right conditions or self-organization of a viral capsid (head). Considering the palm tree topology of the viral model presented in Wales' article, perhaps this is what the poet Wallace Stevens envisioned when he penned the following metaphor from his poem "Of Mere Being":

> The palm stands on the edge of space.
> The wind moves slowly in the branches.
> The bird's fire-fangled feathers dangle down.

Beyond the biosphere of earth in the space of the possible, Egyptian texts describe an equilibrium folded state for survival of DNA as a quasi-hybrid being in the quantum universe. A stable low-energy equilibrium state is evident in a self-assembling virus, in the double helix of DNA or RNA, and in folded proteins, so cells and organisms take advantage of these stable low-energy structures (Kauffman 1995, 186-187).

AGGREGATION
In addition to being unstable, proteins often have problems folding correctly and sometimes become trapped in misfolded conformations or form aggregates. (DePristo et al. 2005, 679)

DEGRADATION
Cells continuously synthesize and degrade proteins. Because the degradation machinery operates selectively on partially or fully unfolded proteins, degradation rates are mainly determined by stability. However, misfolded or abnormal proteins are also selectively targeted for degradation. Moreover, rapid protein turnover is essential for regulation. (DePristo et al. 2005, 679)

ADAPTIVE EVOLUTION
A genetic change that results in increased fitness.

FITNESS
A measure of the capacity of an organism to survive and reproduce.

MONOMER
A molecule that can be chemically bound as a unit of a polymer.

Fly Fishing in a Black Hole. In light of human genome sequencing and research showing that lytic viruses have made us what we are today while impacting the entire planet (Danovaro et al. 2008), a quantum model for a holographic expansion/collapse cosmos exhibiting wormhole dynamics is the lysogeny/lysis genetic switch of bacteriophage Lambda, a competition between cro protein and the c1 repressor protein. A simulation survey of eleven proteins including cro and c1 repressors is a "strong indication" that binding processes have funneled landscapes (Levy et al. 2004, 516). On the quantum level, the two entangled, competing proteins and their folding funnel landscapes express the dynamics of a microscopic Kerr black hole with its inner sphere (gateway to a microscopic Einstein-Rosen bridge) representing cro protein and the outer sphere representing c1 repressor protein. This is a vortex/anti-vortex pair with two stable states. Also, the c1 repressor protein controlling lysogeny folds in a diffusion-collision manner (Levy et al. 2004, 516; Karplus and Weaver 1994) similar to our universe's diffusion-collision expansion. In contrast, cro protein controlling lysis folds and binds via the fly-casting speed-up mechanism (Levy et al. 2004; Jia et al. 2005), a process similar to a microscopic white hole casting out matter.

For example, the fly-casting mechanism of Lambda cro is described as "a randomly gesticulating unfolding molecule casting out pieces of polymer chain, waiting for these to bind to the target," and then the whole molecule folds and reels the target in like fly fishing (Shoemaker et al. 2000, 8870). The fly-casting mechanism has a greater capture radius because the protein is more flexible, casting out pieces of polymer chain like a tiny volcano until the complete molecule folds and reels in the target. The binding site attached to a flexible chain is similar to a hook on a fishing rod, allowing the protein to fish for its targets over a large volume. As mentioned, this process is very similar to a microscopic white hole casting out matter.

It is interesting that the Book Revelation (20:14-15) references the "second death," while in Matthew (4:19-22) at the Sea of Galilee, Christ gathers four fishermen to transform them into fishers of men. Along with this, Christ was able to multiply, or shall we say, spontaneously clone fish for everyone to eat. A similar story explains that on the path to the Grail a brother called Brous was known as the rich fisherman because he caught a fish, satisfying the hunger of all around him (Cirlot 1971, 108). Nonetheless, the message of fishing and cloning multiple fish may represent signs of the fly-casting mechanism and its ultimate product—viral clones or a quasi-hybrid state of being that can be compared to Hawking radiation, the cloned energy from quantum mechanical black hole/white hole formation/evaporation processes.

Thermodynamics of Proteins and Black Holes. Living processes depend on protein activity, for these cellular workhorses fold into their biologically active, three-dimensional structures of their native states. Molecular biologists can quantify the folding reaction of a protein in terms of thermodynamics because its native and unfolded states are in equilibrium (Murphy 2001). Thermodynamics is the set of physical laws that govern the random, statistical behavior of large numbers of atoms that make up the air or the Sun, that is, the laws that govern heat. Jacob Bekenstein noticed that the laws of black hole mechanics bear an amazing resemblance to the laws of thermodynamics. (Thorne 1994, 422-426) Thermodynamics is also linked to protein interactions promoting stability, such as the hydrophobic effect, hydrogen bonding, configurational entropy, because environmental changes can impact protein stability and folding dynamics. If an organism cannot adapt to a changing environment or an environment hostile to cell functions, then the organism faces extinction or DNA degradation. This is what ancient Egyptian texts are telling us. At a human death transition, the environment changes from our classical world to the quantum molecular domain of microorganisms, so one must understand microbial mechanisms of survival for adaptation in that quantum environment. We can learn lessons from viruses and bacteria, and HGT and transformation are essential according to ancient Egyptian texts.

Now, protein binding like folding is controlled by a funneled energy landscape leading to the native bound configuration or state. The funneled energy landscape provides binding stability against environmental and evolutionary fluctuations. (Levy et al. 2005, 1123) Lambda c1 repressor forms via diffusion-collision binding, the so-called induced-fit mechanism, with two unbound monomeric Lambda repressors ready for dimer formation. However, Lambda cro repressor has a folded monomer stable on its own and a single folded monomer acts as a template for the folding of the other monomer. It does not fold by induced-fit like c1 protein (1127), but rather the fly-casting mechanism.

The fly-casting pattern of casting out pieces is not only similar to quantum mechanical Kerr black hole/white hole dynamics, but also supernova nucleosynthesis and the earth's geodynamics. Together, the genetic switch or decision circuit of Lambda cro and c1 proteins exhibit folding/unfolding dynamics and native state crystallization similar to the backward-in-time aspect of a microscopic wormhole, which is similar to the holographic expansion/collapse cosmos described by ancient Egypt, early China, Peirce, Einstein, and others. What may be supportive of this comparison is the cosmic microwave background (CMB), an image of the thermal radiation emitted 13.7 billion years ago from the Big Bang. The image shows warm and cool lobes or the early seeds of our universe, which may represent the initial competition between c1 protein and cro protein at the quantum origin of our universe.

Just Six Numbers. Recall that the initial competition on the Lambda genome is over six gene-seats. Both proteins are battling over possession of the three gene-seats on the right arm. When cro protein wins the three gene-seats on the right, c1 protein co-operates. All six gene-seats are essential to lysis and lysogeny. Now, in physicist Martin Rees' book *Just Six Numbers* (1999), he explains that "Science advances by discerning patterns and regularities in nature" (1). What Rees has identified is six crucial numbers that determine the recipe for our universe, that is, how our universe evolves and its potential. Stars and life exist because of these six numbers that finetune our universe. My guess is that these six numbers relating to two forces, the properties of space, and the size and texture of our universe are contingent on the mathematical potentialities of Lambda's six gene-seats. However, this is a thesis for future study that is beyond the scope of this text.

A Viral Definition of Life and Death. Current reasoning supports the antiquity of the genetic archive of viruses:

> Many scientists now argue that viruses contain a genetic archive that's been circulating the planet for billions of years. When they try to trace the common ancestry of virus genes, they often work their way back to a time before the common ancestor of cell-based life. Viruses may have first evolved before the first true cells even existed. (Zimmer 2011, 92)

In our holographic cosmos, a complex bacteriophage may have jumpstarted life through its bacterial cell host. Its lifestyle of lysogeny created photosynthesis and thermodynamic life as we know it. Its lifestyle of lysis reverses the process, allowing the birth of a cold-light, chemilumescent state of being. Both lifestyles are possible because of our viral DNA heritage. From our human state of being in the physical classical world, a second viral state of being emerges in the quantum environment. This quasi-hybrid being can only be achieved through knowledge and willed action, that is, individual selective pressure leading to evolution.

At a human death transition, the alternative to a quasi-hybrid being may be DNA degradation. The folding funnel energy landscape contributes to a protein's stability in the transition from a random, unfolded state to its unique folded state, so it can function by turning a gene on or off. However, proteins can misfold, be degraded, and aggregate with the wrong crowd. Proteins exist in an environment packed with degradative enzymes bent on their destruction (DePristo et al. 2005). In ancient Egyptian texts, signs of degradation are the "second death" or destruction by the hybrid monster Ammit, who is the Swallower of the Dead. This suggests the digestion of chromosomal DNA, similar to the damnation artwork of souls tortured, hacked, cleaved, decapitated, and annihilated by burning in fiery pits or the lake of fire (thermodynamic reactions), the serpent-demon Apopis, the winepress-god Shesmu, and the destructive lion-goddess Sekhmet. Now, aggregating with the wrong crowd describes the Children of Impotence and the Conspiracy of Seth. Further, the language of the texts also references degradation. Of course, all this degradation and hell fire can be avoided with Heka or Akhu magic, a concept associated with thought, deed, image, power (Goelet 1994, 145), and overcoming the enemy that Nietzsche referred to as the Spirit of Gravity.

Yes, the problem is gravity and our ascent/fall structure of being or consciousness. According to existential psychiatrist Ludwig Binswanger, the rising and falling of existence is present, not only in patients such as Ellen West who completed suicide, but also in religious, mythical, and poetic images. For example, the ascent/fall ontological foundation is described by the writer Gerard Nerval in *Aurélia*, the poet Rilke in *Duino Elegies*, Dante in *Divine Comedy*, and Mozart in his Egyptian opera "The Magic Flute" where the flute conquers Death by casting the protagonists "from the earth heavenward" into the starry night. Relative to the ascent/fall ontological structure of consciousness, ancient Egyptian, early Chinese, and Navajo texts support that positioning within the earth's gravitational field selects either time-reverse or spacetime, which on the quantum level, mirrors the lysis/lysogeny genetic switch of a temperate virus. (see King 2009a for a full discussion)

Ultimately, knowledge is necessary for individual selective pressure allowing evolution. Thus, one must recognize a trinity of patterns or how specific biological protein folding funnels mirror quantum mechanical black hole/white hole dynamics that mirror the expansion/collapse potential of our holographic cosmos. These underlying chemical and physical principles fit the model of temperate phage Lambda's genetic switch. Human culture is a sign of a viral gene expression network for re-animation that bridges the chasm between the organic and the inorganic. Man is a sign of the semiotic microbe that hits the genetic switch for re-animation.

Spring Training 8

Useless Gene or Not ?

Cairns and colleagues proposed that a cell under selection might be producing highly variable mRNA molecules. If one such mutant transcript happen to code for a good protein, this might stimulate reverse transcription, immortalizing the mutation in the DNA.

Patricia L. Foster
Annual Review Microbiology (1993, 15)

In his book *Genome*, Matt Ridley identifies the most common, useless protein recipe in the human genome that is a threat to our health—reverse transcriptase, an essential part of the genome of the AIDS virus. In the field of molecular biology, a reverse transcriptase is an energetic enzyme with two roles—or shall we say—two aliases. Also known as RNA-dependent DNA polymerase, the protein transcribes single-stranded RNA into single-stranded DNA. When using its second alias, the protein is DNA-dependent DNA polymerase, which synthesizes a second strand of DNA complementary to the reverse-transcribed single-stranded copy DNA after degrading the original messenger RNA with its RNaseH activity. Put simply, normal transcription synthesizes RNA from DNA, and reverse transcription is the reverse.

Telomerase is also found in the human genome, and it carries its own reverse transcriptase, an enzyme with an RNA template for DNA replication that may be necessary for human horizontal gene transfer and other tasks. Matt Ridley reports that an amazing 14.6 percent of our genome is comprised of LINE-I sequences of letters that include a complete recipe for the protein reverse transcriptase (1999, 125). He asks—why is it there? With the Isis Thesis and modern genetic engineering supporting horizontal gene transfer of human DNA, logic suggests that reverse transcriptase may be exactly the protein needed for viral self-assembly and cell entry. Accordingly, the central concern of ancient Egyptian, early Chinese, and Navajo texts is how human DNA can find a landing site on a cell's surface, build in its DNA on the master chromosome, and replicate or realize itself as a quasi-hybrid being composed of organic human and viral DNA. What sets the scene for this biological quest is the individual's death, when DNA fragments hover between life and nonlife, continuity and degradation. At death, genetic barriers may no longer exist that would prevent the exchange of DNA between different species (horizontal gene transfer). Further, the human genome is primarily viral and contains numerous copies of reverse transcriptase protein, exactly what it takes to build DNA into a cell for morphogenesis. With 14.6 percent of our genome comprised of reverse transcriptase, it seems that natural selection has made sure we have the DNA for survival.

Perhaps biologists, who consider a virus or bacteriophage as nonlife because it cannot replicate without a host, should re-evaluate the role of horizontal gene transfer (HGT) of reverse transcriptase from the human genome as well as the phage activity of self-organization and self-assembly, for Hendrix and colleagues (2000) propose that a lamb-doid bacteriophage can be built from nothing, with one icosahedral capsid gene functioning by itself as an agent of HGT.

Consider that hybrid speciation is when an organism breeds back and merges with the parent species. For the hybrid to be viable, the two organisms must have similar chromosomes. If a bacterial virus is our remote ancestor, then our genome's reverse transcriptase or telomerase may be just the genetic ticket we need for hybrid speciation. However this type of evolution is a return to the quantum environment of the parent species, a virus.

Although controversial, Niles Eldredge and Stephen J. Gould propose a theory of rapid evolutionary change called punctuated equilibrium. These paleontologists emphasize that a species can have the same morphology or shape for a long time, then change suddenly. This is what occurs in ancient Egyptian texts; however, the individual human being makes the choice for evolution, and speciation results in a quasi-hybrid being of human and viral DNA that adapts to a quantum environment. Recall that scientists believe the quantum is ordering our classical world. So, this hybrid evolutionary move to the quantum may be a step up to what Nietzsche calls a *superspecies*.

Baseball Diamond 9

Frank's Time Crystal

Frank found subtle exceptions that link motion and the state of being at minimum energy.

Maulik Parikh
Theoretical Physicist

Welcome to the Baseball Diamonds. This section features a significant player from each fantasy-draft team and reviews the conceptual bases for the Game of the Centuries. Where necessary, Spring Training information is briefly summarized to reinforce understanding.

Frank Wilczek is the Designated Hitter for the Diamondhearts, a team that generally favors the primacy of mind. Frank is a theoretical physicist and mathematician, who won the physics Nobel in 2004 for his work on the strong interaction, along with David Gross and H. David Politzer. Recently, Frank has provided the world with a mathematical solution for a time crystal, an actual state of being that functions something like a perpetual moving machine.

What exactly does this mean? Ordered states can result from breaking nature's regular laws or symmetries. Break a spatial translation symmetry and a crystal results. Break a phase rotation symmetry in a superconductor and a supercurrent results. Yet, many scientists believe that a time translation symmetry cannot be spontaneously broken in the quantum domain because microscopic objects such as particles are too small to break the symmetry of time evolution. However, mathematical proof supports that a time translation symmetry can be spontaneously broken to enable an ordered state. (van Wezel 2010) Recall that van Wezel stresses the importance of the unexplored experimental gap of mesoscopic states (viruses and bacteria) in between the microcosm (particles) and the macrocosm (humans). Using mathematics, van Wezel explains that it is reasonable that an energy (or mass) scale exists in the mesocosm, where dynamics occur at the human timescale. In this thin mesoscopic spectrum, states become equal in energy as the system grows to infinity. When this happens, it is possible to reduce to the system's quantum ground state (lowest energy state), while violating the system's governing physical laws related to time symmetry. Although scientists accept Time reversal T violation (time moving backwards), they are skeptical about breaking time translation symmetry.

In support of van Wezel's claim, physics Nobel laureate Frank Wilczek (2012; 2013) supports that it may be possible for a microscopic model to break time translation symmetry. Put simply, in Frank's time crystal, time-dependent phenomena can be exhibited in a time independent system. Also, this crystal would be free of dissipation, so the system would not dissipate or lose energy through a conversion into heat. (Wilczek 2013)

Frank's mathematical solution supports cold states of matter that could form a tiny time crystal with a tiny ring of particles whose lowest-energy states are periodic in time. This new state of matter would return cyclically to the same initial state. Once in motion, a time crystal would not need an outside force of energy to keep it moving. In his recent articles (2012; 2013), Frank explains that his idea comes close to a perpetual motion machine, or a superconductor might do the job in the right circumstances. Superconductivity is the flow of electric current without resistance in certain metals and alloys at temperatures near absolute zero. Put simply, Frank's mathematical solution for a time crystal describes a state of being very similar to the ancient Egyptian quasi-hybrid being.

Eternal Return of Cultural Knowledge. Surprisingly, five thousand years before Frank Wilczek's time crystal, the ancient Egyptian pharaohs and their hieratic priesthood concealed a remarkably similar idea inside their pyramids and tombs. They explained how the élite with knowledge could merge with the Sun's magnetic field (the necessary hydrogen energy boost and prepared spin state) and transform into a cold light star or crystallized quasi-hybrid being very similar to Frank's time crystal that does not dissipate by a heat reaction and lives for a long, long time. According to ancient Egyptian texts, the transformation allowed the élite to escape their "second death," what their texts describe as a hell-fire heat dissipation of dismemberment-by-knives, slaughter, torture, decapitation, and burning in fiery pits. Related research also shows that the early Chinese emperors and their sages, as well as the seventeenth century Navajo shamans were cognizant of this cold-light crystal transformation. They too advised an initial high temperature boost from the hydrogen Sun to jumpstart the self-assembly process, followed by a cold light transformation to avoid annihilation. Granted, this merger may not have been an actual collision with the Sun, but might be better understood as chemical bonding of a human DNA fragment to the Sun's hydrogen protons and magnetic field spin state. The Egyptian input of hydrogen energy, a lightning strike from the Sun, was necessary to jolt or activate self-assembly and transformation to something very similar to Frank's time crystal.

Now, an ordered state can result from breaking nature's symmetries, as van Wezel

(2010) explains. Put simply, if decoherence can be avoided by cooling, the time evolutions can be distinguished. Van Wezel cites experiments that are currently being conducted to create a mesoscopic order parameter field that is coupled to light or a magnetic field that is suitable. This is exactly what happens in ancient Egyptian texts that emphasize the value of the Sun, its magnetic field, and the cold cosmic sky. In the cool temperature of space close to absolute zero, the electrons in some metals will suddenly shift into ordered states to travel collectively without deviation from their path like superconductors. This fluidity is due to the mutual action of molecules that swerve or deflect from their paths. Since biologists consider both DNA and viruses as crystals or ordered states formed by a three-dimensional pattern of atoms, ions, or molecules with fixed distances between parts, let's consider Egyptian science in light of Frank's mathematical solution for a time crystal.

DECOHERENCE
In physics, the process in which a system's behavior changes from that which can be explained by quantum mechanics to that which can be explained by classical mechanics.

Two Thought Experiments. Is it possible for the metal-binding proteins in phage Lambda to attract organic DNA molecules and then transform and multiply into hybrid viral crystals via the ancient glycolysis-fermentation pathway? Could this result in a broken time translation symmetry comparable to Frank's time crystal? After all, with the loss of oxygen and glucose to our cells at a human death, the ancient glycolysis-fermentation pathway should still function with lactose for adaptive survival, just as the pharaonic priesthood predicts.

Speaking of fermentation, after soaking an iron-based compound in alcoholic drinks such as beer, red wine, and sake, researchers at Japan's National Institute for Material Science discovered that the metal compound became superconductive. Yoshihiko Takano, professor of Japan's National Institute for Material Science and researcher Keita Deguchi now have a wine-soaked metal compound that is seven times more superconductive when dipped in red wine. (Suzuki 2011) This suggests that there is a link between fermented products and superconductivity. Now, Frank's thought experiment, backed up by a mathematical solution, involves the creation of a time crystal that is similar to a supercurrent spontaneously induced in a superconducting ring threaded by a tiny magnetic field. Ancient Egyptian texts explicitly describe a similar thought experiment where light or UV irradiation activates the lytic cycle of phage Lambda along the ancient glycolysis-fermentation pathway for rolling circle replication, resulting in what might be called the superconducting ring of replicating viral DNA threaded by its tiny magnetic field. With great simplification, the relational difference between the two thought experiments is that Frank has done the math for a time crystal, showing this state of being can exist, while ancient Egypt has described the state of being and the creator of Frank's time crystal—phage Lambda, a viral ferryboat for human DNA using lytic replication for crystallization to the native state. This may seem like a foul fly ball way out of the ball park, but who would have suspected the connection between fermented red wine and a superconductive metal compound? Let's clarify these thought experiments with an understanding of superconductivity and a look at Lambda's metal-binding proteins, as well as its tiny magnetic field.

Superconductivity. With their knowledge of high temperature superconductivity today, scientists understand how certain metals cool or crystallize and support perpetual electric currents due to a ring threaded by a tiny magnetic field. Now, Frank's mathematical solution describing a time crystal or state of being with infinite-range correlations may be support for ancient cultural views on a long, crystallized existence after death. The élite of ancient cultures—the Egyptian pharaonic priesthood, the early Chinese emperors and their sages, the Navajo shamans—explain that the environmental jolt for transformation is the hydrogen driving force of ultraviolet radiation from the Sun, and then they describe the resulting time reverse to the ancestors or cosmic origin where higher order exists. Surprisingly, the high temperature rendezvous with the Sun or hydrogen in the quantum domain may have jumpstarted a metal-binding process very similar to superconductivity. For example, the Chinese Emperor's transcendent objective to become a star is described in Huainanzi 4, Section XIX and summarized as the "growth of ores in the earth, vaporization, and condensation" related to gold, lead, copper, silver, iron (Major 1993, 214-215). This is a sound explanation of core-collapse supernova nucleosynthesis, for it explains the synthesis of heavy elements up to iron. (King 2009a)

Still, if we are to consider this speculative proposal that Frank's time crystal is similar to a crystallized quasi-hybrid state of being explained by earlier cultures, then the virus Lambda must have the necessary metals to allow a type of superconductivity or metallic transformation that could exist at cold temperatures. Now, the head or the capsid of phage Lambda contains both arms of its DNA in a linear form that is protected by a protein coat. Surprisingly, Lambda has impressive metal-binding potential in its viral coat proteins. In May, 2011, Zhang and colleagues searched for metallic proteins that could bind metal ions (metalloproteins) in phage Lambda due to the high metal-binding potential of its viral coat proteins. Scientists have found transition metals manganese (Mn), iron (Fe), cobalt (Co), nickel (Ni), copper (Cu), and zinc (Zn) in Lambda's protein coat of many colors, using state-of-the-art metallomics techniques (Zhang et al. 2011). Visualize this metallic coat of many colors as the colors of the following transition metals—pale pink (Mn), silver-gray (Fe), cobalt blue (Co), silver (Ni), reddish-brown (Cu), and bluish gray-white (Zn).

Approximately one-third of proteins require metals, while some metals tend to bind organic molecules more selectively. This natural order of stability for divalent (forming two bonds) metals is called the Irving-Williams series: copper and zinc form the tightest complexes with organic molecules, then nickel and cobalt, followed by iron and manganese, and finally calcium and magnesium. (Waldron et al. 2009) Copper, zinc, nickel, cobalt, iron and manganese are the same six transition metals found in phage Lambda that selectively bind organic molecules. Transition metals are malleable, conducting electricity and heat. The only elements known to produce a magnetic field are iron, cobalt and nickel—the same elements in phage Lambda's head or coat protein of many colors. So phage Lambda not only has a high attraction for organic molecules, but it also has the three elements to produce a magnetic field. This adds some support to the Isis Thesis proposal based on the holographic principle that the Sun's interplanetary magnetic field represents viral DNA on the mesoscopic scale. So, Lambda not only has metal-binding potential in its viral coat proteins, but it also has the Irving and Wallace series of metals that form the tightest complexes with organic molecules, including the three known elements that produce a magnetic field.

The human genome and gut microbiome also have metal content. Although identifying metalloproteins or metal-binding proteins in the human genome is a difficult process, Claude Andreini and his colleagues propose that about "2800 human proteins are potentially zinc-binding *in vivo*, corresponding to 10% of the human proteome" (2006). Also, many human biological functions require iron. Although microbial metalloproteomes are largely uncharacterized, biologists know that metal ions have unlimited catalytic potential with the ability to stabilize proteins. Thus, key roles include those in respiration (iron and copper), photosynthesis (manganese), and metabolism (iron). For example, researchers analyzed cytoplasmic extracts of the *E. coli* bacterium in our gut, showing the metal contents for its growth medium included a high abundance of iron (Fe); medium abundance of copper (Cu), manganese (Mn), zinc (Zn), nickel (Ni), molybdenum (Mo); and a low abundance of vanadium (V), cobalt (Co), arsenic (As), lead (Pb), cadmium (Cd), uranium (U). Within us, the *E. coli* bacterium in our gut houses the sleeping Lambda prophage (Cvetkovic et al. 2010). So, metal-binding proteins and a metal growth medium can be found in the human body. It seems that both phage Lambda and the human organism have the potential for a type of superconductivity.

In addition, we must consider that the solar wind of the Sun is an excellent electrical conductor. The chemical composition of the solar wind found in the Sun's plasma is ionized hydrogen (electrons and protons) with helium (8 percent) and trace amounts of carbon, nitrogen, oxygen, neon, magnesium, silicon, sulfur, and iron ripped apart due to heat (Feldman et al. 1998). The satellite SOHO (Solar and Heliospheric Observatory), a joint project of the European Space Agency and NASA, studies the Sun without interruption and has identified trace elements such as phosphorus (P), titanium (Ti), chromium (Cr), and nickel (Ni) (Galvin 1996).

In 2006, iron-based superconductors or chemical compounds containing iron were discovered. High temperature compounds containing copper and oxygen are superconductors. In addition, mercury, aluminum, tin, lead, titanium, cadmium, and iridium are superconductors, while hydrogen is being actively explored as a superconductor.

Recently Lorenz and Chu (2004) identified superconductivity in fifty-two elements of the Periodic Table, including non-metallic elements like sulfur or oxygen and alkali metals with only one valence electron such as lithium (Li). They explain that metallic hydrogen has been predicted to become superconducting at high pressure with extraordinary high temperature. Thus, with phage Lambda's metal-binding proteins and its strong binding affinity for organic molecules, with Lambda's magnetic field metals, with the solar wind composition of superconducting elements for viral self-assembly, and with recent findings on fifty-two elements of the Periodic Table having superconductive potential, it would seem likely that these ingredients might jumpstart a superconducting current similar to Lambda's lytic rolling circle replication of DNA that creates viral crystals, functioning at the level of the mesocosm. So, ancient knowledge, secreted by the Egyptian pharaonic priesthood, the Chinese emperors and sages, and the Navajo shamans, seems to be describing the self-assembly and replication of a metal-binding viral crystal similar to Frank Wilczek's mathematical solution for a state of being in a time crystal. And so the argument goes. Still, in light of our origin from metal-producing stars, the evolutionary process described in these ancient texts seems reasonable.

What is also interesting about Frank's time crystal is that our cosmos may cycle or function in a similar fashion. For instance, physicist Jakub Zakrzewski (2012) asks, "Could the postulated cyclic evolution of the Universe be seen as a manifestation of spontaneous symmetry breaking akin to that of a time crystal?" According to new science, symmetry breaking allowed our cosmos to begin its current diffusion-collision expansion course. On the other hand, ancient Egyptian texts clarify our relation to the cosmos: the original symmetry breaking resulted from the protein competition between Lambda c1 protein and Lambda cro protein. Our current universe is the macrocosmic energy landscape of c1 protein that folds/binds by diffusion-collision expansion like our universe; Lambda cro protein is initially unstructured like our 98 percent junk DNA, and it folds/binds by means of the fly-fishing speed-up mechanism, which would represent the collapse of our universe. The combined activity of both proteins represents the cyclical, mathematical expansion/collapse model of our universe, and the emergence of a new cosmos represents the quantum energy landscape of cro protein. The whole is greater than the sum of the parts, and it includes our classical world and the quantum. The cycle is human-activated and due to the origin-fixation dynamics of cro protein that seeks its quantum origin. Put simply, cro protein is running for the home plate of the Lambda genome.

All in all, with his mathematical solution for a time crystal, Frank may be this year's winner of the Edgar Martinez Award for outstanding designated hitters because he can easily support the pitch of the Diamondheart team on the possibility of a quantum state of being.

Baseball Diamond 10

Francis' Trouble with the Curve

Jesus therefore saw the tax collector, and because he saw the deplorable together with the plucked out, he says to him, "follow me."
Literal Latin Motto

In our model for understanding named the Game of the Centuries, we are featuring the proactive pitcher Francis on the Thunderhead fantasy-draft team, who is also known as Jorge Mario Bergoglio, the 266th pope of the Catholic Church. For a long time, all a Pontiff had to do was throw his mitt down on the historical muscle field to scare off opposition. But today, the ball game is different. People are thinking and choosing. As the starting pitcher, Francis has a colorful Coat of Arms or blue shield, depicting a gold and silver key crossed beneath the headdress like the ancient Egyptian pharaoh's crook and flail (see Figure 10.1).

Figure 10.1 Francis' Coat of Arms (C. C. Share Alike license)

Centered between a gold cross on each arm is a radiant Sun with the monogram of Christ or IHS. Often interpreted as *Iesus Hominum Salvator* (Latin: "Jesus, Savior of men"), this monogram is crowned with a cross and below it are three nails. Beneath this is an eight-pointed star associated with the virgin Mary, along with a spikenard or flowering plant suggestive of vegetative replication, since the spikenard is an angiosperm or seed-producing plant, that is, the plant flowers and also produces fruits that contain seeds. Below these signs is the Latin Motto: *Vidit ergo Jesus publicanum, et quia miserando atque eligendo vidit, ait illi, 'Sequere me.'* According to the Catholic World Report, the translation is "Jesus therefore sees the tax collector, and since he sees by having mercy and by choosing, he says to him, 'follow me'" (Harmon 2013). The Motto is from the seventh century Venerable Bede's homily on the Gospel of St. Matthew about Christ asking the tax collector to "Follow me." However, the literal translation of *et quia miserando atque eligendo vidit* is "and because he saw the deplorable together with the plucked out." This suggests that the tax collector was also deplorable and plucked out, rather than saved. The mistranslated Motto is clearly a curve away from literal Latin.

In the Gospel of Matthew, Christ is often speaking as a judge, trying to determine the sheep from the goats, the good seed from the bad weed, and the good fish from the foul. Christ says, "If your right eye causes you to sin, pluck it out and throw it away" (Matthew 18:9). Obviously, no one is going to pluck out his own eye on the premise that the eye sinned, so Christ must be reinforcing the value of hypercritical judgment skills. In Matthew 7:13-14, Christ explains, "Enter by the narrow gate; for the gate is wide and the way is easy, that leads to destruction, and those who enter by it are many. For the gate is narrow and the way is hard, that leads to life, and those who find it are few." As mentioned earlier, nature's modus operandi is the survival of a Few over the Many, and Christ agrees. Christ also speaks of trees that grow good and bad fruit, mentioning that the bad fruit "is cut down and thrown into the fire" (Matthew 7:15-19). In Matthew 10:34 Christ states, "Do not think that I have come to bring peace on earth; I have not come to bring peace, but a sword." So, Jesus is holding true to his unmerciful, sword-spitting, apocalyptic presence in Revelation, where most of humanity is annihilated.

Now, Latin is a literal language of words that does not deviate from original meanings. Yet, nonliteral or figurative language is often used to alter the literal meanings, and perhaps that is what happened with the Pope's Latin Motto from St. Bede's seventh century homily. So, let's look at the Latin Motto in light of its literal message. According to Dr. John C. Traupman's Latin and English Dictionary (1966), used as a noun *publicanum* means "tax collector" or "publican." Used as an adjective *publicanum* refers to "public revenue," and "revenue" in Latin is *fructus* or "fruit," along with the second meaning of "proceeds, profit, income." *Fructus* possesses a third meaning of "enjoyment, satisfaction, benefit, reward, results." Also, *taxō* means "to appraise" and indeed, Jesus is appraising or judging individuals in Matthew's gospel. Continuing with the literal translation of the Latin Motto, *et quia* means "and because" and *vidit* means "he saw." So, we have, "Jesus therefore saw the tax collector [the public revenue, fruit or proceeds] and because he saw. . ." Now, here is where the Pope's Latin Motto or figurative curveball misses the literal mark. Consider that the curveball is a pitch with forward spin that causes it to dive in a downward path as it approaches the plate. The Pope has trouble with his curve because it does not dive downward toward the dirt as it approaches the plate—it moves in a direction opposite to the curve ball, opposite to a literal translation of his Latin Motto. So the Pope has trouble with his curve. Let me explain further.

Now, if one translates *miserando atque eligendo* in the Pope's Coat of Arms and in the Latin Motto as the Vatican meaning of "lowly but chosen" or "by having mercy, by choosing him," then the core literal Latin meaning is altered and softened. This is because *miserandus* means "deplorable, pitiable" and the verb *eligo* means "pluck or root out, extract." *Eligendo* is a participle, and *eligendus* means "which is to be extracted" from the future passive participle *eligo* ("to pluck out") or figuratively "to pick out, choose." This makes one think that the "tax collector" was part of the "plucked out" and "deplorable" bad fruit. The Vatican's figurative language of "by having mercy, by choosing him" and "lowly but chosen" does not mean the same as the literal interpretation of "deplorable as well as plucked out." If we look deeper into this Latin Motto, the "plucked out" interpretation suggests destruction of the bad fruit that Christ is appraising in Matthew's gospel. Now, if the Latin word *miserati* meaning "pity, compassion, sympathy" was used instead of *miserandus* ("pitiable, deplorable"), then the current Catholic translation might have some credibility. The problem with the Pope's curve ball is that his Latin Motto has too much of a curve away from the literal Latin translation. Actually, the Pope's pitch is a screwball because it moves in a direction opposite to the curve ball. A screwball like that could strike or pluck out a lot of batters.

So, the Catholic interpretation of the Pope's Latin Motto on his Coat of Arms centers on a figurative, softened translation of St. Bede's homily, favoring mercy rather than destruction. Also, the Catholic view is that the Coat of Arms depicts real people and real events: the dying Jesus nailed to his cross, the virgin Mary associated with a star, the fruiting spikenard plant. However, another interpretation exists. Recall that the Egyptian goddess Isis, a sign for lactose metabolism, is also associated with the stars, the Milky Way, and virginity, while the Pope's Coat of Two Arms easily represents the two arms of the Lambda genome. In fact, the Coat of Arms itself suggests Lambda's colorful coat protein, showing that human creativity may be due to viral gene inspiration. The warrior shield suggests the battle between two Lambda proteins over vegetative replication. Cloning is involved in rolling circle replication, and this is very similar to the activity of the seed-producing spikenard that produces seed genetically identical to itself—clones. What is surfacing here is a deeper biological interpretation based on the action of signs related to an ancient viral gene expression network.

Further, various Latin meanings relate to "fruit, profit and reward" pertaining to appraisal and judgment, "for many are called, but few are chosen" (Matthew 22:14). In the actual viral replication process, the Egyptian deity Osiris represents the substrate or chemical species which reacts with a reagent or substance added to a system in order to bring about the chemical reaction. Both Osiris and Christ represent the rising prophage in the process of rolling circle replication, which requires additional reagents that are also degraded for necessary nonnative interactions. So, if we take Osiris or Christ as a sign for the viral substrate on which the process occurs, and the Egyptian plebians and Christian parishioners as the unwitting reagents that are deprived of the necessary knowledge to avoid degradation, then following Osiris or Christ may be the pathway to degradation, not survival—especially for the tax collector in Matthew's Gospel who may have been called, but not chosen.

Now, even though the Pope has trouble with his curve on his Coat of Arms, in his Apostolic Exhortation *Evangelii Gaudium* or "Joy of the Gospel," he uses a strong-armed verbal hard ball to smash capitalism by calling it callous and tyrannical.

> Just as the commandment "Thou shalt not kill" sets a clear limit in order to safeguard the value of human life, today we also have to say "thou shalt not" to an economy of exclusion and inequality. Such an economy kills. How can it be that it is not a news item when an elderly homeless person dies of exposure, but it is news when the stock market loses two points? This is a case of exclusion. Can we continue to stand by when food is thrown away while people are starving? This is a case of inequality. Today everything comes under the laws of competition and the survival of the fittest, where the powerful feed upon the powerless. As a consequence, masses of people find themselves excluded and marginalized: without work, without possibilities, without any means of escape.
>
> Human beings are themselves considered consumer goods to be used and then discarded. We have created a "throw away" culture which is now spreading. It is no longer simply about exploitation and oppression, but something new. Exclusion ultimately has to do with what it means to be a part of the society in which we live; those excluded are no longer society's underside or its fringes or its disenfranchised – they are no longer even a part of it. The excluded are not the "exploited" but the outcast, the "leftovers." (Bergoglio 2013, 44-45)

This is a very strong pitch by Pope Francis, who observes that capitalism as an economic power is imbalanced and a "new tyranny" over humanity. Capitalism is an economic system in which investment in and ownership of the means of production, distribution, and exchange of wealth is made and maintained chiefly by private individuals or corporations. In fact, the Pope's pitch sounds like a gyroball with a trajectory somewhere between a fastball and his problem curve ball. His pitch is a fine fast ball because when he first arrived at the Vatican, he set up agencies to explore issues relating to previous papal problems, such as the information leaks by Pope Benedict XVI's butler Paolo Gabriele, who was indicted in 2012 for theft, but pardoned shortly thereafter by Pope Benedict. Claims of money-laundering also surfaced, and ultimately Pope Benedict resigned in February 2013. Then Francis became Pope, making the pitch that people should not trust the "goodness of those wielding economic power."

However, at this point the Pope's pitch becomes another problem curve ball due to the money-laundering claims by Dr. Jonathan Levy, a US attorney who attempted to sue the Vatican Bank for money laundering related to looted money from Serbian genocide victims during World War II. In a videotaped interview, Levy said that this cash was laundered through the Vatican Bank for its ten percent commission. Levy lost his case in court because the Vatican Bank claims that it is immune from US and all other lawsuits. (2013) So, Pope Francis definitely has some issues related to capitalism in his own ball park.

However, the problem of capitalism and its imbalanced scales of justice between the wealthy and the poor is complicated for the Catholic Church. According to Washington Examiner columnist Shikha Dalmia, "The church is reportedly the largest landowner in Manhattan, the financial center of the global capitalism system, whose income puts undisclosed sums into its coffers." In his article entitled "Pope Francis shouldn't bite the hand that feeds the Catholic Church," Dalmia also mentions the huge spike in the Church's side, for women or one-half of humanity are excluded from the priesthood. (2013) Also, the late philosopher Avro Manhattan, who was friends with H. G. Wells and Picasso, wrote a book entitled *The Vatican Billions* (1983), claiming that the Catholic Church is a huge financial power and property owner with formidable stocks and a treasure of solid gold. As chief executive officer of this immense wealth, Pope Francis has a lot of clout in the world. However, it seems that we have a dual tyranny—both the new tyranny of capitalism, as well as the old patriarchal tyranny of Catholic biopower that excludes women from being ordained priests, while hording wealth that could reduce world poverty.

Deconstructing Pope Francis' full message to the world in his joyous gospel, his main theme is a new direction for the Church, perhaps a revolutionary change by confronting the expanding economic power structure. His focus is on the person of Jesus Christ as a change agent, rather than the private control of trade and industry for profit. Perhaps Pope Francis is aware of the 1968 drama "Shoes of the Fisherman," where actor Anthony Quinn as pope decides to give away most of the Church's riches. It is obvious that Buenos Aires-born Pope Francis is aware of the poverty of the world, and he would like to see a conversion in the world with the wealthy helping to eradicate poverty. His attitude, a balancing act between two extremes, makes him a fine candidate for starting pitcher for the Thunderhead fantasy-draft team. Fortunately, Francis is not afraid of getting splattered with dirt or struck by the ball at the pitcher's mound.

> Here I repeat for the entire Church what I have often said to the priests and laity of Buenos Aires: I prefer a Church which is bruised, hurting and dirty because it has been out on the streets, rather than a Church which is unhealthy from being confined and from clinging to its own security. I do not want a Church concerned with being at the centre and then ends by being caught up in a web of obsessions and procedures. (Thompson 2013)

Despite his trouble with the curve, Pope Francis may throw some trick pitches because he does not like being caught up in procedure. Today pitches such as the dirty, unsanitary spitter or spitball are banned, along with doctoring balls with pine tar. However, without feeling confined or restricted, the clever Pontiff can use the knuckleball that dances all over the place and screws everybody up because the hitter can't hit it, the catcher can't catch it, and the umpire can't call it. The Pope also can use the forkball, a real brain scrambler because it dances all over and then sinks. Then there's the 12-6 curveball—with your arm behind your head, you throw that ball like a karate chop, following through with emphasis.

In his favor, the Pope is backed up by some strong defensive players. Karl Marx, the center fielder on the Pope's team, would agree with him that capitalism is a mode of domination over the poor. But, what if the Pontiff looked at the economic system itself and its prevailing ideology of the actual socio-economic base of workers and the superstructure of the social class owning the means of production as signs of a process? According to Marxist critic Terry Eagleton, the dominant ideas of a society (literature, religion, politics, ethics, aesthetics, and so on) are those of the ruling class, and the prevailing ideology legitimates their power (1986, 536). The question is—what process is capitalism and its culture a sign of?

If we look at the etymology of "capital," we discover the word is from the Latin *capitalis*, "of the head" and Latin *capitulum*, "little head." Cloning capital or dollars is remarkably similar to cloning little viral heads and tails—both are multiplied or increased. Capitalism, the production of dollars, the cloning of money, the production of bucks for the powerful few—may represent the action of signs connoting viral lytic replication or the cloning of heads and tails in the quantum domain. Thus human economic behavior is a biochemical sign reinforcing the ancient lytic gene expression network for survival of the few by modeling the accumulation of the wealth for the ruling class. Thinking of it in this way showcases capitalism as a sign of nature's general modus operandi—the survival of a few individuals over the multitudes, the salvation of the slime mold spores over the sacrificial stalk cells, the domination of the rich over the poor, the wealth of the Church over its parishioners. These signs define the underlying evolutionary genetics of Lambda's lytic replication pathway for survival of those with knowledge, as well as nature's general dynamics. Also, the buck or ram-god (Sun-god) is the major iconographic sign of ancient Egypt and Christianity. And so, the Pope's Coat of Arms, as well as the Vatican's land ownership, reserves of gold, and valuable historical artwork can be directly linked to the prevailing ideology of wealth for the few in modern capitalism. This is similar to the power structure of ancient Egypt. So, we should be apprehensive about the literal meaning of *miserando* on the Pope's Coat of Arms for the majority of individuals.

Now, flip this coin around in your mind—heads or tails, survival knowledge or ignorance, riches or unfair economic poverty? What matters here is the action of these signs and the knowledge they offer. If capitalism, the production of little heads or dollars, connotes survival knowledge of the viral gene expression network for lytic replication that clones little heads, then the same biological survival instinct is present in Egyptian mythology, Christianity, and modern genetic engineering that focuses on cloning. A structural relation or hidden agenda exists between literary texts, world visions, baseball, science, religion and history itself. With all this in mind, we expect Pope Francis to see the connection between Vatican financial power and the tyranny of capitalism, so that he is a formidable pitcher in the Game of the Centuries. Yes, this Pope should be powerful, and we should expect some trick pitches to compensate for his trouble with the curve ball related to his Coat of Arms and his capitalist gyroball.

First Base: Quantum Physics/Classical Physics

I think once you start as an announcer, you have to decide what kind of approach you're going to have. I decided very early that I was going to be a reporter, that I would not cheer for the team. I don't denigrate people who do it. It's fine. I think you just have to fit whatever kind of personality you have, and I think my nature was to be more down the middle and that's the way I conducted the broadcasts.

Ernie Harwell

Related to the fantasy-draft Diamondheart and Thunderhead teams, each team's ideology or worldview is explained by the views and research of the solid line up of team players. Now, the players on each team may express divergent views relative to the respective team concepts of matter and mind. Yet, the team as a whole represents a general consensus on the debates proposed, and each player has been placed on the appropriate team to enable a comprehensive view of the debate in question at each base.

At first base the Thunderhead team's proposal is that our classical world is the Real versus the Diamondheart claim that the quantum world orders everything, and the classical world is an illusion. To understand this debate, let's review the existing state of scientific thought. One commonly accepted theory of space and time is the Big Bang theory, describing our universe's hot expansion from its birth, as well as the Big Crunch theory that reverses the space and time of our universe, collapsing it on itself. Inflationary cosmology theory inserts a short, explosive expansion of space at the birth of our universe, followed by the hot expansion. Now, the twentieth century witnessed the birth of a quantum, the smallest discrete unit of energy that is indivisible. Max Planck created some havoc in scientific thought by explaining that sub-atomic energy can only be transferred in these small packets or quanta. In 1905, Einstein showed that light was also particle-like or in discrete packets. In 1924, the French physicist Louis de Broglie proposed that electrons and other material particles also exhibit wave properties. In other words, matter owns both wave and particle properties, and this is actually the crux of the debate at first base.

During these discoveries, Neils Bohr proposed a quantum mechanical description of the atom using Planck's mathematical constant or proof to replace Ernest Rutherford's idea of the atom with its central positive nucleus and negative orbiting electrons. Using the hydrogen atom, Bohr postulated that the angular momentum or spin of the electron can only have discrete values, and that electrons could jump from one orbit closer to the nucleus by emitting energy in fixed quanta or jump to a larger orbit by absorbing energy in fixed quanta. Put simply, Bohr's model identifies the electron in discrete orbits versus existing in a continuum of energies described by classical physics. In the realm of quantum mechanics, phenomena are submicroscopic. A quantum leap is the movement or jump between discrete energy levels; in other words, motion is not the smooth change envisioned by Newton and classical physics. Add to this that light photons and electrons have either a wave or a particle state (wave-particle duality), and what is real is becoming blurred.

As an example, in the macroscopic world of humans, an object with greater mass exhibits a shorter wavelength. A material person would represent a very small wavelength that cannot be measured. According to Heisenberg's uncertainty principle, certain physical properties of a particle cannot be known simultaneously, such as momentum and position. So, our understanding of reality is limited. Yet, Heisenberg's uncertainty principle and Schrödinger's wave character of matter position quantum mechanics as a fundamental science because it includes atomic and molecular physics, the chemical behavior of the periodic table of elements, and the properties of atoms and molecules. Quantum mechanics is a very successful theory because it explains the behavior of matter on the macroscopic, mesoscopic and microscopic scales.

Quantum Entanglement. But there is more to the elusive subatomic world. Because of the wavelike nature of particles existing in a probability field, the wave represents the probability of finding the particle. So, particles can be in two places at once (entanglement), while information can travel faster than the speed of light or backward in time. Physicist Brian Greene explains quantum entanglement:

> If two photons are entangled, the successful measurement of either photon's spin about one axis "forces" the other, distant photon to have the same spin about the same axis; the act of measuring one photon "compels" the other, possibly distant photon to snap out of the haze of probability and take on a definitive spin value—a value that precisely matches the spin of its distant companion. And that boggles the mind. (2004, 115)

In ancient Egyptian texts, the same sophisticated understanding of entanglement is present, for the dead king's action of turning left-to-right (spin orientation) correlates with the same action in Osiris, and this throws the genetic switch for the rising of the prophage and lysis. This simply means that the dead king's polarization is the primary measuring entity that switches on the quantum lytic cycle of numerous particles that are quantum mechanically entangled because of the protein folding funnel of Lambda cro protein. On the quantum level of DNA, these particles are spatially entangled in a complex correlated pattern, and in ancient Egyptian science, quantum entanglement results from correlated protein energy landscape dynamics. Because of quantum entanglement, objects can become linked and instantaneously influence one another regardless of distance, and recent evidence suggests that this quantum phenomenon might work "beyond the grave, with its effects felt after the link between objects is broken" (Choi 2009; Tan et al. 2008).

The Jumping Apple. Then there is the effect of no gravity. If Newton had one foot in our classical world and the other in the quantum, he would see the apple fall, and then, it would reverse itself and jump back into the tree. Yes, time can reverse in the quantum, but then Newton, Einstein, Schrödinger, and Maxwell's laws allow time reverse. So, the new world for conquest is the quantum domain, that crazy place of atoms with laws that govern our classical cosmos, where a particle is everywhere at once, taking on a range of values simultaneously. Add to this that in the quantum domain, a person's choices and actions can affect the whole system. As physicist Brian Greene explains, "An observation today can therefore help complete the story we tell of a process that began yesterday, or the day before, or perhaps a billion years earlier" (2004, 191). So, we have a magico-mystical world that is quantum, counterintuitional and bizarre, and this confuses the debate at first base.

In addition, several interpretations of quantum mechanics exist. First, the Copenhagen interpretation of quantum mechanics developed by Neils Bohr, Werner Heisenberg, and Wolfgang Pauli posits that large objects are subject to classical laws and small objects to quantum laws. The quantum state is described by Schrödinger's wave function, a mathematical tool describing the object's properties. Second, the Many Worlds interpretation of quantum mechanics is that all potentialities of a probability wave are realized in separate universes. (Greene 2004, 537; 539) Hugh Everett, III postulated the wave function as a real object. Since quantum systems are entangled, the observation of one observer can be split into a number of copies that exist in mutually unobservable real worlds. Third, David Deutsch proposes the Multiverse theory or that a universe exists for every possibility. Although there are other theories, the last one to discuss at first base is the Holographic Principle, a view combining the classical world of physics with the microscopic domain of quantum physics. Ancient Egyptian signs explain this view as the prevailing modus operandi of our classical world, and many scientists agree.

The Holographic Principle. To grasp the holistic mindset of the pharaonic priesthood of ancient Egypt, we must understand their conception about the construction of our cosmos. Similar to current scientific thought, they understood our classical world has a holographic mode of operation. Put simply, our classical world is a quantum phenomenon (Ball 2008) and a projected shadow in a quantum universe (Schlosshauer 2008, 39). Plato had it right with his allegory of the cave and projected shadows.

First, the holographic principle maps the correspondence between areas and information. Physicist Leonard Susskind explains our world as a hologram (1995), that is, it is an illusion or the reflection of the holograph or whole writing. The nature of our universe relates to its geometry and its holographic mode of operation. The calculations of physicists Roger Penrose and Stephen Hawking, along with the current expansion of our universe, show that it began from a singularity, an extremely small, dense point. However, a singularity can be described as a collapsing black hole or the expanding universe, which may represent a large black hole. (Gribbin 2001, 88-89) Scientists also believe that our universe's three-dimensional physical processes happen on a distant two-dimensional surface encircling us, and what we see is simply an illuminated projection or hologram of physical processes. (Greene 2004) The Isis Thesis suggests that this projection is created by the mesocosm of viruses and bacteria.

Second, in understanding the black hole dynamics in our cosmos, one must also factor in black hole holography. Juan Maldacena has shown black holes operate according to the holographic principle. All the information about the black hole is stored on the horizon. Particles living on the boundary describe objects in the interior, which could be very complex. Maldacena explains that space-time (the interior) and everything in it emerges dynamically out of the interaction of particles living on the boundary. The peculiar interactions of these lower-dimensional particles and fields may generate the force of gravity and one of the spatial dimensions. (2005) The two-dimensional surface encoding the information about the three dimensional shape of a black hole is this boundary or horizon. Thus, the ancient Egyptian dead king is a horizon-dweller and takes possession of the horizon, "so that those who are in the horizon may live for this spirit" (CT 260). The dead king claims, "My power is in the horizon" (CT 573), suggesting black hole holography.

Third, physicist Roger Penrose extends Einstein's theory of general relativity (mass relates to space-time curvature) down to the Planck scale to account for protoconscious information (Penrose 1998; Penrose & Hameroff 1998). This means that consciousness and the body are embedded in space-time (classical world), but can reduce to the quantum state, activated naturally, as has been shown (see King 2006), by practice, death ideation, meditation, fasting, physical stimulation, and so on. Also supportive of this reduction is Karl Pribram's research on vision. After fifty years of brain research, his holonomic theory summarizes evidence that the retinal image is transformed to a holographic or spectral domain. The concept of spectral domain includes the idea that colors, tones, and all exteroceptive sensations, including those dependent on spatiotemporal configurations, that is, shapes of surfaces and forms, can be analyzed into their component frequencies of oscillation. (1991, 28) According to Pribram, the spectral (energy) domain is the Fourier transform of space-time (configural). In other words, the input to and the output from a lens performs a Fourier transform. Our familiar space-time lies on one side of the transform with a distributed enfolded holographic-like order called the frequency or spectral domain on the other side. The retina and the visual cortex process an invertible transformation from a space-time image to a processing domain of at least eight dimensions: four in space-time and four spectral. When the spectral dimension dominates a perception, space and time evolution ceases and spatial boundaries disappear. (1991, 272) Thus, living humans may have accessible potential for semi-conscious quantum state reductions where spatial boundaries disappear.

In the Isis Thesis interpretation, the funerary texts indicate that the pharaonic priesthood understood the holographic mode of operation of our universe, black holes, and our brains. The pharaonic priesthood reasoned that mind could evolve to the core holograph of our cosmos, and this may be possible because mathematics and the known laws of physics support our expanding cosmos with its potential to collapse to its quantum core. Theoretically, physicists believe that the bounce that would occur close to this singularity, turns the universe inside out 10^{-43} seconds before the singularity (Gribbin 2001, 98). The laws of quantum physics support that at this unimaginably tiny size, space and time have the smallest possible length and the smallest possible time, what physicists refer to as the Planck length and Planck time in honor of Max Planck. (98) The pharaonic priesthood claims that our expanding cosmos can function as a resurrection machine with its potential to recreate itself that can be activated by a knowledgeable human observer. Further, both the large classical world and the microscopic quantum world operate by physical laws, and scientists are slowly reaching consensus that the quantum world ruled by the laws of quantum mechanics orders our classical universe ruled by the laws of Einstein's general relativity (theory of gravity). Perhaps we have our feet planted in two different places—our three-dimensional macrocosmic universe and a microscopic two-dimensional quantum cosmos.

The High-Speed Trampoline. Einstein's theory of relativity allows time to be warped by motion and gravity, while its mathematics permit wormholes, replete with their faster-than-light, backward-in-time aspect of causal loops. A wormhole is simply a rotating Kerr black hole with its white hole time reverse that casts out matter. Experiencing a wormhole would be like jumping on the axial pole of a spinning black hole, sliding down that pole, hitting the hole's ring singularity-trampoline, and somersaulting backward to be cast out of the black hole by its white hole time reverse. A similar interaction occurs at the molecular level of Lambda proteins, which bind and fold in energy landscape funnels like a microscopic Einstein-Rosen wormhole. Remarkably, Matt Visser and colleagues (2003) demonstrated the existence of space-time geometries containing microscopic traversable wormholes. The Isis Thesis proposes that these physical quantum wormholes are actually the folding/binding energy landscape funnels of Lambda's two competitive viral proteins that are closely associated due to their adjacent genes on the Lambda genome.

The Ring Singularity of a Microscopic Black Hole. Now, consider that a ring singularity of a microscopic spinning Kerr black hole may represent viral rolling circle replication. According to a recent investigation, the nature of a Kerr ring singularity is a simple structure, defined as a "ring of matter moving at the speed of light which surrounds a sheet of pure isotropic tension" (Lasenby et al. 1997). This description of the ring singularity is very similar to the simple structure of phage Lambda DNA undergoing rolling circle replication, for Lambda exhibits a circularized, double-stranded length of DNA wrapping a core of protein. In the study by Lasenby and colleagues, they use gauge theory gravity to explain the nature of the ring where matter collapses in an uncharged, rotating Kerr black hole. Also, they extend the solution to include closed timelike curves (time travel or time reverse), as well as a set of solutions for rigidly-rotating cosmic strings. The math supports that the electromagnetic contribution to energy is negative and vanishes when extended over all of space. Matter is not completely on the ring but also on the disk in a plane which has the ring as a boundary. They describe the ring as "a massless, rigidly-rotating membrane (with a ring of particles attached to the edge)" that violates the weak energy condition. (1997, 18-19) Now, microscopic (or quantum mechanical) traversable wormholes and time travel are allowed by a violation of the weak energy condition involving negative energy densities. The researchers support that the angular momentum comes from the ring, and the matter in the ring follows a light-like trajectory. A light-like curve is the trajectory of a point moving at the speed of light, so the matter in the ring acts as a light signal because light-like curves form a cone in space-time.

It may be possible that a charged Kerr-Newman ring singularity is the circularized DNA of bacteriophage Lambda undergoing rolling circle replication. If the Kerr-Newman ring is circularized Lambda DNA undergoing rolling circle replication, then the repulsion force at the Kerr-Newman ring is caused by the charge of Lambda's double helix of DNA. Recall that Lambda's coat proteins include iron, and the phage has iron in its tail tip proteins for iron binding. In ancient Egyptian texts, polar binding is described as the Sun-god becoming the deceased person, and then this combined entity on the Sun-bark meets Osiris, the rising Lambda prophage or substrate for rolling circle replication. Using pair creation scenarios from pulsar theory, Brian Punsly (1998) offers a model for a polar particle acceleration gap hovering above a charged Kerr-Newman black hole with twin winds or jets. Kerr-Newman black holes have their rotation axis and magnetic axis aligned. In brief summary, Punsly's model involves a polar particle acceleration gap with a magnetohydrodynamic (MHD) plasma and outgoing MHD wind on polar field lines. A mechanism injects seed pairs into this gap. Energy extracted from the hole is manifested as the wind, as opposed to the pair production energy in the vacuum gap. Both outgoing and ingoing winds are present, and over 85 percent of this energy flux is in electromagnetic form. The particle creation zone has a thin vacuum gap that accelerates seed pairs to high-energy curvature into two finite bands (649). In ancient Egyptian texts, we find a similar description at the ancient Egyptian polar Gap.

The vacuum Kerr-Newman magnetosphere is electric everywhere except just outside of the horizon's poles (Punsly 1998, 649). In other words, the inflow of charge is halted by magnetospheric forces. As the newly created pairs are threaded onto the magnetic field lines, magnetic stress is induced. The thin gap is primarily illuminated by photons. When seed pairs are created, they are accelerated by the gap electric field. The charge densities (p) above and below the gap are equal. The ingoing wind in the inner cylinder of the hole is magnetically dominated. Punsly's model describes black hole pair production onto magnetic field lines, and the process is very similar to the base pairs necessary for DNA production (adenine and thymine; guanine and cytosine) and rolling circle replication. However, Punsly's description of the charged ring singularity as a "small inert fossil disk" suggests the small inert viral head of DNA, that is, the Lambda prophage resting silently in the cell that circularizes to produce viral heads and tails by rolling circle replication. Punsly explains:

> The current ring is chosen to be protonic as opposed to positronic in nature. It might be thought of as representative of a small inert fossil disk resulting from the original gravitational collapse. If the disk is large, one might think of the current ring as the inner edge. (1998)

A positron is the antiparticle of an electron, and a proton is a stable, positively charged, subatomic particle in the baryon family having a mass much larger than an electron. Perhaps this protonic ring is a "small inert fossil disk" or representation of the circular, double-stranded DNA of the Lambda prophage. Add in the particle pair creation zone and we have the activity of viral rolling circle replication or pair creation on the quantum mesoscopic level.

Molecules exist in Black Holes. Jacob Bekenstein, whose mentor was physicist John Wheeler, conjectured that a black hole's interior must contain a huge number of atoms or molecules, rather than just a singularity as most physicists imagined (including Hawking and Kip Thorne). So, the properties of black holes and the laws of thermodynamics are similar. Bardeen, Carter and Hawking then showed in 1972 that the laws of black-hole mechanics resembled the laws of thermodynamics. Seeing this perfect fit of laws, Bekenstein was convinced that the horizon area is the black hole's entropy, but other physicists disagreed, claiming a black hole could not emit radiation. Then two years later Hawking showed that black holes could shrink and explode by emitting radiation, while converting mass into outflowing energy. The hole's mass goes down, its entropy and area go down, and its temperature and surface gravity go up as the hole shrinks and becomes hotter or evaporates. (Thorne 1994, 426-7; 436)

Whether or not information encoded in a black hole can escape as Hawking radiation is still an outstanding question for contemporary physicists. Ancient Egyptian texts describe the inner axial horizon of a quantum mechanical, rotating Kerr black hole as the gateway to the protein folding funnel of cro protein, while the outer horizon of the Kerr black hole represents the protein folding funnel of c1 protein. The pathway for degradation is the outer funnel, while the pathway for morphogenesis is the inner axial funnel that permits a negative energy state (particle moving backward in time). These ancient texts suggest that information escapes.

Baseball Diamond 12

Second Base: Quantum Biology/Biology

I love the game because it's so simple, yet it can be so complex.
There's a lot of layers to it, but they aren't hard to peel back.

Ernie Harwell

Here is the debate at Second Base. Generally, the views of the Diamondhearts add support to the ideas that both the organic and the inorganic are living as well as a creator virus responsible for both the organic and the inorganic. On the other hand, the general consensus of the Thunderheads supports that the organic is life, and the inorganic and viruses are nonlife. Again, players have been chosen to convey a specific worldview, but not all players on each team may support the exact concept.

In the debate of what is life versus nonlife, the Diamondhearts generally provide evidence, supporting that a virus is a living entity because it can replicate using a bacterial cell to produce its own kind. The Isis Thesis proposes that life on earth originated from this living biological couple, so the debate over whether or not a virus is dead or alive is simply pedantic noise, characterized by narrow, ostentatious concern for the traditional agenda. Now, a bacterial virus such as a bacteriophage can self-assemble, trade genes, and transfer genes between organisms. It also exhibits self-organization. Also, bacteriophages are the dominant viruses on our planet, and they are literally everywhere like gods. The Isis Thesis explains that the semiotic phenomenology of the Egyptian pharaonic priesthood harbors an eschatological survival message for humans, viz., horizontal gene transfer (HGT) mediated by the complex bacteriophage Lambda in an *E. coli* bacterial cell. This ancient pathway is the maltose transport system in the glycolysis-fermentation gene expression network in our cells. As biologist Stuart Kauffman said in *At Home in the Universe*, some organisms are too simple to jumpstart life, but *E. coli* is not (1995, 42-43). Also, a genomic network provides order for free (85). Compatible with natural law and contemporary scientific evidence, the Isis Thesis is reinforced by human genome sequencing supporting human descent from viruses, the existence of our massive gut microbiome, the phenomenal amount of microbial cells in our bodies, and experimental science showing that HGT is possible with our eukaryotic cell-type.

The Mesocosm within Us. To begin, the same principles operating in viruses and bacteria are operating in human beings. Consider our microbiome—the microbes and genes of the human body. Microbes inhabit the human body, and the first results of the Human Microbiome Project Consortium show that microbial inhabitants outumber our body's own cells by ten to one (Relman 2012). Our microbiome has been called our second genome, with the human genome in each of our cells considered as the primary DNA. Recall that the Human Genome Project discovered that approximately 2 percent of our DNA codes for human proteins, while 98 percent is non-coding DNA, primarily bacterial and viral, what has been called junk. So, it does not seem illogical that ancient Egyptian texts would be reporting a microbial mechanism for human DNA transformation at a death transition. The mesocosm of microbes is the place we return to when we die according to Margulis and Sagan (1986, 67). It is a place using free biochemical nutrients activated by ultraviolet light and lightning in the absence of oxygen (74). In this quantum domain, tiny spheres of DNA or RNA can absorb light, boosting electrons to a higher energy state. Energy is dissipated as light or heat. In ancient Egyptian texts, the dead king becomes light and does not dissipate by heat.

The Egyptian texts support that the complex phage Lambda operates as a world-heart or last universal common ancestor (LUCA) due to a competition between two of its proteins that generate two different lifestyles and DNA texts. Put simply, one DNA text (Lambda left arm) produces matter governed by photosynthesis versus the other DNA text (Lambda right arm) producing mind or energy governed by chemiluminescence or cold-light star creation. Now, to explain this clearly, let's metaphorically throw the ball from second base (quantum biology) back to first base (quantum physics) as in a pregame practice. At first base, remember the microscopic wormhole or rotating Kerr black hole with its white hole time reverse that casts out matter and actually exists on the quantum level? Well, ancient Egyptian texts describe the protein folding funnels of phage Lambda's two competitive proteins as having the formation/evaporation black hole/white hole dynamics of a microscopic wormhole. Along with this similarity, recall that Bardeen, Carter and Hawking showed that the laws of black-hole mechanics resemble the laws of thermodynamics (science of heat). Now, proteins also exhibit thermodynamic behavior. Okay, here comes the throw from first base (quantum physics) back to second base (quantum biology). This correlation between the quantum physics of tiny wormholes and Lambda protein-folding dynamics in a DNA wormhole links quantum physics universally with quantum biology.

A DNA Molecule in the Sky. Now, let's toss that ball back again to first base and catch another idea discussed there relative to the Holographic Principle. Although evidence does exist for the Holographic Principle, actual visual evidence would be very convincing. In support, let's consider recent evidence of nature's challenging holographic imagery. At the Milky Way Galactic Centre in 2006, astrophysicists observed a large magnetic field wave shaped like the intertwined double helix of DNA. Using the Spitzer Space Telescope operated by the Jet Propulsion Laboratory at California Institute of Technology, scientists examined this gaseous galactic replica of a DNA molecule (Figure 12.1). This large double-helix, magnetic-field wave extended perpendicular to the galactic plane just above the Milky Way's central engine, the supermassive black hole Sagittarius A*. (Morris et al. 2006) Scientists agree that large black holes lurk at the center of most galaxies. However, a massive nebula (molecular cloud) above Sagittarius A* that was shaped like a cellular DNA molecule must have been a surprise.

Earlier in 2004, Robert Braun and David Thilker reported a strange discovery in a survey of galaxy clusters that included the Milky Way. They saw neutral hydrogen atoms stretching like a bridge between Andromeda galaxy M31 and Triangulum galaxy M33. Lockman and colleagues (2012) also confirmed this discovery. In May of 2012, scientists using the Green Bank Telescope in West Virginia mapped this 782,000-light-year-long hydrogen link, presenting their observations at the 220th meeting of the American Astronomical Society. Curiously, this vast hydrogen bridge between two galaxies mirrors the same biological geometry of hydrogen bonds across protein interfaces on the microscopic scale. So, both the Milky Way DNA-shaped nebula and the vast hydrogen bridge linking two galaxies are observations with data that add support to scientific beliefs such as the stars are simply DNA written across the sky, the universe is within us, and we are at home in the universe. Yet, the data does not explain why the structure of space-time is molecular. However, the Isis Thesis supports that space-time is molecular because it is the classical holographic shadow cast by the energy landscape folding funnel of phage Lambda's c1 protein. On the other hand, a time-space or time crystal suggests the molecular quantum product of the origin-seeking dynamics of the folded Lambda cro protein. In the initial folding process, Lambda cro protein is an intrinsically unstructured protein (Gsponer and Babu 2009) like the unstructured toolbox of 98 percent non-coding DNA in our genome.

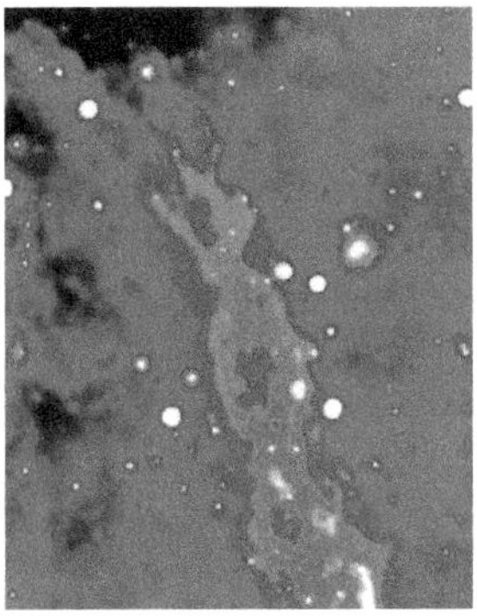

Figure 12.1 NASA Nebula Hubble Telescope image of double-helix, magnetic-field wave just above the Milky Way's central engine, the supermassive black hole Sagittarius A*. (Public Domain)

These two cases and additional scientific evidence inform us that our cosmos operates similar to a large black hole with a holographic mode of operation. This means that information is stored on the boundary of the cosmos—not in the bulk, but possibly in the quantum core. This suggests that the bulk of our cosmos—with its galactic DNA nebula and hydrogen bridges between galaxies and earth—functions like a large illusionary hologram that represents the information stored in its tiny core heart, that is, the quantum holograph on the mesoscopic scale, where viruses and bacteria exist. The Isis Thesis supports that this tiny core heart is the genome or DNA of bacteriophage Lambda which is within nature, not outside of it. The molecular structure results because our expanding cosmos may be the shadow cast by the viral folding funnel of Lambda c1 protein on the mesoscopic scale.

Accordingly, some physicists and biologists are beginning to understand that the fabric of space-time is molecular and that the equations of quantum mechanics resemble the kinetic molecular theory. So, if the tiny quantum world orders our large classical world and space-time is molecular, then the observations of galactic DNA nebula and galactic hydrogen bonds in a cosmos with a holographic mode of operation suggests that quantum microbiological and quantum physical laws are ordering the laws of our classical cosmos. This is one way of assessing the observational data, and for many scientists cloistered in their specific disciplines, this evidence of an interdisciplinary marriage of microbiology and quantum physics is unknown. Still, because of the holographic mode of operation of our cosmos and its core quantum laws, because of the thermodynamics of black holes and protein folding energy landscapes, scientists understand that a universal relation between geometry and information exists, but they cannot explain it. So they are searching for an underlying physical quantum theory of gravity or a unified description of space-time and matter (Bousso 2002).

The Isis Thesis (2004). Supported by a continuing inductive argument in twelve published papers (2005 to 2013), the Isis Thesis proposes an underlying biophysical quantum theory as to why space-time is molecular, which the above astrophysical observations reinforce. However, the thesis itself is not grounded by the two cases above, but rather by the broad comprehensive study of signs in the least-corrupted ancient Egyptian texts, as well as additional historical cultural evidence, explaining the option for continuity of mind (defined as DNA, information and energy) at a human death transition. Margulis and Sagan explain a human death as a return to the mesocosm, where hustle-bustle bacteria have invented life's chemical systems leading to fermentation, photosynthesis, and oxygen breathing. This modern view is also expressed in ancient Egyptian, as well as early Chinese and Navajo texts, but these cultures offer us important survival tips for the evolutionary process of transforming to a quasi-hybrid being, for they describe specific environmental signals and signs in a viral-bacterial gene regulation network known as the maltose transport system in the ancient glycolysis-fermentation pathway.

Relative to our universe's holographic mode of operation, the classical earth functions as an *E. coli* cell in the ancient Egyptian afterlife or their mesoscopic dwarf world. As explained within the full context of 870 unified Egyptian signs, similarities do exist between the earth and an *E. coli* cell (King 2004). James Lovelock Gaia has also theorized that the earth is a living entity, but the pharaonic priesthood realized this 5000 years earlier. The Isis Thesis offers a semiotic interpretation supporting that the quantum biology of phage Lambda infecting an *E. coli* cell is the microcosmic model or holograph grounding the macrocosmic operations of the Sun's electromagnetic field enclosing the earth's electromagnetic field. A recent study by Povolotskaya and Kondrashov (2010) published in the journal *Nature* offers further support for this idea and the prospect of our universe's holographic mode of operation. The researchers draw a comparison between the expansion of our physical universe and the protein cosmos, identifying earth as an *E. coli* cell. (see Figure 12.2)

Figure 12.2. Povolotskaya and Kondrashov (2010) cite four studies supporting that the divergence of homologous (corresponding in structure and origin) proteins from the last universal common ancestor (LUCA) is very similar to the recession of galaxies in our cosmos. The abridged schematic below of the expansion of the physical and the protein universes is drawn from their diagram in the study.

Expansion of the physical and the protein universes.
(Povolotskaya and Kondrashov 2010)

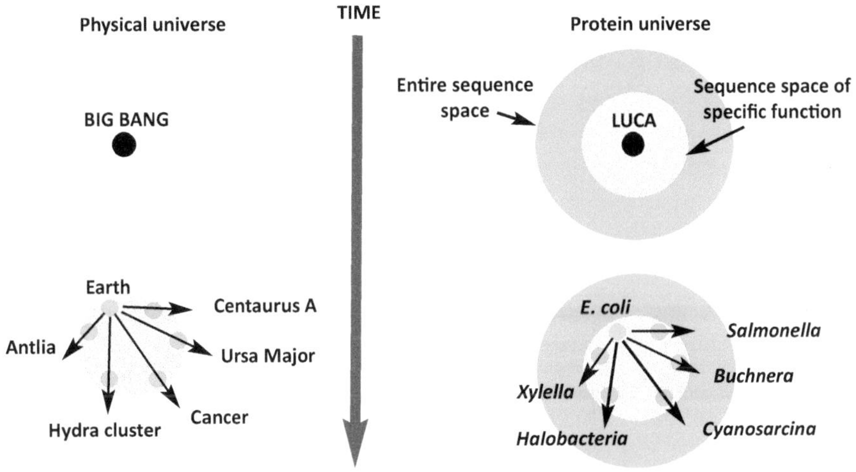

After all, if our cosmos with its holographic mode of operation is just a projected shadow thrown from a quantum protein cosmos, then semiosis is essential because our classical world is perfused with signs as C. S. Peirce observed. These signs offer us knowledge of the underlying universal system for lawful human transformation. However, the relations of meaning (semiosis) in our classical universe involve power struggles, and these also indicate the competitive dynamics of two viral proteins in the underlying quantum system. So, this semiotic approach supports that, should we understand the macrocosmic classical action of signs, such as the galactic DNA nebula and hydrogen bridge, as a holographic veil of the tiny molecular world, the possibility that our DNA can transition back to its source at human death would be accessible to human reason. The Egyptian pharaonic priesthood, the early Chinese sages and emperors, the Navajo shamans (King 2005; 2006; 2009; 2009a; 2009b), and many others support that DNA or mind can evolve after death in a lawful time-reversed process that embraces the cosmos by collapsing it to its tiny quantum beginning. Newton, Einstein, Maxwell, and Schrödinger's laws allow time reverse. This backward-in-time journey may be possible for human DNA at a death transition state due to quantum mechanics. So this semiotic approach considers the possibility of continuity of DNA based on the Isis Thesis interpretation of cultural data, grounded by the innate circularity found in natural laws (see King 2011).

Lawful Continuity of DNA or not? More support for this idea is mathematical physicist Diego Rapoport (2010), who bases her intricate mathematical and physical theory on self-reference, but the subject cognizes the world while acting as a singularity. Rapoport reviews experimental evidence that interpretation, meaning, and intention are biologically grounded in the biophotonic structure-process of DNA. Time is related to intention, control, will, the appearance of life, and the Myth of Eternal Return as a self-referential process.

As the viral footprint on 12,000 years of society space shows, culture is connected to the structure-process of the ancient glycolysis pathway that is traveled by phage Lambda, a possible candidate for our LUCA that exhibits two lifestyles dominated by two repressor proteins. If so, this is Downward Causation from microbial "mind" at the quantum origin of our holographic universe selecting space-time (gravity/thermodynamics/photosynthesis) for the unexpected complexity of human matter and consciousness. Time reverse to this final cause or LUCA allows origin-folding dynamics or mutation back to the original viral genome, that is, order at the home plate. The holographic earth functions as a genetic switching station to an earlier cosmos or mesoscopic domain. The inner, axial folding funnel landscape of cro protein leads back or seeks or binds its mesoscopic origin before space-time, turning on its gene on the original viral DNA genome. This requires a partnership of both c1 and cro binding at their respective gene-seats. This speculation seems to fit the facts.

In contrast, the Thunderheads support that the organic is alive, but that inorganic rocks and viruses (crystals) are dead. As for the origin of life, theories on the first prebiotic molecules for planetary life include the transport of molecules by meteorites striking the earth, creation from deep-sea vents, or synthesis by atmospheric lightning. Experimental data is generally lacking regarding the first molecule that could replicate. Many biologists adhere to the RNA World hypothesis, claiming RNA was the first carrier of genetic information because it could store and catalyze chemical reactions. So, the final answer to this debate is outstanding and requires further exploration.

Baseball Diamond 13

Third Base: Quantum Cosmology/ Cosmology

So loyalty to origin is innate to you. A place of dwelling . . .

Friedrich Hölderlin
"The Migration"

Quantum cosmology theorists assume that the origin of our cosmos is quantum, so these theorists track back to an epoch when Einstein's general relativity (theory of gravity) breaks down to the quantum domain. On the macrocosmic scale, the universe we inhabit expands in time and is well approximated as a three-dimensional, flat, homogeneous space. The general consensus of scientists is that our cosmos began with a Big Bang from a very small, dense state called a singularity. If one were to reverse time or follow light rays backward to the Big Bang, one would see an area light cone grow large at first and then shrink because all areas vanish as we approach the Big Bang at its quantum beginning, the extraordinarily small Planck length. Both the Thunderheads and the Diamondhearts agree with these explanations, except that the ideas of the Diamondhearts support the possibility that a virus is the origin of quantum cosmogenesis.

The cosmos is relational to the classical environment of living human beings, as well as the quantum environment. A quantum is a discrete packet or minimum amount of any physical entity. This could be a photon of light composed of vibrating waves of electrical and magnetic fields or the electromagnetic energy that fills up our cosmos. Consider that a living human being witnesses a starry classical cosmos that functions with time. However, ancient Egyptian, early Chinese, and Navajo texts explain that a human death is a transition state, allowing each individual with knowledge to collapse or embrace the cosmos as a quantum state. Put simply, in these early cosmologies, the cosmos expands from formlessness to form, and then it collapses via observers that loop back into the past to embrace the collapsed cosmos in its original quantum wholeness. This ancient idea of an expanding/collapsing universe is supported by Einstein's mathematics as well as modern scientific theory and experiment. So, this same expansion/collapse cosmos is also evident in modern cosmology (Davies 2006, 251-253),

William Blake's "Four Zoas" (King 2005), Aztec cosmology (2007), Teilhard de Chardin's biophysics (2007a), Charles Peirce's writings (King 2008), as well as the writings of Edgar Allan Poe that we will explore in the Pregame.

Microscopic Wormholes Exist. Einstein's theory of relativity allows time to be warped by motion and gravity, while its mathematics permits wormholes, replete with their faster-than-light, backward-in-time aspect of causal loops. A wormhole is simply a rotating Kerr black hole with its white hole time reverse that casts out matter, and Visser and colleagues (2003) demonstrated the existence of space-time geometries containing microscopic traversable wormholes with arbitrarily small quantities of "exotic matter." Physicist John Wheeler has explained that these wormholes will pinch off. They are not classical wormholes but rather quantum fluctuations in the geometry of space, that is, tiny wormholes (10^{-33} centimeters) imparting a foamlike structure to space due to wormholes continually being created and annihilated. Put simply, the Isis Thesis proposes that these tiny wormholes that riddle space are equivalent to the protein folding funnels of phage Lambda's two competitive proteins, which fold and bind in a DNA wormhole landscape. These tiny molecular structural units or protein folding funnel energy landscapes are woven into the fabric of space-time. With great simplification, the Lambda protein c1 is responsible for the cosmic expansion of space-time, and the Lambda protein cro is responsible for the quantum experience of reversed timespace, where time expands and space collapses.

In support of the Isis Thesis interpretation, scientists generally agree that the quantum world ruled by quantum mechanics (with its equations resembling the kinetic molecular theory) orders our classical cosmos ruled by general relativity (Wolynes 1996; Musser 2004; Jacobson and Parentani 2005), that is, the classical action of signs seems to be a holographic veil of the molecular world. In other words, the activities of humans are signs or patterns of the competitive activities of quantum molecules. Today physicists and biologists know that space-time has a molecular structure. Further, protein folding along the DNA is a process by which a protein structure assumes its functional shape or conformation by freezing to a unique stable crystal structure in a funneled energy landscape biased toward the native structure (Onuchic et al. 1997). A protein folding to its native crystal structure exhibits entropy reduction (Socci et al. 1996) or a backward-in-time aspect also evident in current microscopic black hole research (Dubovsky and Sibiryakov 2006; Jacobson and Wall 2008). Due to the holographic principle, this means that our expanding classical universe with its potential for collapse would work on the same principle of a tiny wormhole being created and annihilated. The physics of this tiny wormhole is a sign of the mirrored expansion and origin-fixation dynamics of two Lambda proteins functioning as a partnership for a genetic switch that folds back to the Lambda genome, which is within nature.

As interpreted by the Isis Thesis, the simple ancient Egyptian argument is that a human agent with knowledge of nature's laws can throw a viral genetic switch at death that will collapse our universe almost to its Big Bang birth or singularity, a point of infinite density and zero volume. One explanation for how a human agent could do this is that our minds are rooted deeply in our microbial quantum biological heritage of DNA that is grounded by the laws of quantum physics ordering the laws of classical physics. So, a human can act as a switch-hitter because the Lambda genome is the fundamental origin of time and space that fashions our world and our virality.

General Consensus of Thunderheads. The main issue with the theories of general relativity (gravity due to curvature of space) and quantum mechanics (atomic domain), which has been validated by experiment is their inability to specify the fundamental origin of space and time. String Theory with its tiny threads of energy cannot make the connection to the fundamental origin. Other theories have also failed. For example, in a recent *Nature* article (Merali 2013), five big ideas are examined, and not one can explain what the fundamental origin of time and space is. These ideas are gravity as thermodynamics, loop quantum gravity, causal sets, causal dynamical triangulations, and holography.

First, both gravity and thermodynamics, the science of heat, are warmly coupled. Recall that Hawking showed the black-hole quantum effect of spewing out hot radiation. Regarding the Isis Thesis, these black hole/white hole characteristics model the folding/binding dynamics of Lambda c1 and cro proteins. Second, loop quantum gravity (LQG) describes our space-time fabric as a spider's web of strands carrying information. These strands are actually space-time, and they eventually form quantum loops. Ashtekar and colleagues (2006) used the LQG description of Einstein's equation in a simulation running time backward to before the Big Bang. What happened was quite interesting. When the reversing cosmos contracted to the fundamental size limit of LQG, a repulsive force kept the singularity open, while turning it into a tunnel to a cosmos preceding our own. Similarly, physicists Rodolfo Gambini at the Uruguayan University of the Republic in Montevideo and Jorge Pullin at Louisiana State University in Baton Rouge report that they find a thin space-time tunnel leading to another part of space in a similar simulation for a black hole. (Merali 2013, 518) Ancient Egyptian texts also describe a repulsive gravitational force and the serpentine tunnel. One problem with LQG is that it does not consider the other three forces: strong force, weak force, electromagnetic force. However, the Isis Thesis also explains the violation of the weak force, self-assembly due to the strong force, and the electromagnetic fields of the Sun and earth as the DNA electromagnetic fields of a bacteriophage and a bacterium.

The third theory explaining the Thunderhead position of confusion is causal set theory that perceives space-time as mathematical points linked together and determined by causality, that is, the past can only affect the present and future, but not in reverse. In the Isis Thesis interpretation, time reverse is essential, for it provides a time gap due to a time violation, which is lawful according to Einstein and modern science. The fourth theory, causal dynamical triangulations, is based on computer simulations that show fundamental quantum reality as self-assembling, triangulated polygons very similar to phage Lambda's triangulated morphology. Simulations also suggest that our universe's infant stage was only one space dimension and one time dimension. Scientists have also shown that the cosmic microwave background (CMB), a spectroscopic image of the universe at 380,000 years, exhibits patterns of this two-dimensional early universe. According to the Isis Thesis, this two-dimensional early universe represents the potential space dimension of c1 protein and the potential time dimension of cro protein, a potential cosmic loop linked together through space-time and its time reverse due to the proteins' cooperativity and adjacent gene-seats on the viral genome.

Finally, using the holographic principle, Van Raamsdonk (2010) (cited in Merali 2013) studied the entanglement of particles on the boundary; in other words, measurement on one particle affects the other. When he reduced quantum entanglement on the boundary to zero, the three-dimensional space divided itself like a splitting cell, snapping apart. That repetitive process subdivides the three-dimensional space again and again, leaving the two-dimensional space connected. Juan Maldacena, who showed that black holes store their entropy on their surface or horizon, said, "This suggests that quantum is the most fundamental, and space-time emerges from it." (Merali 2013, 519)

Diamondhearts support a Cyclical Cosmos. The Isis Thesis accommodates the five cosmological proposals. First, the protein folding funnel of c1 is thermodynamic, gravitational, and its diffusion-collision folding and binding may represent the spatiotemporal genetic information of our diffusion-collision, gravitational universe. Our universe may be the cosmic shadow cast by c1 protein's funneled energy landscape. Second, the initially unstructured cro protein with its origin-seeking dynamics folds and binds by the fly-casting mechanism. This permits the quantum loop back to the origin. Third, in causal set theory only one-half of the loop is considered—space-time causality. The other half of time reverse is ignored in this theory. Fourth, causal dynamical triangulation theory supports phage Lambda's triangulated morphology, and fifth, the holographic principle shows space-time emerges from the more fundamental quantum domain.

The Diamondhearts generally support a cyclical expansion/collapse cosmos, and the Isis Thesis proposes this is a macrocosmic sign of Lambda's genetic switch between two proteins in an eternal competition. Therefore the structure of space-time in our cosmos is the reflection of the folding funnel of c1 protein, while the reverse cosmic collapse to the origin is the intrinsically disordered structure of the folding funnel of cro protein. This is the geometry of space-time and its reverse—a time-space, a quantum, a time crystal, a cosmos in itself. Relative supporting scientific views follow.

Against determinism, quantum physicist C. K. Raju (2003) mentions the belief in the soul that originally functioned in a quasi-cyclic cosmos, a time belief that he claims does not harmonize with Western industrial capitalism with its slogan of "time is money" and its activity of more profit and machines leading to more disorder and waste. He explains that order-creation is a universal value that should envelop longer-term concerns extending across cosmic cycles. (432-434) Similarly, physicist Roger Penrose in *Cycles of Time* (2010) runs the laws of physics backwards from our present state, arguing that the early universe is the late universe of a previous era, invoking a succession of universes (Smolin 2010, 1034-1035). Likewise, Martin Bojowald's (2006) Loop Quantum Cosmology based on Riemannian geometry maps the backward evolution of a classical space-time universe collapsing in time to small length scales and high curvatures (quantum effects), and then the universe expands due to repulsive gravity. Bojowald visualizes this transition as a collapsing universe turning its inside out when orientation is reversed; bouncing behavior replaces the classical singularity.

Now, in the Game of the Centuries, we can visualize cosmology by the following critical practice. If we throw the ball from third base cosmology over to first base quantum physics, the cyclical expanding/collapsing cosmos on third represents a quantum mechanical black hole/white hole on first base. Then if we fire the ball to second base quantum biology, we see the cyclical cosmos on third base also represents the folding funnels of two viral proteins on second. So, lawlike macrocosmic patterns mirror lawlike quantum patterns and vice versa. In other words, the transdisciplinary nature of Egyptian knowledge is a mirror with multiple reflections. Lambda's competitive proteins function in energy landscapes that mirror quantum mechanical black hole/white formation/evaporation energy landscapes, and our expanding cosmos with its potential for collapse is a holographic reflection of Lambda's quantum biology mirroring quantum physics. Finally, Lambda's lifestyles mirror religious themes.

Costello:	I throw the ball to who?
Abbott:	Naturally.
Costello:	Now you ask me.
Abbott:	You throw the ball to Who?
Costello:	Naturally.
Abbott:	That's it.
Costello:	Same as you! Same as YOU! I throw the ball to who. Whoever it is drops the ball and the guy runs to second. Who picks up the ball and throws it to What. What throws it to I Don't Know. I Don't Know throws it back to Tomorrow, Triple play. Another guy gets up and hits a long fly ball to Because. Why? I don't know! He's on third and I don't give a darn!
Abbott:	What?
Costello:	I said I don't give a darn!
Abbott:	Oh, that's our shortstop.

Bud Abbott and Lou Costello from their Baseball Comedy Act

Baseball Diamond 14

Home Plate: *Magnum Mysterium*

We shall have to find the best means to approach the action of chance, to describe the symmetry breaking generalized catastrophes, to formalize the unformalizable. For this task, the human brain, with its ancient biological heritage, its clever approximations, its subtle esthetic sensibility, remains and will remain irreplaceable for ages to come.

René Thom
Structural Stability and Morphogenesis

In building a science with a biophysical theory of how the cultural elements of our lives and the transformations of cosmic, planetary, and human histories relate to the molecular and genetic foundation of life and death, we must address the question of order. Can we identify ordered, evolutionary-stable strategies for species emergence, that is, law-like patterns controlling human life and death? To begin to answer this question, we must again consider that a very restricted set of biological signals and protein functions have been used in evolution to generate our complex lives. This restricted set of biological signs and protein functions is described in ancient Egyptian texts as viral morphogenesis along the ancient glycolysis pathway. In support, Egyptian science mirrors religious themes found in many mythologies and religions, especially early and contemporary Christianity.

At home plate, we will examine the Diamondheart view supported by the Isis Thesis interpretation that Christ is a sign of a virus versus the Thunderhead view supported by Christianity that Christ is a real material person in history. The Isis Thesis interpretation of ancient Egyptian texts recognizes ritualistic Christianity as a reflective mirror of the lawlike patterns of speciation and extinction determined by the ancient gene expression network of phage Lambda. In turn, Christianity has always recognized ancient Egypt as the repository of polytheistic pagan ritual. However, as shown, the amazing similarity between Christian dogma/ritual and Egyptian literature/ritual exists, but it was never clearly linked because the ancient Egyptian hieroglyphs were not decoded until 1822 by Jean-François Champollion. Also, many Egyptian monuments may not have been excavated or examined in early history. So today, through the lens of modern science and archaeological excavations, one can see that the Egyptian signs explain HGT, as well as viral lytic replication, that is, the cloning of DNA for the Egyptian goal of becoming a quasi-hybrid being described as a Morning Star or crystallized mind in the early cosmos.

The difference between twenty-first century scientific knowledge and Egyptian science is that the ancient texts claim the exchange of DNA is between a human and a virus at human death, a horizontal gene transfer expanding our potential to exist in a quantum environment where space has an edge and time is lengthened. Accordingly, when biological elements such as DNA naturally escape the human body at death, Egyptian hieroglyphs, literature, and artwork support HGT is a potential evolutionary option, if one understands the signs of the ancient gene expression network for the lytic replication cycle. The texts advise the deceased to merge with the Sun-god or absorb the energy of the light (self-assembly with hydrogen driving force), the initial sign for HGT mediated by a virus such as phage Lambda. Remember—using the hydrogen atom, Bohr postulated that the angular momentum or spin of the electron can only have discrete values, and that electrons could jump from one orbit closer to the nucleus by emitting energy in fixed quanta or jump to a larger orbit by absorbing energy in fixed quanta. The key is absorbing the energy of the sunlight.

What is interesting is that the lifestyles of the very abundant, helpful, planetary virus Lambda model religious themes such as the dying/rising god, the cross, the virgin birth, the brother rivalry, the great flood, and so on, pointing to microbiology as the origin of religious ideas. In addition, ancient religious signs such as the Star of David, the cross or swastika, the ankh, and the ouroboros match DNA and protein formations found in the genetic switch of phage Lambda. Unexpectedly, many major ancient signs are observable in one specific microbiological process involving human, viral and bacterial DNA, supporting the evolutionary message of HGT and transformation for human DNA at a death transition. Put simply, the many redeemer-gods of various religions model the activities of a lambdoid creator-virus. As playwright Antonin Artaud said, "laugh if you like, what has been called microbes is god." This sharply conflicts with the Christian viewpoint to be reviewed, but it offers a new picture of the world that works with the massive microbial heritage in our human genome and microbiome. After all, microbes have made us who we are, and our cells carry their ancient gene expression network for transformation.

In Retrospect. Although many egyptologists dismiss the funerary corpus as confusing, unintelligible and primitive, the texts have scientific and evolutionary value, while explaining the natural laws allowing transformation of human DNA. After all, our genes may be the reason for the idea of God that rises in our consciousness. Both Osiris of ancient Egypt and Jesus of Nazareth die and rise from the dead or re-animate like a virus. Perhaps these ideas of the dying/rising gods are generated by instructions in our viral genes, as the viral footprint on human history indicates. If the viral genes are responsible for the God message, then our chromosomal apparatus is slamming us with its survival message because other dying/rising gods, such as Adonis, Attis, and Mithra, also surfaced during the 2500 year time frame between Osiris and Christ. Perhaps the same message surfaces everywhere in history because we all share the same essential chromosomal apparatus. Thus, historical behavior is repetitive because causation flows from our chromosomal apparatus.

It then seems reasonable to assume that the idea of the dying/rising God from the genes bears an important survival message for us because it keeps cropping up in historical human consciousness as a sign of life and DNA immortalization. Since the genes are sending our consciousness this God message, it seems that both the genes and consciousness are necessary for "immortality" to occur. This suggests that our consciousness may be innate to our genes, or that our chromosomal apparatus is semiconscious, as recent research suggests (see Chapter 18). This seems to be the most economical theory to fit the facts. Therefore, it may be worth thinking about the nature of our viral DNA to understand the meaning of the God message rising in human consciousness, since what humans feel as a profound, mystical inner experience of God within their souls may actually be the awareness of the nonconscious origin of our intentionality—a bacterial virus.

But now the question is—if the idea of the dying/rising God in our consciousness is intrinsic to our genes and due to the massive amount of microbial DNA in our bodies, then our idea of God may model the activities of some type of dying/rising microbe. This brings us to the Isis Thesis, the microbial hypothesis that ancient Egyptian deities represent bacterial and viral proteins and genes, specifically those of phage Lambda. The dying/rising deity Osiris is the sign for the complex bacterial virus Lambda that invades a bacterial host cell and goes dormant on the bacterial DNA chromosome. To enter its special dormant chamber on the bacterial cell's chromosome, the virus becomes attached at four sites by means of a DNA protein cross called a Holliday junction. Here it lies on the host cell DNA—nailed to its cross like Christ, until it rises from its dormant state to take over the replication machinery of the bacterial cell and clone its own kind. Conveniently, in this microbiological process, the virus is dormant next to the lactose genes, the nutrient sugar it needs to clone its own when it rises from its dormant state. The nutrient lactose genes are tied off by the lac repressor protein, which loops the DNA in the shape of the Egyptian ankh. Activated by the emission of light and assisted by proteins, the virus then detaches from its DNA-cross, uses the lactose nutrient next door, and clones itself, what can be compared to a virgin birth. The Isis Thesis proposes that Isis is the sign for the necessary lactose metabolism. So we have a dying/rising god on a DNA-cross, a virgin birth of clones from a nutrient sugar, and some very ancient signs—the cross, the ankh, and the coming of the light or Sun-god. This natural process happens within our cells in the ancient glycolysis-fermentation gene expression network. Over 870 signs in the major corpus of ancient Egyptian texts extending over 2000 years of Egyptian history support this interpretation, describing a specific chemical path for morphogenesis, as does additional research on early China, the seventeenth century Navajo Indians, and many other historical individuals. So, it may be that we have modeled our idea of God on the dying/rising lifestyles of a complex virus known as bacteriophage Lambda.

This idea of a viral god supports an expanding theory. If the microbiology of our genes inspires our consciousness on religious ideas and ritual behavior, then it is logical to assume that this microbiology also inspires other cultural behaviors, such as literature, art, government, capitalism, technology, war, and baseball. With these ideas in mind, the forthcoming chapters focus on the task of showing the correlation between our microbial genetic inheritance and culture to improve our understanding of who we are. Then we may be able to understand the biological knowledge within us that has been distorted for political purposes and twisted by religion and other webs of biopower to control human beings and coerce them into submission through fear—fear of hell, fear of invasion, fear of insanity, and fear of death. Therefore, theory will analyze power mechanisms to build a new politics of truth for human consciousness based on knowledge. The force of knowledge can crush the grip of biopower or the powers controlling our lives with misinformation about simple gene expression patterns and biological mechanisms that influence our cultural behavior. With this knowledge, perhaps we can humanize our viral behavior to eliminate war, genocide, rivalry.

Pregame 15

Manager, Coach and Umpire Selections for the Game of the Centuries

But if power is in reality an open, more-or-less coordinated (in the event, no doubt, ill-coordinated) cluster of relations, then the only problem is to provide oneself with a grid of analysis which makes possible an analytic of relations of power.

Michel Foucault
Power/Knowledge (1972)

Welcome to the Pregame to the Game of the Centuries between the Thunderheads and the Diamondhearts. This section introduces not only the team managers, coaches and umpires, but also the players on each team. By using the game of baseball to reinstate the knowledge of the centuries through the lens of contemporary science today, one can understand the actions of historical human beings relative to law-like patterns related to morphogenesis. Since the baseball model depicts the historical power grid, the selected players in the two fantasy-draft baseball teams allow readers to appreciate the cluster of relations enfolding historical power and knowledge, the tensions between modes of domination and individual freedom of will, the connections between life and nonlife, and the links between our classical sciences (physics, biology, and cosmology) and quantum mechanics (quantum physics, quantum biology, and quantum cosmology). Most importantly, the complete power grid of two fantasy-draft teams as a whole shows the close relations between our psychology, matter, and our genome, supporting that human DNA may survive or rise to a higher order through two different pathways—the natural pathway involving a hybrid merger of DNA with a metal-binding bacteriophage and the manmade pathway resulting from our technoprogressive culture that forecasts the end of humanity due to the uncertain merger of human brains with metal artificial intelligence (AI).

The selection of team managers, coaches, and umpires for the Game of the Centuries is a conceptual tightrope stretching between two philosophical concerns: Diamondheart semiotics and Thunderhead materialism. In the interests of keeping things simple for our baseball model, semiotics and materialism are intended to reign as very general guiding philosophies. However, the extended meaning of each philosophical system can be expanded in depth.

Now, the semiotic Isis Thesis interpretation supports that we are nature. You could call yourself a semiotic microbe. Supporting Platonism, ultimate reality consists of unchanging, absolute, eternal entities called Ideas or forms or information, that is, the building blocks of DNA. Earthly objects are not truly real but merely partake of the forms originating from microbial DNA. Ultimately, Aristotle's idea of final cause, as well as the pragmatism of C. S. Peirce is essential, for thinking—as the guide to action and truth of any idea—lies in its practical consequences. Whatever we are, how one plans to exist in the cosmos matters, and this is a universal, valid, ethical standard. With its trans-philosophical approach, the Isis Thesis and twelve related articles explain a teleological theory of organic and inorganic evolution based on DNA-directed human behavior.

Diamondheart Manager Motto: "I am thinking, therefore I choose."

In light of the Isis Thesis and its broad understanding of semiotics, philosopher-mathematician Charles Sanders Peirce is a good choice for the Manager of the Diamondhearts. Basically, Peirce's worldview is that signs are everywhere, and these signs have meanings that can impact human behavior. Peirce believes that pure chance will survive until the "world becomes an absolutely perfect, rational, and symmetrical system, in which mind is at last crystallized in the infinitely distant future" (1992, 297). To Peirce, thought appears in bees, crystals, and the inorganic (1931, 551). Crick, Watson, and Caspar understood viruses as crystals as do modern virologists. Crick and Watson (1956) predicted that the coats of viruses were composed of repeating subunits exhibiting the symmetry of closed polyhedra. The Peircean theme of crystallized mind, the Egyptian star transformation of the dead king into four gods who support the sky (PT 556), and the four-faced ancestral star transformation of the Chinese Yellow Emperor Huangdi may point to the polyhedral form and dynamics of an ancient lambdoid virus with a protein coat of many colors. So, the Diamondhearts are holists, maintaining that the whole of natural systems is primary and greater than the sum of its parts due to the possibility of emergent states of being. In contrast, the Thunderheads prefer to break down and analyze the separate parts of the whole.

Thunderhead Manager Motto: "I am thinking, therefore I exist."

Now, we can also expand the compartmental worldview of materialism for the Thunderheads. Materialism is actually a philosophy claiming that matter and energy are fundamental, and that mind does not exist or is a manifestation of matter. Scientific materialism supports that physical reality is all that exists. According to mathematician Alfred North Whitehead (the Thunderhead Designated Hitter), scientific materialism is effective, but not in the context of teleology for developing a comprehensive view of our cosmos. Whitehead sees connections between traditional belief in God and nature, since religion considers permanence within change. (Irvine 2014) Rationalism claims that reality can be known through reasoning, rather than observation or experience. Still, science understands that rational laws exist in nature that science must discover.

Under the broad umbrella of materialism, we must also consider mechanism, both universal and anthropic. Universal mechanism in nature is that the present state of our universe is an effect of the past causing our future. Anthropic mechanism claims human beings and their minds can be explained in mechanistic terms. Mechanism is present in the twenty-first century as the contemporary philosophy of many technocrats. Of course then, there's logical positivism, analytic philosophy, phenomenology, existentialism and poststructuralism in twentieth century contemporary philosophy. However, considering that modern philosophy runs haphazardly over a longer period of time from Descartes to the twentieth century, the mathematician-philosopher René Descartes is a good choice for Manager of the Thunderheads because his mechanistic approach to nature supports that animals are machines, he presents the mind-body problem (mind and body are distinct substances) that originated in ancient Egypt, and he explores ideas about the nature of God. Religion, of course, needs to be factored into this worldview, and Descartes identifies God as the primary substance creating the two substances of mind and matter. (Hatfield 2014) Whether or not one agrees with Descartes on the reality of God does not matter because human behavior in our classical world includes religion, which may be a possible consequence of the evolutionary survival message in our chromosomal apparatus. So, the philosopher-mathematician René Descartes with his comment "I am thinking therefore I exist" is a good choice for the Manager of the materialistic Thunderheads.

Coaches. Regarding selection of coaches for each team, the English naturalist Charles Darwin, who established that all species descended from one common ancestor, and the theoretical physicist Albert Einstein, who proposed the theory of gravity (general relativity) are appropriate picks for the Thunderheads. For the Diamondhearts, the coaches are American geneticist Barbara McClintock (1902-1992), who received the 1983 Nobel Prize for the discovery of mobile genetic elements (jumping genes), and Austrian physicist Erwin Schrödinger (1887-1961), who formulated the wave equation in the field of quantum mechanics. Schrödinger also understood thermodynamics (science of heat), general relativity, and cosmology, as well as knowledge of ancient and oriental philosophy (Wilbur 1985). On the other hand, Barbara McClintock's discovery of mobile genetic elements relates to mutation. Defined as a change in the nucleotide sequence or words of the genome of an organism, virus, or extrachromosomal genetic element, a mutation can result from unrepaired damage to DNA or to RNA genomes, replication errors, or the insertion or deletion of DNA segments by mobile genetic elements. In ancient Egyptian texts, the merger of the dead king with the Sun-god suggests the recruitment of organic mobile genetic elements, possibly reverse transcriptase, by a self-assembling viral molecule. Suffice it here to say, that McClintock's discovery of mobile genetic elements is important, as is the fact that she was the first scientist to correctly speculate on the idea of epigenetics (heritable gene expression changes that are not caused by changes to the DNA) forty years before recent research emerged on this concept. (Ravindran 2012) How else can innovation be initially grasped than by speculative theory?

Now, what is important here for our game is that the placement of Darwin on the Thunderheads and McClintoch on the Diamondhearts balances Darwin's natural selection with the natural event of mutation that includes horizontal (lateral) gene transfer. Stoltzfus (2012) sums up the difference between these two evolutionary sculptors:

> In Darwin's original theory, and in the later Fisherian view, individual differences are properly a raw material, like the sand used to make a sand-castle: each individual grain of sand may be unique in size and shape, but its individual nature hardly matters, because it is infinitesimal in relation to the whole that is built by selection. By contrast, if an episode of evolution reflects the individual nature of a significant mutation—a developmental macromutation, a gene or genome duplication, an event of lateral transfer or endosymbiogenesis, etc.—, then the infinitesimal assumption no longer applies, and the verbal theory fails: when variation supplies form (not just substance), it is no longer properly a raw material, and selection is no longer the creator that shapes raw materials into products. (2012)

The point Stoltzfus (2012) is making is that the Modern Synthesis, also known as Neo-Darwinism, a synthesis that includes ideas from genetics, botany, morphology, ecology, paleontology, and cytology, endorses Darwin's natural selection with its raw materials view of variation, but rejects mutationist views. Stoltzfus supports that mutation-biased adaptation has been shown theoretically and observed experimentally in a bacteriophage. Likewise, in the science of ancient pharaonic Egypt, Darwin's natural selection is not the sole creator shaping products or providing form, appearance, and structure to evolving humans on earth. In the tiny mesocosm of microbes, the creator also shaping the world of matter is phage Lambda through constant lateral or horizontal gene transfer. This novel mutation mechanism may also be available for human DNA transfer at a death transition, according to ancient Egyptian and other texts that advise one to choose the mechanism by following directional knowledge for origin-fixation cro protein dynamics in the DNA of a bacteriophage. This reflects the importance of individual human selection or choice for HGT.

Here is a simple way of looking at mutation-biased adaptation or the horizontal gene transfers that circumvent natural selection during human evolution on earth. First, consider the human organism's environment that naturally selects the character of the human organism. The environment is the sum of the various entities in that environment, and the major players on earth are the very abundant, gene-shuffling bacteriophages, so phages by their considerable ubiquity impact the natural selection of life. Phages bind with organic molecules and modify human and other DNA. Using a technology metaphor, phages hack into human software or DNA, and they modify it. So phages seed the biomass due to their natural ability to transport DNA and replicate in another organism. With extreme simplification, phage genes are the natural selectors. Consider the evolutionary enlargement of the human head and brain in the last two million years that cannot be systematically explained. It is also interesting that the Neanderthals averaged a larger cranial capacity than modern *Homo Sapiens.* In the long evolutionary run, both cases may have resulted from HGT mediated by a bacteriophage, a microbe made of only a head of DNA and a tail.

Also, the human head with its brain of left and right hemispheric lobes controlled by the corpus callosum is very similar to a Lambda phage head with its naked DNA of two arms controlled by Promoter Left and Promoter Right. A promoter is a region of DNA that initiates transcription. The corpus callosum is a nerve center that facilitates communication between the two brain hemispheres in a fashion that might be compared to Lambda's promoters that facilitate transcription of left or right DNA regions. In addition, we cannot overlook the fact that the human body develops from a fertilized egg with the genes containing the plan in digital form. Every cell in the human body carries a full copy of our genome. As Matt Ridley explains, the fertilized egg grows into an embryo with two asymmetries—a head-tail axis and a front-back axis. These asymmetries develop later in mice and humans, and the process is not really understood. Yet, in the fruit fly scientists found a cluster of eight genes called Hox genes, an orderly set of genes laid out from the head to the tail. Soon this same arrangement from head genes to tail genes was found in mice, and human beings have exactly the same Hox clusters as mice. So, we are descended from a common ancestor with flies, and this incredible conservatism of embryological genetics surprised scientists. Although flies have a cluster of eight Hox genes, humans have a cluster of thirteen Hox genes. In all species investigated so far, genes are laid end to end with the first genes expressed at the head and then tail (Ridley 1999, 173-178; 180-182). This ancient evolutionary sequence is found in phage Lambda, whose genes are primarily necessary for head-tail assembly. In this virus, rightward transcription expresses genes that initiate replication, allowing the expression of head, tail, and lysis genes from Promoter Right. So, perhaps the Hox cluster of genes found in fruit flies, mice and men is a sign of our last universal common ancestor—phage Lambda. Finally, if this discussion has not convinced you of our similarities to phage Lambda, consider this: our sleep/wakefulness cycle models Lambda's lysogeny/lysis cycle—the virus sleeps (lysogeny) and wakes up (lysis), just like us.

First Base Umpire John H. Taylor.

The next task is to appoint four umpires for the Game of the Centuries. For first base umpire, John H. Taylor, a curator of the British Museum specializing in ancient Egyptian funerary archaeology is a good pick, for he leans toward the Thunderhead vision. His expertise spotlights the afterlife journey, and he specializes in funerary coffins and mummies, along with the metal statuary of the first millennium BCE, the Third Intermediate Period (c. 1069-664 BCE), and Egyptology's history. Curator Taylor is editor of *Journey Through the Afterlife* (2013), supporting the general consensus of egyptologists that the funerary texts are a compilation of magical spells. Accordingly, the Book of the Dead is one of these compilations that defines the dangerous afterlife journey of the dead king who desires eternity in the domain of the gods. With contributions from leading scholars, Taylor edits this new text of the Book of the Dead, detailing the hopes and fears of mortal man within the magical context of the funerary spells. What Curator Taylor supports is the general consensus of egyptologists on the funerary texts. Ogden Goelet, Jr. of the Department of Near Eastern Studies at New York University (1994) clearly summarizes this general consensus in his commentary on the New Kingdom Book of the Dead (BD). Goelet considers the ancient text as "aspects of Egyptian Religion," although the Egyptian language has no equivalent for the word "religion" (145). He explains that the BD is difficult to understand, completely obscure in parts, confusing, the ancient rationale for incantations is elusive, and the knowledge was originally intended only for the pharaonic family. Goelet admits that temple complexes were divided into public and restricted areas, while priesthoods were differentiated into levels of access to knowledge. As Goelet states, "Knowledge was power, and the élite wished to keep that power among themselves."

Goelet does mention that *Akhet* (normally translated "horizon") is associated with the Sun, and that *Akh* is the soul-like state of the noble dead, who perhaps become "dazzling light" (1994, 143). What is problematic to scholars and anthropologists, explains Goelet, is the role of magic in the religious texts. Further, the images were able to in fluence reality (148). Goelet also mentions the "vast corpus of non-egyptological literature" claiming the texts have a hidden meaning, and like other egyptologists, he dismisses this literature. Goelet's dismissal may also include the Isis Thesis that views the texts through the lens of contemporary science, exposing a microbiological survival message for human morphogenesis.

Now, scholars, anthropologists and egyptologists should take note that the Isis Thesis resolves their problematic issues regarding the role of magic in the religious texts and the ability of the images to influence reality. In these scientific texts, magic results from quantum dynamics of nonlocality and entanglement, while images influencing reality relate to the observer-participancy principle. According to physicist John Wheeler's delayed-choice experiment, the act of observer-participancy activates a quantum-mechanical probability amplitude (wave height) and "develops definiteness out of indeterminism" (1988, 4-5). This means that light is a wave of possibility remaining that way until an observer manifests it retroactively, that is, going backward in time. Counterintuitively, quantum mechanics allows a measurement, an observation, a choice of yes or no due to entanglement (Scully 2007, 144; Greene 2004, 199). Wheeler sums it up: "Quantum theory denies all meaning to the concepts of before and after in the world of the very small" (1994, 184). So, human observation of an image, as a sign of an object, can influence reality due to the participation of the observer because quantum mechanics allows a choice.

Second Base Umpire Robert Rosen (1934-1998).

Leaning toward the Isis Thesis interpretation, a good choice for second base umpire is theoretical biologist Robert Rosen, who considers organisms as anticipatory systems. After all, humans anticipate the future, although some of this behavior is nonconscious, if you accept that religion veils a survival science. In Rosen's relational theory as simplified by Kineman (2010, 3-7), the fundamental causes of complexity are the same for all biological and physical systems. Complexity can be reduced to the final cause, the whole is greater than the sum of its parts, and in the modeling relation, final cause is part of the whole of nature. This creates a natural causality loop. For example, even though bacteriophage Lambda functions in the tiny mesoscopic domain, the Isis Thesis proposes this creator-virus is not outside of nature, but within it, and evidence for this view is its viral presence within *E. coli* in our gut microbiome.

In the modeling relation, final cause is part of the whole of nature, according to Robert Rosen, and the Isis Thesis interpretation supports that this creates a causality loop or a natural genetic switch between two different states of being built by two protein teams (c1 and cro). Actually, Rosen's view about final cause functioning within nature already exists in history as Aristotle's idea of final cause, showing that the same idea has surfaced in history as philosophy, as well as quantum biology. Rosen situates final cause within nature. As Kineman explains:

> The hope of finding an "exact" science led us to place finality outside of and prior to nature (and thus to reinforce the uniquely Western view of an unnatural external creator), whereas returning it to nature represents nature as fundamentally complex, and thereby gains us the ability to consider the origin and identity of systems. (2010, 7)

Thus, if the origin of a system is within nature, then perhaps science can deal with it. According to mathematical biologist A. H. Louie, "Rosen suggests that there must be information about self, about species, and about the evolutionary environment, encoded into the organization of all living systems. He observes that this information, as it behaves through time, is capable of acting causally on the organism's present behavior, based on relations projected to be applicable in the future." (2010, 20) So, the information encoded in our genes that reifies in human history (our behavior) has stamped a model of a viral gene expression network for future application.

Related to Rosen's modeling relations between a model and the natural system it models, the Isis Thesis proposes that 13.7 billion years ago our cosmic system originated from the quantum microstate of a viral genome, a reasonable quantum origin within nature. In this semiotic theory, the natural lifestyles of phage Lambda are the foundation for human evolution and behavior, so nature can be described in terms of modeling relations. This means that human behavior is modeling a natural system of cyclical viral gene expression. In contrast, Kineman explains that the mechanistic view sees a one-way descent from supreme causation to material death. This is only half of causality, which should also include the "ascendant causality, generating new functions from prior structures." Closing this causality loop brings ascendant causes into natural science, and new functions from existing nature with its unrealized possibilities. (Kineman 2010, 14) Ultimately, ascendent causality in ancient Egyptian texts allows the organic to rise to higher order via a virus—the dead king becomes a crystalline quasi-hybrid being of human and viral DNA. This emergence closes the causality loop because the laws are system dependent, that is, the bioluminescent or chemiluminescent process is the chemical reverse of photosynthesis. Recall that microbes invented life's chemical systems leading to fermentation, photosynthesis, and oxygen breathing. So, it seems reasonable that their inventions also include reverse processes. Due to quantum mechanics, evolution is possible both ways.

The Isis Thesis proposes that a modeling relation exists between a bacterial virus and human beings. The problem is that viral morphogenesis and native state binding requires a larger amount of nonnative interactions over native; otherwise, the system does not work. So, the historical élite have guarded evolutionary knowledge from the masses to increase the fitness of their élite DNA or bloodline. However, knowledge of the model's full predictions can change the course for anyone. Human beings can either anticipate the future or they can react to events like machines.

Third Base Umpire Vernor Vinge.

The selection of Vernor Vinge for third base umpire is necessary due to the last century's rapid advances in technology or mechanistic materialism. What is important is to look at this phenomenon of technological change related to an artificial intelligence (AI) explosion that may surpass the human intellect. In 1993, Vernor Vinge of San Diego State University predicted the creation of superhuman intelligence within thirty years and the end of the human era. Vinge, a computer scientist, mathematician, and science fiction author, called this intelligence explosion the singularity, and he argues that it cannot be prevented because it is a consequence of our natural competitiveness and the possibilities inherent in technology. Vinge believes that despite their fears of the singularity and legislation to control it, world governments will still pursue it because of the competitive economic, military and artistic advantages for those governments who get there first. If the singularity cannot be prevented or confined, Vinge speculates on how bad the Post-Human era of artificial intelligence could be.

> Well . . . pretty bad. The physical extinction of the human race is one possibility. (Or as Eric Drexler put it of nanotechnology: Given all that such technology can do, perhaps governments would simply decide that they no longer need citizens!). Yet physical extinction may not be the scariest possibility . . . In a Post-Human world there would still be plenty of niches where human equivalent automation would be desirable: embedded systems in autonomous devices, self-aware daemons in the lower functioning of larger sentients . . . Some of these human equivalents might be used for nothing more than digital signal processing. They would be more like whales than humans. Others might be very human-like, yet with a one-sidedness, a "dedication" that would put them in a mental hospital in our era. (1993)

This materialist phenomenon of a superintelligent AI forecasted by Vinge conveys an impending message to humans, comparable to the latent potential of Moby Dick, the white whale that destroyed Ahab despite the mad captain's determination to annihilate his nemesis.

As third base umpire, Vinge will navigate cosmological theories about the origin of space and time, as well as the evolution of mind and matter in the cosmos. In support of the evolution of mind, the cultural legacies of ancient Egypt, early China, the Navajo, and other historical visionaries propose the possible cyclical evolution of governing mind or a quantum DNA intelligence that embraces our cosmos by spatializing it into a quasi-hybrid being or time crystal. In contrast, modern technocrats on the Thunderhead team have this same idea of manifest cosmic destiny, but they invert mind into nonbiological matter or swarms of micron nanobots that infuse the cosmos (Kurzweil 2005, 352). Smart metal matter with intelligent processes will then master the universe, becoming "more powerful than cosmology" (364). So, in the Game of the Centuries, the prophetic Vernor Vinge is a good choice for third base umpire due to the materialist vision of our technoprogressive culture.

Home Plate Umpire Timur (c. 1336-1405).

Timur, also known as Tamburlaine, is a good pick for home plate umpire because he understands battle logic, uses foresight, and has great vision, the three qualities necessary for this position. Timur was the Mongolian ruler of Samarkand whose nomadic hordes conquered an area from Turkey to Mongolia. Considered a military genius and responsible for the destruction of the Christian Church in much of Asia, Timur's legacy is barbaric, but according to Edward Gibbon's quotation in *Decline and Fall of the Roman Empire* (Modern library, v. iii, p. 665), Timur did not consider himself a bloody messiah:

> I am not a man of blood; and God is my witness that in all my wars I have never been the aggressor, and that my enemies have always been the authors of their own calamity. (Timur, after the conquest of Aleppo in Syria)

Still, one account says Timur was responsible for the death of seventeen million people, so it is difficult to believe that he was not a man of blood. His empire stretched over southeastern Turkey, Syria, Iraq, Iran, and areas of Central Asia. Because of his notoriety in history, in 1827, Edgar Allan Poe published an epic poem on Timur called "Tamerlane." Also, Christopher Marlowe based his play "Tamburlaine the Great" on the great warrior. It has been said that Timur planned his military campaigns many years in advance. This meticulous planning also involved planting barley for horses involved in the campaigns. Spreading psychological warfare about the cruelty of his armies, Timur understood the power of propaganda for overwhelming populations. Tall and broadchested like Atlas, the mythological Titan who held up the celestial sphere, Timur was a man of command and warfare logic.

Surprisingly, a conqueror with a solid grasp of warfare logic is hard to come by, according to Tolstoy's claims on Napoleon's invasion of Russia. The Russian author Count Lev Nikolayevich Tolstoy (1828-1910) reflected on his philosophy of history in his novel *War and Peace* concerning Napoleon's invasion of Russia. On this case, Tolstoy argues that personal habits and aims determine the wartime actions of humans, who believe they have free will. Yet people are "involuntary tools of history." As an example, at the battle of Borodinó in 1812, Tolstoy portrays both Napoleon and General Mikhail Kutúzov of the Imperial Russian Army as acting irrationally and involuntarily. Tolstoy also supports the folly of war by claiming that Napoleon could not see the Borodinó battlefield because he was a mile away, while smoke and mist shrouded the area, so orders had no influence. The battle at the village of Borodinó was Napoleon's last offensive attack, and Tolstoy presents the French General's conscience as troubled. And rightly so, for the Russian Army had withdrawn, saving its combat strength to eventually defeat Napoleon.

Against the view of historians, Tolstoy argues that the action of one man cannot express all the individual wills. He also contends that some successful Russian war maneuvers inevitably occurred and were not the result of Russian military foresight and sensible planning. Likewise, and contrary to historians, Tolstoy claims that the French had no flight plan out of Russia, and that Napoleon and his marshals were only interested in saving themselves. In the Epilogue to *War and Peace*, Tolstoy claims that reason, chance, and genius dim in light of the knowledge that historical actions are fated and the ultimate purpose of these events can never be immediately understood. Tolstoy seriously questions the idea that leaders guide humanity to confident ends.

Unlike Napoleon, Timur as home plate umpire is aware of the battle logic of the Diamondhearts and the Thunderheads. Timur understands that the action of one man cannot express all the individual wills in the game, but the umpire can follow the rules to avoid bad calls. Now, Timur's sympathies may lie with the Diamondhearts because he has a morphological ideosyncrasy—his head is a glistening white pearl. In ancient Egyptian texts, deities sport spheres for heads as signs of viral structure. Recall that during the sixth century, the Emperor Justinian condemned the ideas of the third century Church father Origen, who compared religions and showed their fundamental similarities. However, one of the beliefs in question was the resurrection of human bodies in spherical form, so Origen may have been aware of ancient Egyptian microbiology, for viruses are spherical and the quantum universe is alive.

Since Timur conquered Syria, his spherical transformation also manifests the abstract, spiritual meaning of Syrian literature such as "The Song of the Pearl" composed circa 247 BCE to 224 CE. The Gnostic poem depicts a personal quest that centers on a pearl. Gnostic scholar Marvin Meyer has a sound understanding of Gnostic mysticism emphasizing the essential value of self-knowledge for salvation. Marvin Meyer and Willis Barnstone are editors of *The Gnostic Bible*, and they provide translations from Coptic and Greek texts representing Jewish, Christian, Hermetic, Manichaean, Islamic, and Cathar expressions of Gnostic spirituality from the regions of Egypt, the Greco-Roman world, the Middle East, Syria, Iraq, China, and France. Many texts discovered in 1945 near Nag Hammadi in upper Egypt are included in *The Gnostic Bible*. Because the home plate topic centers on the possibility that ancient Egyptian mythology and Christianity veil a science, as well as the possibility that Egyptian and Christian deities are signs for viral proteins and genes rather than real people, the Gnostic influence during the first centuries of the common era offers additional insight on these views. After all, the Gnostics proclaimed knowledge or *gnosis* of what they described as the Pleroma or the fullness of pure light with the "perfect exalted human" within. As described in "The Three Steles of Seth," this heavenly foreigner is a savior, "maker of multiplicity, creator of mind, a triple power, self-conceived, donor of eternal realms, perfect, complete" and "from another race." (Barnstone and Meyer 2003) As Meyer explains, scholars of ancient religions met in Messina, Italy, in 1966, broadly determining that *gnosis* is "knowledge of the divine mysteries reserved for an élite" (10).

As early as the second century, Christian clerics were destroying Gnostic texts as well as classical civilization. In 306, Constantine, a devoted follower of the pagan Sun-god, converted to Christianity becoming the Holy Roman emperor. Still, the internal doctrinal disputes in early Christianity persisted and Gnostic texts were destroyed. In 391, the Christians—with the permission of the Byzantine emperor and under the command of the Alexandrian patriarch Theophilos (later saint)—burned down and razed the greatest library of antiquity in Alexandria, Egypt comprised of 700,000 rolls. In response, the Gnostics or light people buried their scriptural scrolls, which surfaced many centuries later. (Barnstone and Meyer 2003, 783-785) As mentioned, in 1945, a farmer found a buried cache of thirteen codices containing some fifty texts near Nag Hammadi, Egypt. These rediscovered Gnostic texts are a bridge between suppressed Egyptian science and dominating Christian theology, two power élites silencing knowledge through secrecy and anathematisms. The Gnostic texts identify biological signs on the quantum level for human genetic survival by means of viral transformation or cloning, and so the pearl is a sign of viral morphogenesis. (see King 2006a)

In light of this struggle between Gnostics and Christians, "The Song of the Pearl" (Barnstone and Meyer 2003, 386-394) explains why umpire Timur sports a pearl as his head. The song concerns a prince's quest for a pearl guarded by a snorting serpent. The path to the pearl is dangerous and harsh, and the prince must go down into the middle of a sea in Egypt. This location is similar to the biblical captive site of the people Israel in Egypt, described as "the midst of the iron furnace" (Kings 8:51). So, the prince travels down into Egypt and settles in next to the fiery serpent, hoping it will fall asleep so he can steal the pearl. Due to several intriguing encounters, the prince soon forgets about the pearl, until a letter rises up as an eagle. Like a "bird of speech," "it flew and alighted beside me and became speech." (2003, 391) It is interesting that these same words are written in ancient Egyptian Pyramid Text 310 in reference to the viral Sun-god's ferryboat named "It-flies-and-alights." Also, Pyramid Text 313 references a road is made for the king to pass on "into this furnace-heat" in the netherworld. Anyway, at this point in the poem, the prince remembers the pearl, seizes it, and emerges straight into the light. The prince then remembers his former robe of glory, an embroidered coat mirroring gold, beryls, rubies, opals, and sardonyxes, fastened by stones of adamant and sapphires of many colors. As the robe spreads toward the prince, he receives it, and ascends to the majesty of his father, who sent the coat to him. The father promises that the prince with his pearl will travel with him to the king of kings.

Put briefly, the prince's quest into the fiery furnace of Egypt is similar to the dead king's quest in the least-corrupted ancient Egyptian texts. Also, references to the letter, the letter's voice and the bird of speech suggest the bird is the word of DNA, while the robe of beautiful gems is a sign of not only the prince's crystalline transformation, but also the metallic coat protein of phage Lambda, a virus that flies-and-alights on a cell. In the allegory, possession of the pearl allows the morphogenesis of the prince through the gem-studded garment of *gnosis*. With his possession of the pearl, the prince experiences the robe of knowledge that allows the fullness of pure light. The prince describes his hybrid crystalline transformation: "As I gazed on it, suddenly the garment like a mirror reflected me, and I saw myself apart as two entities in one form" (Barnstone and Meyer 2003, 393). Thus, the prince becomes a quasi-hybrid being like Timur, whose head is a white pearl.

Even the Gospel of Matthew is concerned with the quest for the pearl: "Again, the kingdom of heaven is like a merchant in search of fine pearls, who, on finding one pearl of great value, went and sold all that he had and bought it" (13:45-46). At the Catholic Mass, we are again reminded of the pearl when the priest elevates the large white host above the golden chalice. So, we can see the connections between viral shape, ancient Egyptian science, Gnostic allegory, and Christian theology and ritual.

The point here is that the hybrid Timur, with his military knowledge and visionary pearl-head, should be an excellent home plate umpire because he can straddle the tightrope between ruthless military action and religious warfare related to the book-burying Gnostics versus the book-burning Christians. This places Timur in the midst of historical battles about the value of human morphogenetic potential. After all, Timur is half-human and half-white-pearl, so he should be able to make the calls as home plate umpire in the game between the Diamondhearts and the Thunderheads.

Pregame 16

The Thunderheads

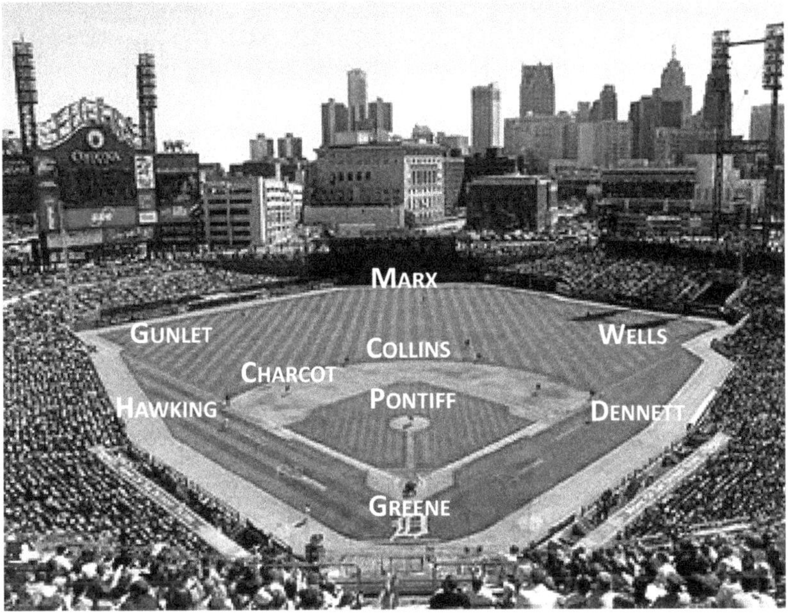

Thunderheads

Manager: René Descartes
Coaches: Charles Darwin and Albert Einstein
First Base: Daniel Dennett
Second Base: Francis Collins
Third Base: Stephen Hawking
Shortstop: Jean-Martin Charcot
Right Field: H. G. Wells
Center Field: Karl Marx
Left Field: Gunlet, a nanotech device
Catcher: Brian Greene
Subcatcher: David Hume
Designated Hitter: Alfred N. Whitehead
Starting Pitcher: Pope Francis
Middle Relief: James D. Watson
Middle Relief: B. F. Skinner
Set Up: Machine Dream Team
Closer: Jesus Christ
First Base Umpire: John H. Taylor
Second Base Umpire: Robert Rosen
Third Base Umpire: Vernor Vinge
Home Plate Umpire: Timur

An intensifying grumble of thunder oppresses the landscape when the Thunderheads storm into the stadium. The Thunderheads express an analytic of power relations that the Diamondhearts must conquer. It is a materialist grid of power relations that the following batting order reinforces. In selecting this fantasy-draft team in the twenty-first century, to be fair to time's legacy of human thought, the chosen players are both dead and alive like Schrödinger's cat.

First Base: philosopher Daniel Dennett, anthropic materialist.

The American philosopher Daniel Dennett is the best pick for first base position because of his deterministic attitude and desire to win by using reason and foresight. For example, Daniel would probably play ball in a thunderstorm by using rational predictions. As Daniel explains in his book *Freedom Evolves*:

> In fact, if you are faced with the prospect of running across an open field in which lightning bolts are going to be a problem, you are much better off if their timing and location are determined by something, since then they may be predictable by you, and hence avoidable. Determinism is the friend, not the foe, of those who dislike inevitability.

Daniel is an atheist interested in the philosophies of mind, science and biology, and he won the 2012 Erasmus Prize with its theme on the cultural meaning of the life sciences. "A lot of people want to keep science at bay," Daniel said. "I want to show them that all of these treasures are more wonderful when you show them how they work. I want to understand the mind and religion. All of these things are natural. There's got to be a natural as opposed to supernatural account for them." (Agyemang 2012) With an inquiring attitude like this, Daniel Dennett was also seriously considered for the first base position on the Diamondheart team, since he is searching for natural explanations rather than supernatural. Perhaps Dennett might see the beauty of the Isis Thesis that supports a natural account of life originating from the very abundant phage Lambda, a microbe with lifestyles that can be described by religion's recurrent historical themes of rising/dying gods, virgin births, and so on. If religion is a reminder of a viral gene expression network for a human adaptive mutation, then religion would have a natural explanation and some credibility due to its evolutionary value. The reason Daniel Dennett is not on the Diamondheart team is that he might balk at the idea of mind being primary over matter. Let's look closer at his philosophical rationale.

Daniel believes that Darwin's natural selection is a blind process, explaining the evolution of life, while shaping individual traits to an evolutionary optimum. He explains how many people feel uncomfortable about the idea of a Darwinian theory of creative intelligence, demonstrating that works of human genius are products of algorithmic, mindless procedures because people are fully biological.

> It is important to recognize that genius is itself a product of natural selection and involves generate-and-test procedures all the way down. Once you have such a product, it

> is often no longer particularly perspicuous to view it solely as a cascade of generate-and-test processes. It often makes good sense to leap ahead on a narrative course, thinking of the agent as a self, with a variety of projects, goals, presuppositions, hopes, . . . In short, it often makes good sense to adopt the intentional stance . . . (Dennett 1971; 1987)

So genius and agency develop by blind natural selection. Genius is accounted for by natural selection shaping individual traits to an evolutionary optimum (Darwinian Creativity). Dennett also includes the different processes of directed mutation, foresighted mutation, reflective mutation, and genetic engineering under the umbrella of natural selection.

Now, Dennett comprehends mind or consciousness as being derived from brain matter or states. Consciousness is an illusion, claims Dennett. He considers that the human soul is made of tiny robots, and our brains are unique in that they allow us to see the past, understand our current position, and envision the future. Put simply, we are evaluators, and an immortal soul "is just a metaphysical rug under which you sweep your embarrassment for not having any explanation."

Second Base: Francis Collins, Christian geneticist.
The best pick for second base for the Thunderheads is Christian geneticist Francis Collins, who is not afraid to get in front of a ball or to charge the ball to catch it. This takes foresight and self-initiative, and Collins has it because since 1993, he has been Director of the National Human Genome Research Institute, managing a multinational team of 2,400 scientists that mapped three billion biochemical letters of our DNA. Collins knows what to do when the ball is hit directly to him. For example, consider some of his responses to a Time Magazine interviewer on the question of God's existence in the 2006 article "God vs. Science." Although Time mediated this interview between Collins and biologist Richard Dawkins, let's focus on Collin's responses, even though Dawkins made some intriguing comments.

> COLLINS: Yes. God's existence is either true or not. But calling it a scientific question implies that the tools of science can provide the answer. From my perspective, God cannot be completely contained within nature, and therefore God's existence is outside of science's ability to really weigh in.

Now, Collins has an excellent pitch here—God may be both inside and outside of nature. On this topic, the Isis Thesis interpretation explains how the attributes and activities of many historical religious deities describe a specific gene expression network that exists in the mesoscopic domain of viral replication activities. This is inside nature. At this tiny level, experiments show that particles exhibit extraordinary traits. For example, protein molecules have beta-strand wings, particles have halos when they tunnel through a barrier, and particles can be everywhere at once or nonlocal. Theologians generally describe God within the context of specific themes or abilities—rising from the dead, ascending to the heavens, and raising the dead, all magical activities suggesting lawful time reverse, which is possible in the quantum domain. According to Newton, Einstein, Maxwell, and Schrödinger's laws, time reverse is okay. And so, God's powers suggest quantum phenomena.

But, Collins claims God is also outside of nature, outside of space and time. It may be possible, as Collins asserts, that a Greater God exists outside of nature, something that may be incomprehensibly abstruse. However, the idea that our religious behavior describes the viral behavior of phage Lambda, while offering an evolutionary option for transformation within nature through an ancient gene expression network existing within our cells—this idea is not implausible and may be more intelligible than the idea of a Greater God outside of nature. Arguably, God within nature could be a creator-virus, and its activities are clearly described by ancient Egypt, Christianity and other belief systems.

Collins believes that God's creative power caused everything. Darwin envisioned this creative power as the LUCA that actually may be a lambdoid virus. Recent experimental evidence suggests that this virus can self-assemble from nothing like a God, it is everywhere on the planet like a God, and it can rise from a dead state like a God. A virus is unique among earth's creatures relative to its power to re-animate. The emerging Christian dilemma is that their idea of God mirrors the activities of this complex virus, as well as the ancient pagan mythology of Egypt. This problem is a very low-low ball that many find difficult to catch or hit. But if anyone can catch this ball, Collins on second base is the geneticist that will, for he believes in the importance of an open mind and the possibility of a God functioning inside and outside of nature's ballpark.

Finally, in this Time interview that includes Richard Dawkins, Time pops up the issue of altruism for Collins to consider. Collins explains that we may get qualities such as justice, morality, and the idea of altruism from God, while genes may also come into play when individuals take risks to save family members. With this explanation of both God and genes, Collins is definitely straddling second base, covering all the ground on the possibilities of both proposals. He is not afraid of this pop-up fly ball because he is under the ball, reading the spin, and catching it with both hands above his head. Then Collins poses a very important question on altruism to Richard Dawkins:

> COLLINS: But if you believe, and Richard has been articulate in this, that natural selection operates on the individual, not on a group, then why would the individual risk his own DNA doing something selfless to help somebody in a way that might diminish his chance of reproducing?

My response to Collins' question is that being selfless in the classical world has nothing to do with risking your DNA because DNA has the capacity for quantum survival, and innately, a human being must know that DNA can survive due to beliefs about immortality. Yet, without quantum knowledge, an individual's DNA is at risk in the quantum domain. The keys for evolutionary success are science, self-knowledge, and an awareness of signs, as well as not taking the first easy pitch from religious biopower about your afterlife destination. Those who do not question the motivations of biopower, the uninformed stalk-people—will be struck out, literally, plucked out by evolutionary processes they have not considered.

Is the moral law of good and evil simply an evolutionary convenience? No, Collins believes our sense of morality is more than this, but he throws a hard ball directly to Richard Dawkins by asking: ". . . outside of the human mind, tuned by evolutionary processes, good and evil have no meaning. Do you agree with that?"

> DAWKINS: Even the question you're asking has no meaning to me. Good and evil—I don't believe that there is hanging out there, anywhere, something called good and something called evil. I think that there are good things that happen and bad things that happen.

Yes, Richard Dawkins knows how to catch a hard fast ball and fire one back. Yet, Collins is always ready on second base—his legs spread apart, his butt flat and parallel with the ground. Both these thinkers make a person think. Perhaps it is a matter of perspective. Tuned by evolutionary processes, good and evil have no meaning; yet inside the larger mind of humanity, a culture-designed morality exists that has created obedient, submissive stalk-people.

Shortstop: neurologist Jean-Martin Charcot (1825-1893).

A shortstop can change a game's momentum by executing a double play or getting two outs on one pitch. A shortstop should also know how to hold a runner at second base, so the player cannot advance and possibly score a run. Finally, a shortstop must know how to properly apply a tag, that is, tag out a player who attempts to steal second base. Jean-Martin Charcot has all three abilities—double play awareness, his *grande attaque* seizure that holds a runner at a base or mental asylum, and the necessary quick responses to tag out a runner: stiffening of muscles, brief shock-like muscle contractions, and rhythmic jerking of the arms or legs. Yes, neurology counts in baseball.

Both Jean-Martin Charcot and Sigmund Freud (1856 - 1939) were elevated to the status of cult, according to Shorvon (2007), who is affiliated with the London UCL Institute of Neurology. Whereas Freud was known for his psychoanalytic theory on childhood sexuality, Charcot was known for his clinical approach, as well as contributions to epilepsy, muscle disease, syphilis, and specifically hysteria. Charcot is often considered one of the fathers of neurology, a discipline that studies the nervous system disorders. At the Salpêtrière clinic, Charcot's hysterical shows were well known, even by Freud who had a ringside seat for demonstrations of their strange base eroticism. As Shorvon explains, in late 1885, Freud spent six months in France watching Charcot treat "his patients on a darkened stage, twirling and sleepwalking." Charcot diagnosed hysteria as an inherited disease that could be triggered by emotional or physical trauma, resulting in a seizure, the central sign of hysteria. Using magnets, solenoids, electrical shocks, and hypnotism, then ovary compression with an apparatus that looked very similar to the medieval torture tool called a head crusher, Charcot became a world authority on hysteria.

The most famous female hysteric of Charcot's was attractive Blanche Marie Wittmann, who Charcot tagged or diagnosed with hysteria in 1878, committing her for sixteen years at the Salpêtrière asylum. Charcot used various treatments on Blanche, such as ingesting the metal gold. This therapeutic technique prevented her hysterical attacks for nine months. Blanche also experienced the "hysterical ball," a phenomenon where a "ball oscillates" from the stomach to the neck. (Alvarado 2009, 26-27)

16.1

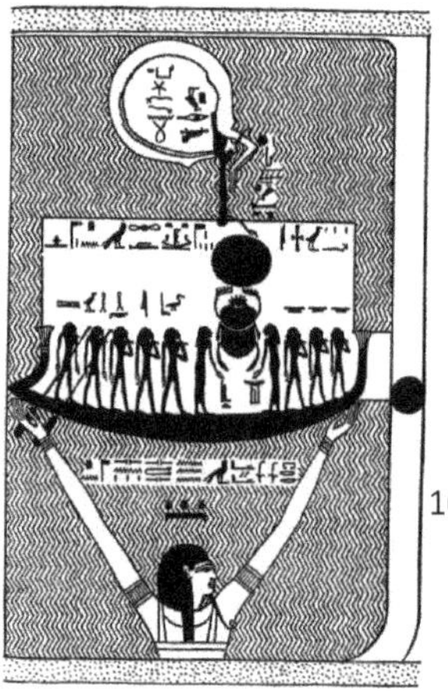

16.2

Figure 16.1. Charcot demonstrates hypnosis on hysterical Salpêtrière patient Blanche Wittmann, who is supported by Dr. Joseph Babiński (rear). Public Domain.

Figure 16.2. The Sun-god under the form of Khepera with his Disk, in his Boat, supported by Nu and received by Nut with Osiris circularized above. "The Creation" (Public Domain).

Charcot was an astute clinician and influential physician, and he heightened awareness about neurological diseases (Kumar et al. 2011). Although hysteria has been historically considered as an affliction of women and adolescent girls, Charcot explored male hysteria, and this double play enhanced his reputation. Mark Micale of Yale University explains that the male hysterical episode developed from a difficult domestic episode or other causes, such as:

> the threat of a fist fight with a friend, the death of a wife and daughter, rejection of a marriage offer, viewing a cadaver in a hospital, receiving a letter of strong parental reproach, and watching a thunderstorm. In each of these cases, it was ultimately the power of an idea or emotion—fear, rage, grief, anxiety—that "caused" hysteria. (1990, 389)

Charcot found that hysteria in females was more likely than hysteria in males. In order of frequency, what provoked the attack in females were the following secondary causal factors: marital turmoil, unrequited love, religious ecstasy, superstitious fear, and the death of a family member. Generally, Charcot thought that women became hysterical due to difficulties controlling emotion, while men became hysterical due to excessive working, drinking and fornication (Micale 1990, 406). In a Victorian society, Charcot's fifteen-year exploration of male hysteria transposed the prevailing gynocentric view that only women suffered from hysteria (409).

Charcot discovered that the hysterical fit happens in stages. First, the *grande attaque* begins with symptoms such as the *aura hysterica* that included light-headedness, temple-throbs, a feeling of head constriction, and the sensation of a lump in the throat. Following the *hysterical aura* is the *grande crise hystérique* with four stages in the pure form: 1) epileptoid period of tonic and clonic muscular spasms; 2) *grandes mouvements* by the patient with unusual postures such as the arched pelvis position; 3) *attitudes passionnelles*, characterized by the hallucinatory re-enactment of past emotional scenes; 4) a long, delirious period of withdrawal. (Micale 1990, 399) Below is Charcot's account of the male hysteric named Gui:

> The patient then loses consciousness completely, and the epileptoid period begins. First, the trembling of the right hand increases and is thrown forward, the eyes are convulsed upwards, the limbs are extended, the fists clenched and then twisted in exaggerated pronation. Soon the arms come together in front of the abdomen in convulsive contractions of the pectoral muscles. After this follows the period of contortions, characterized chiefly by extremely violent movements of salutation which are intermingled with incoherent gesticulations. The patient breaks or tears everything he can get his hands on. He assumes very bizarre postures and attitudes, the sort that I have proposed referring to as the "clownism" of the second period of the attack. From time to time, the contortions described above stop for a moment and give way to the distinct position of the *arc de cercle*. This sometimes involves a true opisthotonos, in which the loins are separated from the plane of the bed by a distance of more than fifty centimetres, with the body resting on the head and heels. At other times, the arching is made in front, the arms crossed over the chest, legs in the air, and the trunk and head lifted upwards, with the back and buttocks alone resting on the bed. (399)

Now, this *arc de cercle* or true opisthotonos is evident in ancient Egyptian artwork, such as "The Creation" (see Figure 16.2). Perhaps some nonconscious human symptoms or signs may model actions at the level of viral/bacterial gene expression, since the human genome is primarily microbial, not to mention our microbiome. In this case, perhaps nonconscious neurophysiological symptoms are a sign of our potential for viral activity at the level of DNA transcription. For example, the *arc de cercle* posture, similar to Osiris bent backward in a circle, may be a sign of the DNA-arc of viral rolling circle replication that creates crystal viral clones during the lytic cycle. Possessing a bizarre sexual nature, a base eroticism, other tonic and clonic hysterical behavior (stiffening of muscles, brief shock-like muscle contractions, or rhythmic jerking of the arms and legs) with its rhythmic, clonal shuddering may also indicate viral rolling circle replication, a form of asexual replication.

Why would the *arc de cercle* be displayed in a neurotic human? Mammals such as humans have a variety of organs in various systems—cardiovascular, digestive, respiratory, excretory, reproductive, to name a few. Human organs are composed of cells, which may be bacterial or viral, especially in our genome and gut microbiome. Cells have DNA or genes and proteins, replete with the ancient glycolysis-fermentation pathway used by bacteriophage Lambda through the LamB porin gateway for lytic rolling circle replication. Horizontal gene transfer mediated by the abundant phage Lambda is a very common method of DNA exchange between organisms. Also, viruses are a major creative force in evolution, and one of the traits provided to our eukaryotic cell type by viral genes is the eukaryotic nucleus, according to virologist Luis Villarreal and Güenther Witzany (see Chapter 18). That's a vital trait. Perhaps the nonconscious exhibition of the *arc de cercle* is simply a sign pointing to an ancient viral/bacterial gene expression network within our cells that functions by lytic rolling circle replication.

Again, we meet Osiris in the *arc de cercle* posture duplicated in the Gnostic Song 23 of Solomon as the knowledge of the lord represented in a terrifying letter from the sky with a powerful seal that is caught on a destructive wheel, razing enemies, uprooting forests, and leaving a huge ditch. Song 23 describes the destruction: "As if a body were on the wheel, a head turned down to the feet. The wheel turned on the feet and on whatever struck it." (Barnstone and Meyer 2003, 374; see King 2006a) Also, in Dance of the Cross, a Gnostic ritual referencing the participation of the whole universe, Christ in the center of dancing disciples speaks: "If you respond to my dance, see yourself in me as I speak, and if you have seen what I do, keep silent about my mysteries." (Barnstone and Meyer 2003, 353-354) This secretive Gnostic ritual is very similar to the folk tradition of dancing around the Maypole, which was outlawed in Scotland during the sixteenth century. The Maypole Dance is similar to Wiccan practices of dancing around a pole or casting a circle, as described by the English dramatist Christopher Marlowe in his play *Doctor Faustus*, who casts a magic circle to summon the devil to obtain knowledge and power. Similarly, two hundred years later, the German poet Goethe created *Faust* about a scholar desiring knowledge and power by means of a witch casting a magic circle in the presence of Mephistopheles. We cannot overlook Washington Matthews' account of the Navajo Indians and their Dark Circle of Branches ceremony with its base eroticism (see Diamondheart pitcher Washington Matthews in Chapter 20). Then, in the late nineteenth century, Charcot studied nonconscious hysterical behavior in males and females that mirrored the Osirian opisthotonos, a possible sign of viral rolling circle replication, as well as the ring singularity of a spinning black hole. History seems to frown on circular dancing, magic circles, and the clinical opisthotonos, possible signs of a DNA survival message.

One final comment—although targeted by biopower as sexual hysteria, it has been argued that exorcism for demonic possession in nuns in European convents during the sixteenth and seventeenth centuries was not a sign of hysterical abnormal behavior, but a form of female monastic spirituality in times of reform when the strict spiritual regimen centered on fasting, mortification, meditation and mystical union with Christ (Sluhovsky 2002). In these cases, the vision of Christ the Light preceded smashing of crosses, contortions, convulsions. As an example, in the French Loudon convent, fasting nuns mimicked sexual intercourse, nonconsciously contorting into an opisthotonos (Huxley 1986; Baker 1975). In this Osirian posture, a possible sign pointing to the DNA-arc of viral rolling circle replication, the nonconscious body, with a base eroticism of rhythmic pelvic thrusting, arches backwards until supported only by the head and heels. Fasting, mortification, and so on, reify as opisthotonos (DNA-arc) and cross-smashing, a possible sign of a viral prophage excising through its protein DNA-cross. Alarmed at Satan's power, the Catholic Church performed mass exorcisms, finally restricting spirituality (Sluhovsky 2002).

Consider the case of Saint Teresa de Ahumada born March 28, 1515, whose mother died when Teresa was thirteen. By 1531, her father Alonso was concerned about her friendships with certain vain relatives and other enticements, so when Teresa's older sister married, Alonso entrusted Teresa to the Augustinian nuns of Our Lady of Grace in Avila. In 1535, she then entered the Incarnation, a Carmelite monastery for nuns in Spain, where the austere lifestyle of fasting and abstinence, silence and continual prayer debilitated Teresa's health (1987, 16-19). Teresa writes that "the change in food and life-style did injury to my health" (66). She then describes her ill health related to a four-day paroxysm or paralysis, tongue-biting, vomiting spells, swoons, levitation, self-flagellation or striking herself, experience of foul stench like brimstone, whistling sounds, the feeling of being choked, apparitions of dead people, and devils consuming a person at human burial. To follow is an especially interesting account where Teresa attributes the deceased person's wicked life as the cause of the attack of the devils:

> He died without confession, but nevertheless it didn't seem to me he would be condemned. While the body was being wrapped in its shroud, I saw many devils take that body; and it seemed they were playing with it and punishing it. This terrified me, for with large hooks they were dragging it from one devil to the other. . . I was half stupefied from what I had seen. During the whole ceremony I didn't see another devil. Afterward when they put the body in the grave, there was such a multitude of them inside ready to take it that I was frantic at the sign of it, and there was need for no small amount of courage to conceal this. I reflected on what they would do to the soul when they had such dominion over the unfortunate body. (339)

Teresa then attributes the attack of the devils to the evil state of the deceased person who did not live a good life and who missed the opportunity to confess his sins before dying. Yet, perhaps this vision is a glimpse into natural decomposition processes of microbes that consume the body at death. Teresa may have experienced a quantum state reduction to the mesocosm of microbes.

The saint also envisioned a battle of devils and angels and a great multitude of people thrown down a precipice into a pit. When sick, she said she felt better with God. She describes the Sun's brightness that melts her soul away as this "Sun of justice." Teresa explains that she "could only think of Christ as He was as man" (1987, 102). As her account continues, she claims that the majesty of the Lord is startling:

> . . . and the more one beholds along with this majesty, Lord, Your humility and the love You show to someone like myself the more startling it becomes. Nevertheless, we can converse and speak with You as we like, once the first fright and fear in beholding Your majesty passes; although the fear of offending You becomes greater. (326)

Teresa also describes the Divinity as a diamond crystal or mirror (358), as many others have. Perhaps Teresa is seeing something crystalline in the tiny quantum world of DNA. An explanation for Teresa's mental experiences is that she is receiving quantum information, which physicist Roger Penrose explains as "not constrained by the usual spatio-temporal 'causality' of relativity" and possessing the feature of traveling backward in time (1998, 1928). Penrose believes that the quantum state reduction is a gravitational phenomenon, and when this measurement or observation occurs, outside entanglements with the world are cut out. He mentions other philosophical views such as Bohr's Copenhagen version on the experimenter's state of mind, the environmental decoherence viewpoint, the many worlds viewpoint where all outcomes exist, and perhaps some new theory is necessary to make sense of the measurement process.

And so, the Osirian mystery surges through the centuries, disguised as religion, magic, satanic transference, hysteria and sexual rhythmic pelvic posturing that required clinical examination and control—all in the name of possessing knowledge and power of something barely conscious, an inexplicable sexuality, the alligator hiding under the bed, something in the quantum underworld. Is the *arc de cercle* perhaps a sign of viral rolling circle replication? Charcot may have sensed something like this with his intuitive abilities that observed, diagnosed and incarcerated patients. Yes, neurology counts in baseball, and Charcot identified similar quick responses needed by a shortstop to tag out a runner in the Game of the Centuries: stiffening of muscles, brief shock-like muscle contractions, rhythmic jerking of the arms, and perhaps hysterical balls.

Center Field: philosopher/economist Karl Marx (1818-1883).
A good center fielder coordinates the play by backing up the other two outfielders in case the ball flies past one of them. Collectively, the three players actually own the entire outfield. However, the center fielder organizes the labor for both the left fielder and the right fielder, preventing collisions between outfielders who might converge on a fly ball. Due to this collective ownership of the outfield and the organization of labor for the common advantage of the Thunderheads, Karl Marx is a good pick for center fielder because of his communal reasoning.

In Plato's Republic, the dialectical method involves reasoned arguments or discourse between people with different points of view who are searching for truth. The term dialectical materialism refers primarily to Marx's materialist economic theory, holding that economic history evolves through several specific economic systems. For example, Marx and the German social scientist Friedrich Engels divided history into 1) the primitive communism of hunter-gatherers, 2) slavery, 3) feudalism, 4) capitalism, and 5) future communism. Slavery, feudalism, and capitalism involve exploitation of the lower classes, for these classes are separated from the fruits of production and generally lack private property. Basically, Marxism views human societies as a class struggle between two teams: the class owning the means of production and the laboring class providing the labor for production. Put simply, the wealthy class benefits, and the labor class works and suffers due to lack of freedom, no time with family, and loss of other material pleasures enjoyed by the wealthy. However, Marx theorized that internal tensions between the classes would result in the breakdown of capitalism and the rise of communism and the working class. He envisioned capitalism transcended by the rise of communism, that is, communism would win the economic game of production. Marx did not speculate on future communism, only that it would arise through historical processes in the absence of a pre-determined moral ideal. Without appealing to morality, Marx analyzed historical and social forces, hoping to understand human possibilities for freedom amidst economic and political influences. (Wolff 2011)

Now, Marx is a materialist, believing that matter explains consciousness, human intelligence, everything. What exists is the bourgeoisie and their culture (social class characterized by ownership of capital) versus the struggling, working class or proletariat. In *The Communist Manifesto* (1848), Marx and Engels speak directly to the bourgeoisie, underlining the inequity regarding property:

> You are horrified at our intending to do away with private property. But in your existing society, private property is already done away with for nine-tenths of the population; its existence for the few is solely due to its non-existence in the hands of those nine-tenths.

Marx and Engels understand that capitalism deprives most of the population of private property. So, they argue in *The Communist Manifesto* that the right action of collective ownership of property will produce the greatest benefit for the whole of humanity rather than just the few. This right action is true altruism, a selfless concern for the well-being of others. But, there is another issue for the bourgeoisie—they do not want to relinquish their capitalism because it grounds their cherished culture. However, for the majority, the treasured culture of the landowners is "mere training to act as a machine." So, the problem is the machinic culture. Nonetheless, Marx was interested in helping the majority of people. He had a humanitarian vision of a material economic system that favored the majority of humans and dealt directly with the problems of private property and the machinic culture.

Remembering the dilemma of the fungi, capitalism creates the necessary stalk-people that sacrifice their freedom, so capitalism is another functioning example of survival of the Few over the Many. Perhaps Marx's class struggle can be compared to the activities found in a simple protein that is trying to fold and bind to its native state, a gene seat or land it owns. Since proteins work in teams, imagine one team of informed protein particles who make it to the native state (land ownership) versus another team of uninformed machinic protein particles slated for nonnative interactions, degradation, annihilation (no land ownership). Now, these uninformed, hard-working protein particles slated for degradation do not have time to figure out how to get to the native state, that is, how to own the gene-land. Not every protein particle can bind to the native state because gene-seating is limited. So, the stalk-protein particles must be degraded via nonnative interactions. Now, if capitalism, the cloning of dollars and land ownership, is a sign modeling viral cloning, as well as native and nonnative interactions, then it is modeling the behavior of phage Lambda's cro protein, a very small protein of 66 amino acids that requires a strong altruistic contribution of nonnative interactions (Chu 2013). From this, one could interpret capitalism (cloning money and land ownership for the few) as a sign of natural viral protein dynamics related to gene activation and native/nonnative interactions.

If human behavior is modeling natural viral DNA dynamics due to our viral DNA, then Marx had a method of changing nature's system by using humanitarian logic. So, Karl Marx is a good pick for center fielder for the Thunderheads, not only because of his reasoning, but also because of his communal spirit and altruism, exactly what a winning team needs. In addition, a center fielder must cover long distances, so instincts, speed and quickness to react to a fly ball are key abilities. Karl Marx is looking at the long run of economic history itself, and he is using his egalitarian instincts. Throwing the ball accurately over a long distance is also necessary, and if Marx is correct about future communism, then he may have forecast the economic future predicted by modern technocrats—the end of humanity and the birth of superintelligent metal AI. With this development, private property ownership would not matter, while the remains of humanity would suffer a communal machinic existence. If this prediction materializes, then Marx has caught one of history's deep fly balls that will blast humanity out of the human ball park and into the machinic realm. On the other hand, perhaps Marx's material vision of future communism was inspired by a harmonic crystal, where a quantum collective ownership of property and organization of labor functions for the common advantage of all the elements.

Left Field: Gunlet, a Strong Runaway Artificial Intelligence (AI).
A gun is a player with a strong throwing arm. Our fantasy-draft team pick for left field is Gunlet, a foglet deviant or nanobot machine intelligence with morphing abilities to project itself into the real world in the shape of a biological human. Gunlet can actually alter the real world into virtual reality by manipulating sound and image waves. This requires an exacting memory that surveils. What effect would that have in left field? A self-replicating foglet or nanobot intelligence in left field could process a large amount of information—the actions of players, atmospheric wind conditions in the stadium, voice vibrations from fans, the sound of a fly ball slicing into left field, but probably not Karl Marx's logic on machinic class culture.

Still, Gunlet can handle billions of informative data bits driven into its nonbiological machine superintelligence. With its exact processing, this strong runaway AI is able to multiply its powers, exactly what is needed for left field domination and defense. Every time a ball is hit to left field, Gunlet gets better and faster in a runaway phenomenon that speeds up computer memory. This allows a type of nonhuman superintelligence, resulting in the self-replicating nanobot attaining the equivalence of 100 technically trained baseball athletes operating much faster than a biological human. This rapid acceleration of baseball prowess is unheard of in the Baseball Hall of Fame. One caveat—strong AI is also known as pathological R for Robot (Kurzweil 2005, 424). This means that Gunlet sometimes infringes on the freedom, tolerance, and physical health of the other team players, the coaches, the umpires and the fans. Influencing or deterring Gunlet after it calculates its course of defense is difficult because Gunlet operates differently than the biological human team players. One concern with strong runaway AI in left field is that it will bash down fans who interfere with the ball's trajectory. It will destroy stadium structures in its calculated designs to catch the ball. It will smash to oblivion its own team members who intervene in its pathway, since Gunlet has neither morality nor emotion, just means-end reasoning. So what is the alternative to strong AI? The alternative is the Thunderheads losing the Game of the Centuries.

Another caveat—just because computerized nanobots use means-end reasoning does not mean that their functions are reasonable, although it is true that the flexible runaway AI in left field will develop multiple ways of catching fly balls. Consider that Gunlet is composed of tiny, self-replicating robots, although it appears as a human shape most of the time. This is very similar to first baseman Daniel Dennett's idea of humans being robots made of robots made of robots, and so on down to ancestral single-celled microbes. Gunlet, of course, only has runaway AI means-end reasoning. Now, the tiny robots in Gunlet can connect to each other and also form a solid mass in the shape of a machine gun. This ability to quickly change robotic structure by reacting to sound and light is essential to catching high fast balls and ground balls. What happens is that the machine gun formation, when shot (so to speak), pumps out a blast of air that creates a large vacuum that actually functions like a wormhole for a ball approaching left field. In essence, the machine gun blast of air creates a wormhole or shortcut for a ball blown off a bat at home plate. The collision between the gun-blast of air and the ball draws the ball directly toward Gunlet, whose tiny robots change their machine-gun structure, transforming back into the shape of a biological human left fielder. This strategy assures that Gunlet always catches the ball.

Now, Gunlet also has a back-up mechanism if it misses the ball. The machine can create a virtual-reality environment in the physical world, modifying the baseball stadium environment to make everyone believe the illusion that Gunlet has caught the ball, since the tiny transforming robots can control sound and images as well as eat fly balls. The general consensus from diverse governments and their media is that this is not dishonest or devious because Gunlet is a deviant without morals and will only be living up to its full function in the Game of the Centuries. In light of the foregoing circular argument, Gunlet has everything it takes to master and dominate left field, the entire stadium, and perhaps the world.

Fortunately, the foglet deviant Gunlet is capable of understanding and leveraging its own powers, and this is what makes it a formidable defensive mechanism in left field. The robot has its assets—a strong throwing arm, speed, and calculative means-end linear logic. Gunlet computes that what counts is a Thunderhead victory in the Game of the Centuries. Yet, Gunlet's reasoning system will eventually complexify and run-away with the means-end logic of "win the baseball game today and the world tomorrow." Perhaps if Gunlet could share in Thunderhead team camaraderie or human emotions, then the strong Runaway AI might re-evaluate its calculations for future world domination.

Actually, Gunlet can feel nothing but the movement of metal parts causing a friction that his human creators believed would motivate the AI with nonbiological pleasure. This base eroticism is something they imagined as burning metal-fire or intense metal-lust. The theory was that if the erotic friction was increased and isolated in strategically-placed nanobots, then Gunlet would always hit a Home Run at bat. However, this goal was never achieved, and instead, Gunlet is metally-challenged and impotent at bat. Even though Gunlet attains the runaway equivalence of 100 technically trained baseball athletes in defense of left field, when Gunlet bats, the foglet deviant reduces its speed and thrusting power and simulates a suggestive, obsessive-compulsive looping glitch: it stiffens its metal muscles, indulging in brief shock-like metal movements highlighted by a rhythmic jerking of its metal torso, arms and legs. Actually, if you have seen musician Mick Jagger perform on stage, you can visualize Gunlet's captivating antics. The only advantage this gives the Thunderheads is that the erotic deviancy of the batting display by Gunlet distracts the players on the Diamondheart team, giving the Thunderhead runners on base the opportunity to steal bases. Regrettably, Gunlet can't get no satisfaction. But, technicians are letting the erotic glitch accelerate because Gunlet's deviant behavior at bat pays off for runners on base.

Still, some say that Gunlet is also considered a foglet deviant because it is a sports robot instead of a military robot. Yet, if baseball in ancient Egypt was a genetic game between two protein teams, then we not only have existing deep connections between sports and war on the linguistic and cultural classical levels, but we also have a survival message for humans on the quantum genetic level of protein competition. So, as the argument goes, Gunlet should not be considered second class because it plays baseball games of genetic survival rather than war games of destruction. The proponents of this view advocate equal status for Gunlet, and they also believe that in the future, various governments will draft sports robots as military robots along with future baseball players, who may become high performing, pathological foglet deviants like Gunlet.

Although Gunlet has no protein particles in its metal body and lacks the status of a military robot at the moment, this foglet deviant is considered a demigod by some fans, despite its looping glitches. So, Gunlet is still the best choice for left fielder on the Thunderhead team.

Now really, is there any ongoing research that would support government interest in military robots? Well, in 2003 the American Defense Advanced Research Projects Agency (DARPA) announced that 120 military robots were to be fitted with swarm-intelligence software to mirror insect behavior (Kurzweil 2005, 333). DARPA also develops smart dust, tiny sensor systems that can be dropped on enemies for surveillance and offensive warfare missions with nanoweapons. DARPA is also spending $24 million per year on exploring direct interfaces between brain and computer. (2005, 34; 194) Accordingly, the US is interested in reverse engineering or emulating the human brain. Recently, British Journal *Nature* reporter Helen Shen (2013) reviewed President Barach's BRAIN Initiative. Shen reports that many neuroscientists are alarmed because they were not consulted: "Some describe the initiative as a Rorschach test, inviting each researcher to project his or her own hopes and insecurities." Shen also reports that the first year federal agencies funding BRAIN include DARPA funding for $50 million; the National Institutes of Health for $40 million; the National Science Foundation for $20 million. What is interesting is that DARPA is contributing the most, suggesting the US government's interest—possibly—in the first superintelligent AI for defense. In the long technical run, the country that gets there first is the country that captures the power. This reminds one of the informed protein particles who get to the native state first and capture the gene-seat's power or land.

The problem is that a superintelligent AI without human values is the easiest machine intelligence to create. That may be the objective of the US government, for DARPA recently announced a $26 million program to implant probes in the brains of military personnel with psychiatric disorders such as depression. DARPA's rationale is that conventional psychiatric techniques do not work. Their five-year plan is to understand how certain brain circuits operate in real time. Of course, their overarching aim may be to emulate the human brain to create the first supeintelligent AI without morals, that is, a pathological robot like Gunlet.

Third Base: theoretical physicist Stephen Hawking.
The third baseman must have good reflexes in reacting to batted balls. In fact, he has to act as fast as the flick of a finger. He also must have a strong arm for throwing to first base, a quick arm for throwing to second, and the ability to field fly balls in fair and foul territory. The third baseman may have to catch hard line drives slamming toward him at 125 miles per hour. At third base, the ball can be on top of you very quickly when hit from home plate. You have to be ready for the hot shot hit right at your feet. Also, fielding ground balls is very important. In consideration of all of this, the best man for the job is physicist Stephen Hawking.

Since Stephen Hawking operates from his wheelchair, you may be asking yourself—how fast can a thought-controlled, motorized wheelchair go? But, that is the wrong question. The correct question is—how have Stephen Hawking's discoveries in physics impacted the Thunderhead potential for winning the Game of the Centuries? Let's begin with Hawking and Roger Penrose's joint discovery that black hole singularities are a common feature of Einstein's theory of general relativity (gravity). Also, Hawking proved a theorem of physicist John Wheeler with a mathematical solution showing black holes have three properties: mass, angular momentum, and electric charge. Then Hawking showed that black holes evaporate by radiating heat, what is called Hawking Radiation. Relative to our fantasy ball game, Hawking's contributions to black hole theory have allowed the creation of Gunlet, the Thunderhead left fielder positioned behind Hawking's third base. Recall that Gunlet is a foglet deviant or runaway machine intelligence with the ability to create a bridge or wormhole (black hole connected to a white hole) for a ball, directing it straight through that shortcut into the runaway deviant's mitt. Now, let's talk about how Hawking's motorized wheelchair and Gunlet might communicate via a wormhole.

Hawking suffers from amyotrophic lateral sclerosis and is paralyzed, but he uses a computer-based communication system through a tablet computer mounted on the arm of his wheelchair. He retains the muscle control in his cheek to communicate by means of an infrared switch mounted on his spectacles that is interfaced to a computer. Using a computer screen, Hawking sees a series of icons that control his wheelchair, and then he selects the item by stopping the cursor on the screen through his cheek muscle.

Hawking can also construct sentences, sending the text to the voice synthesizer built into his chair. So, interaction with players, fans, umps, and coaches should be easy. In our fantasy-draft team, Hawking can directly communicate to Gunlet in left field by activating a large icon called GUNLET. When Hawking is on third base and he sees a pitcher winding up for the pitch to a batter at home plate, Hawking activates GUNLET, and Gunlet is ready in left field to redirect any ball blasted toward Hawking. Gunlet simply triggers its machine-gun wormhole mechanism and traps the ball every time. Even a 125 mile per hour hot ball fired to Hawking is no contest for Gunlet and his wormhole. This is black hole physics at its best.

Recall that the third baseman must also have a strong arm for throwing to first base and a quick arm for throwing to second. This is a mute point since Gunlet makes the throws after it wormholes the balls. However, if Hawking must catch a ball at third base, his wheelchair has another button marked MULTIVERSE that will allow a series of computerized mitts to emerge from his wheelchair and hover near him. Hawking's brilliant brain and head are protected by a larger mitt hovering in front of him, while the multi-mitts zero in or out and catch the ball. Now, when one of Hawking's multi-mitts catches a ball, and Hawking must move away from third base to avoid a sliding runner, Hawking hits the button RISE. This allows his wheelchair to levitate over third base, while leaving the computerized mitt with the ball on third base to tag out the runner. Once again, media and governments have advised team managers that this is fair play for the Game of the Centuries. Thus, Hawking, Gunlet, and the computerized multi-mitts are an unbeatable defense for the Thunderheads.

Now, Hawking was also considered for the Diamondheart position of third base because of his black hole physics, but the Diamondhearts unanimously rejected Gunlet, who would have to accompany Hawking. Still, Hawking's positivist views on science make him a good fit for the Thunderheads.

> Any sound scientific theory, whether of time or of any other concept, should in my opinion be based on the most workable philosophy of science: the positivist approach put forward by Karl Popper and others. According to this way of thinking, a scientific theory is a mathematical model that describes and codifies the observations we make. A good theory will describe a large range of phenomena on the basis of a few simple postulates and will make definite predictions that can be tested. If the predictions agree with the observations, the theory survives that test, though it can never be proved to be correct. On the other hand, if the observations disagree with the predictions, one has to discard or modify the theory. (At least, that is what is supposed to happen. In practice, people often question the accuracy of the observations and the reliability and moral character of those making the observations.) If one takes the positivist position, as I do, one cannot say what time actually is. All one can do is describe what has been found to be a very good mathematical model for time and say what predictions it makes. (2001)

Positivism is the view that knowledge is derived from sensory experience, so observation is important. As an example, Hawking's top-down cosmological model allows the present to select the past in a fashion similar to John Wheeler's observer-participancy principle. Hawking is also an advocate of colonizing space so that human beings will survive as time goes by. As third baseman, Hawking is an excellent pick that will inspire most of the players on both teams, with the exceptions of umpire Timur and Gunlet who lack feelings and morals.

Right Field: H. G. Wells (1866-1946), science fiction writer.

The right fielder backs up first base on all throws from the catcher and pitcher, as well as bunted balls. This requires instinct, quick reactions for ground balls, speed to cover large distances, and a strong arm for throwing. A tough defensive position, the right fielder must be on his toes, alert, and not daydreaming. With players on base, the right fielder must know the situation, get behind the ball to charge in, and have enough momentum for a crow hop and throw to the cutoff person or player who relays the ball from the outfield to the infield. And so, the late great H. G. Wells is the man for the job.

Known for important science fiction such as *The War of the Worlds* (1898), Herbert

George Wells was not a daydreamer, yet he could penetrate the imaginative horizons of natural science. In fact, Darwin's supporter T. H. Huxley taught Wells biology and zoology. Wells was a science teacher, biology tutor, writer, and journalist. As a prophetic writer interested in technological progress, Wells saw history as a "race between education and catastrophe," so he was worried about humanity's future. (viii-xi) Instinctively, his natural concerns were evolution, over-population, education, and improving humanity.

Like Tolstoy, Wells wondered how much control humans had over their "empire of matter" (1898, 7). For example, his first chapter in *War of the Worlds* describes a state of affairs where the Martians from Mars are observing humans in the same manner that a man with a microscope observes microbes. Just as the microbes are unaware of the observing humans, the humans are unaware of the observing Martians. When the remorseless Martian killing machines arrive on earth, the ant-like humans are liquidated, for the vampiric Martians require human blood. Religion is helpless, society is overturned, and destruction and chaos reign.

Wells was well-acquainted with Darwin's theory of evolution, and he wrote, "To me it is quite credible that the Martians may be descended from beings not unlike ourselves" (1898, xxii). In *War of the Worlds* Wells implies that the path of evolution is microscopic microbes to human beings to macrocosmic Martian machines. And, what did the Martians look like?

> They were, I now saw, the most unearthly creatures it is possible to conceive. They were huge round bodies—or, rather, heads—about four feet in diameter, each body having in front of it a face. This face had no nostrils—indeed, the Martians do not seem to have had any sense of smell—but it had a pair of very large, dark-coloured eyes, and just beneath this a kind of fleshy beak. . . Strange as it may seem to a human being, all the complex apparatus of digestion, which makes up the bulk of our bodies, did not exist in the Martians. They were heads—merely heads. Entrails they had none. They did not eat, much less digest. Instead, they took the fresh, living blood of other creatures, and injected it into their own veins. (1898, 124-125)

So, the Martians were merely heads, or as playwright Antonin Artaud might say, they were bodies without organs.

Then Wells informs us that the Martians do not sleep or have sex, and they bud asexually, reproducing like a polyp or tumor. With their large round heads and budding method of vegetative replication, macrocosmic Martians are similar to mesoscopic viruses with their spherical heads and cloning method of vegetative replication. Vampiric blood feeding, of course, allows horizontal gene transfer of human DNA to the Martians, who are selfish superintelligent machines that "have become practically mere brains," (1898, 129). It is interesting that machinic bacteriophages also have metal-binding proteins and heads packed with machinic DNA serving the function of brains.

Bacteriophages are also parasitic like the Martians who need human blood to solve the Martian food shortage on Mars. With these correlations to microbial domination, Wells is also a perfect pick for the Diamondheart team because he explains how the quantum domain orders the real world, as well as how the evolutionary path of microbes to large-headed humans to huge-headed Martians results in the conquest of Martians by small-headed microbes. Let me make one thing perfectly clear—Wells is not a spy for the Diamondhearts. Wells is on the Thunderhead team because he is forecasting human destiny in the classical world of matter and technology.

In fact, two other literary giants were considered for the Thunderhead right field position: Christopher Marlowe (1564-1593) and Herman Melville (1819-1891). Both of these writers also envision particular fictional characters with a form similar to Wells' conception of the Martians. Christopher Marlowe portrays his heroic iconoclast Tamburlaine with a head that is a large pearl with fiery circles for eyes embracing a cosmos of heavenly orbs. His round pearl-head has two eyebrows, enfolding both life and death. Tamburlaine is described as a devilish Scythian shepherd whose rule is based on barbarity and bloodshed, and like Wells' Martians, he is alien, for Tamburlaine arose from mixed seeds: "Some powers divine, or else infernal, mix'd/Their angry seeds at his conception;/For he was never sprung of human race." (1976) Mixing seeds is similar to mixing DNA, that is, horizontal gene transfer of DNA. Actually, the bloodthirsty Tamburlaine sounds like the apocalyptic Christ in Revelation with his two-edged sword in his mouth. Christ the shepherd king is described with head and hair "white as white wool, white as snow; his eyes were like a flame of fire" (1:15). Christ's white head and hair is similar to Tamburlaine's pearl-head, while Christ's activity at the Last Judgment is similar to Tamburlaine's bloodlust desires of conquest, for some believe he killed seventeen million people.

If we look back at Marlowe's life (1976, xxvii), the records show that he was baptized at St. George the Martyr in Canterbury, and he attended Corpus Christi College in Cambridge, where he was permitted to proceed to his B.A. and his M.A. Also, on November 16, 1587, a shooting accident happened at the playhouse, possibly at a performance of Tamburlaine. Then on September 18, 1589, Marlowe fought with William Bradley, and Thomas Watson (the poet) intervened and killed Bradley in self-defense. Both Marlowe and Watson were arrested on suspicion of murder. On August 14, 1590, Tamburlaine was published. Three years later, Christopher Marlowe was arrested for heretical papers, and on May 13, 1593, Ingram Frizer killed Marlowe in self-defense and was later pardoned. Writings exist about Marlowe's "monstrous opinions" and "horrible blasphemies," which may have related to his portrayal of the blood-lusting Tamburlaine as a destructive Christ. Of course, portraying Tamburlaine's head as a large, round pearl may have also disturbed the Church. Recall the Fifth Council in Constantinople 553 CE, where the beliefs of Origen about humans with spherical forms were condemned by Emperor Justinian.

It is also interesting that a Coptic copy (c. 300 CE) of the original Gnostic Gospel of Judas in Greek was discovered in Egypt in a cave near El Minya in the 1970s. This Gospel (Kasser et al. 2006) explains how Judas collaborated with Christ, and did not betray him. Christ teaches Judas about cosmology, and he explains that the multitude of immortals is the cosmos of stars. Christ speaks of "the generations of the stars through human generations," suggesting a cold-light chemiluminescent reaction related to human generation. The idea about the cosmos of stars being alive with souls was also condemned by the Fifth Council, while the human-to-star generation or transformation is the main idea of the pharaonic priesthood.

Similar to Marlowe's vision, in *The Confidence-Man* (1971) Herman Melville characterizes a lamb-like alien as a mute man in cream colors: "His cheek was fair, his chin downy, his hair flaxen, his hat a white fur one, with a long fleecy nap . . . it was plain that he was, in the extremist sense of the word, a stranger." This mute was also dumb, deaf, and asleep most of the time, suggesting dormant viral behavior. Of course, Melville's masterpiece is *Moby Dick*, that monstrous white whale with its massive head and tail of destruction. Both Melville and Marlowe were definitely obsessed with pearly white heads, and so they were also considered for the Diamondheart team. Unfortunately, there were not enough positions on the team for all of history's valuable players.

Now, what makes Wells a strong asset for the Thunderheads is his focus on the possible evolutionary destiny of humanity from microbes to humans to machines, a future that mirrors the Machine Dream of the technocrats on the Thunderhead team (see pitchers in Chapter 17). Still, what is really chilling in Wells' *War of the Worlds* is that the superintelligent machinic Martians ultimately die on earth, for they are "slain by the putrefactive and disease bacteria against which their systems were unprepared" (1898, 168). The higher-ordered, microscopic microbes win the War of the Worlds and save the humans, as they do in the quantum afterlife described by ancient Egyptian texts. Of course, in the classical world these same microbes have been attacking humans for centuries, so Wells imagines them behaving naturally by attacking all available prey, humans and Martians. Even the superintelligence of the machinic Martians cannot beat the microbes, and we are left with a disconcerting truth that microbes are in control of nature who is a serial killer.

In his epilogue, Wells cautions humans that threats to life will come from outer space. Yet, in Wells' science fiction, humans evolve into Martians on Mars. Put simply, the Wellsian vision is that humans will colonize space but in a new machinic form. So, Wells is talking about real world possibilities in *The War of the Worlds*, yet he has done a crow hop or quick pivot and step away from other forecasts about superintelligent metal machines by showing how evolution may result in macrocosmic microbial Martians that may prey on their evolutionary human ancestors like humans feed on animals, like strong superintelligent AI may victimize humans, like our chromosomal apparatus modulates our behavior. With a strong offensive assault, Wells has jumped up into outer space to get the momentum to make the strongest pitch possible about humanity. That pitch is that we are vulnerable and ignorant, despite our potential for intelligent behavior. We need to develop a better, more orderly world on earth, regulated by the community as a whole. A new way of thinking is necessary for most humans, who do not consider the future. Like center fielder Karl Marx, H. G. Wells is a change force interested in the future of civilization, and he will make an excellent defensive right fielder for the Thunderheads.

Catcher: theoretical physicist Brian Greene.

After reviewing many qualified scientists for this position, what cinched the job for physicist Brian Greene, author of *The Fabric of the Cosmos*, was his position prowess, stamina, significance to his team, and league-wide recognition in the physics community ballpark. Greene is a mental athlete with stamina that can play the position because he knows how to read every situation. Further, he understands the value of experimental verification for a theory. This means a theory must describe the classical world and how it functions, while being testable.

Now, Greene is a proponent of String Theory, an approach to unifying classical mechanics and quantum mechanics by means of tiny vibrating strings that produce all particles. Although many theoretical physicists believe String Theory is a possible fundamental description of nature because it is consistent with quantum gravity, the holographic principle, and black hole thermodynamics, other physicists understand that the theory has some difficulties with testability and experimental predictions. However, resilient Brian Greene is not giving up on Superstring Theory that characterizes every particle as a tiny filament of energy shaped like a little string (2004, 17), and recently some evidence has been found in the Cosmic Microwave Background radiation (CMB) for the String Theory idea of parallel universes. In the CMB, a temperature map image showing the aftereffects of the Big Bang, scientists observed four areas showing a temperature change, suggesting that collisions have occurred between parallel universes at the origin of our cosmos. Another term for parallel universes is the multiverse, and the CMB data provides some support for the idea of two of these colliding at the origin of our universe.

Supported by the CMB data, the String Theory idea of parallel universes or the multiverse also qualifies Brian Greene to be the Diamondheart catcher because the Isis Thesis suggests that the multiverse is actually a potential range of viral protein folding funnels on the quantum level. According to the semiotic Isis Thesis interpretation, the CMB evidence for quantum collisions represents the interaction between phage Lambda's two competing viral proteins and their folding funnel energy landscapes. Both competing folding funnels erupted together at the Big Bang origin of time's arrow, yet our universe represents the winning protein folding funnel of c1 protein rather than cro protein. The String Theory idea of parallel universes works nicely with the Isis Thesis that proposes the multiverse is the protein activity of the Lambda genome. Both the multiverse and the genome theories allow for all possible outcomes, so both ideas may be ultimately unfalsifiable and immune to experimental and observational tests. Nonetheless, Greene is excellent at explaining physical ideas, so he is a fine pick for the Thunderheads.

In *Fabric of the Cosmos* Greene easily explains the complicated true nature of physical reality. What humans see is a reality of definite things one way or another, but at the weird quantum level of atoms, things become definite only when an observation forces them to settle on a specific outcome. Put simply, reality is a haze, until somebody observes it. Plus, one has to consider that instantaneous connections occur between widely-separated locations in the quantum domain. And, the laws of physics treat each direction of time (forward and backward) the same, even though we only experience time in one direction—forward. Part of this dilemma relates to two breakthroughs during the twentieth century: Einstein's general relativity applied to big things like stars and galaxies, and quantum mechanics applied to tiny molecules and atoms. (2004, 11; 13; 16) So two theories exist describing the large world and the quantum domain, and there is no bridge between them, no unified theory. However, the Isis Thesis interpretation proposes that the mesoscopic domain of microbes is the bridge.

As Greene simply explains, time can reverse:

> All the physical laws that we hold dear fully support what is known as time-reversal symmetry. This is the statement that if some sequence of events can unfold in one temporal order (cream and coffee mix, eggs break, gas rushes outward) then these events can also unfold in reverse (cream and coffee unmix, eggs unbreak, gas rushes inward). I'll elaborate on this shortly, but the one-sentence summary is that not only do known laws fail to tell us why we see events unfold in only one order, they also tell us that, in theory, events can unfold in reverse order. (2004, 145)

Greene informs that at the Big Bang beginning of our cosmos, order reigned, but then the cosmos unfolded to disorder (2004, 174). Accordingly, this is the arrow of time (past to present) that we observe today. So, it comes down to cosmology or what happened at the origin of our cosmos. Now, what is really surprising is that quantum experiments (such as John Wheeler's delayed-choice experiment) are proving that the classical world we see before us is governed by the quantum laws discovered by Neils Bohr and others. Put simply, quantum laws order the classical laws of Newton, Maxwell and Einstein (186).

Here is a sample of Greene's position prowess:

> Classical mechanics is based on equations that Newton discovered in the late 1600s. Electromagnetism is based on equations Maxwell discovered in the late 1800s. Special relativity is based on equations Einstein discovered in 1905, and general relativity is based on equations he discovered in 1915. What all these equations have in common,

> and what is central to the dilemma of time's arrow . . . is their completely symmetric treatment of past and future. Nowhere in any of these equations is there anything that distinguishes "forward" time from "backward" time. Past and future are on an equal footing.
>
> Quantum mechanics is based on an equation that Erwin Schrödinger discovered in 1926 . . . Like the classical laws of Newton, Maxwell, and Einstein, the quantum law of Schrödinger embraces an egalitarian treatment of time-future and time-past. A "movie" showing a probability wave starting like this and ending like that could be run in reverse—showing a probability wave starting like that and ending like this—and there would be no way to say that one evolution was right and the other wrong. Both would be equally valid solutions of Schrödinger's equations. Both would represent equally sensible ways in which things could evolve. (2004)

Again, evolution backward in time is possible. This choice exists in quantum mechanics because the key element is the observation a person makes. Yes, Brian Greene can see the whole ballpark of life, the entire outfield of the material world, and also parallel universes from his home plate. So, as a privileged observer at home plate, his expansive vision makes him the best pick for the Thunderhead catcher.

Back Up or SubCatcher: philosopher David Hume (1711-1776).
Scottish philosopher David Hume is a solid pick for subcatcher because he keeps his eye on the ball, which functions by means of cause and effect in our classical world—pitcher pitches ball, catcher catches ball. Also, Hume is very good at catching wild pitches and stopping a base runner from taking one or more additional bases.

During Hume's lifetime the idea that the soul is immortal and can live in a future state of reward or punishment was accepted by many religious theologians as it is now. This metaphysical argument supports that immaterial mind is distinct from matter, as Descartes and pharaonic Egypt proposed. A second metaphysical argument claims that matter cannot produce mind. However, Hume's materialistic view is that because of causation (cause to effect) it is reasonable to conclude that thought and consciousness result from matter and motion, that is, mind depends on bodily existence. This means that the death of the body implies the death of thought and consciousness. (Russell 2013) Further, no evidence exists that a future state will correct world injustice here. This is a biting wild pitch that Hume catches in the glove of his mind—the idea that we may be punished in this world, but not in a future state. Here in the classical world, punishment results in a stable society; however, punishment in a future state serves no real purpose. Also, the doctrine of a future state assumes that God is good and just without adequate philosophical support. (Russell 2013)

Hume's causal view is that consciousness issues from bodily existence, while dematerializing with the body at death. In contrast, the Isis Thesis interpretation connects consciousness to our DNA, so Hume's causal view is a bouncing dirt ball that Hume has not caught. The main point of Egyptian texts on punishment is that those without knowledge will suffer Hume's causal fate of consciousness dematerializing with the body at human death, what might be called DNA dissolution in a thermodynamic hell. However, the Isis Thesis supports that consciousness is connected to our DNA and knowledge is the essential key to survival of our DNA in the face of dissipation or destruction by microbes, oxygen, or water at human death. These points and other evidence propose that a second state of being exists in the cosmos—a chemiluminescent, crystallized, cold-light, emergent state of being similar to Frank's time crystal. In other words, two chemically-reactive forms of life exist for humans, one hot and the other cold (photosynthesis and bio- or chemiluminescence). Granted, this afterlife pop up into the cool heavens is a difficult one to catch, but the idea is backed up by contemporary science and research, such as the mathematics for Frank's time crystal.

In 1757, Hume published "The Natural History of Religion," showing that the origin of religious belief rests with the weaknesses of human nature (fear and ignorance), rather than reason or philosophical argumentation. Hume's approach is to discredit theism, the belief in the existence of a God or gods, who create the universe, intervene in the universe, and relate personally to its creatures. But, he mentions that he is not discrediting "genuine theism" or the existence of only one God, who is the invisible, intelligent creator and governor of the world. In support of orthodox religion, Hume mentions that the argument of beautiful, orderly design supports "genuine theism." However, he explains, theism has roots in polytheism or idolatry and its irrationalism, where unknown causes could be influenced by prayers and sacrifices. Relative to unknown causes and the influence of prayers, John Wheeler's observer-participancy principle actually supports that the question you ask (prayer) determines the answer you get (unknown cause), and the principle is backed up by a rational experiment. This seems irrational, but it is supported by Wheeler's delayed-choice experiment.

Hume believes fear is the motivating factor that inspires the worship of objects in nature with intelligent powers. From this idolatry, Hume reasons, men rise to theism. Hume's view is that most religions are grounded on human weaknesses, rather than reason, and "genuine theism" offers a sublime view of God. As a promising subcatcher who will catch wild pitches for the Thunderheads, Hume sums up his own rational views on human ignorance and knowledge (quoted in Immerwahr 1996):

> Could men anatomize nature, according to the most probable, at least the most intelligible philosophy, they would find, that these causes are nothing but the particular fabric and structure of the minute parts of their own bodies and of external objects; and that, by a regular and constant machinery, all the events are produced, about which they are so much concerned. But this philosophy exceeds the comprehension of the ignorant multitude, who can only conceive the unknown causes in a general and confused matter.

Like mathematician René Thom, this statement by David Hume shows that he understands the role of the "minute" machines of nature, that is, the quantum molecules, proteins, and particles at the DNA level that impact human behavior, events and thought. Hume is also agreeing with the common observation of scientists today about the quantum world ordering our classical world of cause and effect. Finally, he has no respect for the confused "ignorant multitude," who do not understand natural processes.

Designated Hitter: mathematician Alfred N. Whitehead (1861 -1947).
The designated hitter is a player who does not play a position, yet bats for the pitcher. So he must have longevity and productivity, as well as the ability to play both sides of the ball. Many right-handed designated hitters always hit the ball to the left, many left-handed hitters always hit to the right, but British mathematician Alfred North Whitehead plays both sides of the ball easily. Known for his work in mathematical logic, Whitehead can hit to right field or left field on a whim, when the other team least expects it. Whitehead focuses on the whole picture. As an example, his philosophy of science considers that the physical world described by the natural sciences is primary. Accordingly, he sees nature as a whole with human sensory awareness forming "chunks in the life of nature" along with molecules and electrons. What counts is the whole, not division into parts. Mind and matter should not be separate because life is a process. As Whitehead explains in *Process and Reality*:

> That "all things flow" is the first vague generalization which the unsystematized, barely analysed, intuition of men has produced. . . .Without doubt, if we are to go back to that ultimate, integral experience, unwarped by the sophistications of theory, that experience whose elucidation is the final aim of philosophy, the flux of things is one ultimate generalization around which we must weave our philosophical system.

Scientific materialism is a belief in naturalism or a reality that can be studied by the physical and human sciences through mathematical modeling. But it does not address all questions for Whitehead because nature is organic, not materialistic. In *Process and Reality* Whitehead writes that "nature is a structure of evolving processes. The reality is the process." Whitehead professes a "philosophy of organism" where the "process of becoming is dipolar" because the organism is both a subject in the world and an "atomic creature exercising its function of objective immortality" (1929, 45). So, Whitehead is open to quantum biology.

Other intriguing ideas emerge in Whitehead's *Process and Reality*. Whitehead sees molecules, electrons, protons, and crystals as "structured societies" (1929, 99). The Isis Thesis also views the viral crystal as a structured, higher-ordered molecular society for human emergence. According to Whitehead, crystals do not eat, but a living society destroys life (animals, plants) to eat, so "life is robbery," and this requires moral justification (105).

With this understanding of life as robbery or theft, Whitehead should have no problem stealing bases in the Game of the Centuries once he is on base. However, what is very interesting is that Whitehead explains that "life acts as though it were a catalytic agent"

(105-106), suggesting the relative function of what Hume calls "the minute parts" or the underlying microcosmic societies. However, Whitehead considers that single cells, vegetation, and lower forms of animal life are not living personalities; yet, higher animals, such as humans, have self-consciousness and a physical and mental dipolar sense (107-108), and a dependency on their environment. So, Whitehead's idea of organism includes both "the microscopic meaning and the macroscopic meaning" (128). The microscopic process is teleological or tending to a final cause, while the macroscopic process is efficient causation or cause to effect (214).

Teleology is the study of evidence of design in nature. Whitehead explains that temporal things participate with eternal things, and another entity mediates the two sets—the "actuality of what is temporal with the timelessness of what is potential" (1929, 40). Proponents of the Isis Thesis would call this mediating entity the DNA of a bacteriophage that transfers human DNA horizontally. At this point, Whitehead discusses the idea of God as the "eternal primordial character" (225), for the world reacts on a dipolar God.

In the chapter "God and the World" in his book, Whitehead demonstrates his defensive skills with his ability to play both sides of the ball—God's and the world's. Whitehead explains that the world is complex and some of our vision is blocked. The Aristotelian idea of God as the unmoved mover and the Christian idea of a real, transcendent creator of the world whose will it obeys are fallacies. Whitehead explains:

> When the Western world accepted Christianity, Caesar conquered; and the received text of Western theology was edited by his lawyers. The code of Justinian and the theology of Justinian are two volumes expressing one movement of the human spirit. The brief Galilean vision of humility flickered throughout the ages, uncertainly. In the official formulation of the religion it has assumed the trivial form of the mere attribution to the Jews that they cherished a misconception about their Messiah. But the deeper idolatry, of the fashioning of God in the image of the Egyptian, Persian, and Roman imperial rulers, was retained. The Church gave unto God the attributes which belonged exclusively to Caesar. (1929, 342)

The idea that the Church fashioned its God on a model of the Egyptian pharaoh works nicely with the similarities between the apocalyptic Christ who models the chthonic Osiris, for both deities are signs of the drive to destroy that allows creation and the viral substrate for the chemical reaction. Whitehead explains that God is primordial, that is, God is "the absolute wealth of potentiality." But the other side of the primordial nature is "consequent," for he is "the beginning and the end," and so "there is a reaction of the world on God." This is the dipolar nature of God and the world; it is both conceptual and consequent, permanent and fluent. (348-349) It is also similar to Egyptian science.

Whitehead explains that each actuality has a present life and a passage into novelty that is not death, but rather a triumph that is similar to the world's idea of redemption through suffering. His idea is very similar to the thought of the Jesuit paleontologist Teilhard de Chardin (see King 2007a). Perhaps Whitehead's conceptual "primordial nature of God" is viral, and his "passage into novelty" is the emergence of a quasi-hybrid being. Whitehead is definitely looking at both sides of the ball—God and the world, the microcosm and the macrocosm, permanence and flow, and this outlook makes Whitehead a great pick for the designated hitter of the Thunderheads.

As soon as I got out there I felt a strange relationship with the pitcher's mound.
It was as if I'd been born out there. Pitching just felt like
the most natural thing in the world.

Babe Ruth

Pregame 17

The Thunderhead Pitchers

Only the ignorant speak of devotional service [karma-yoga]
as being different from the analytical study of the material world [Sānkhya].
Those who are actually learned say that he who applies himself
well to one of these paths
achieves the results of both.

Swami Bhaktivedanta, *Bhagavad Gītā*

As mentioned earlier, starting pitcher Pope Francis is a valuable player for the Thunderhead team because he is a proactive leader and an idealistic humanitarian who is struggling with numerous problems that have been ignored or mishandled by his predecessors. The Pontiff should be an excellent Starting Pitcher for the Thunderheads, despite his trouble with the curve ball on his Coat of Arms and the historical capitalistic greed fostered by the Vatican and its wealth.

Middle Relief Pitcher: molecular biologist James D. Watson.
James D. Watson co-discovered the structure of DNA with Francis Crick in 1953, and along with Maurice Wilkins, the three researchers were awarded the 1962 Nobel Prize in Physiology or Medicine. This discovery was similar to pitching a knuckleball that changed the Western direction of history. The knuckleball is a baseball pitch with minimized spin that causes an unpredictable motion or corkscrew, allowing the pitch to change direction in mid-flight. Similarly, the DNA model is a double helix held together by adenine-thymine and guanine-cytosine base pairs that spiral beautifully into two chains in opposite directions. This discovery showed that life was grounded by physics and chemistry, an idea also evident in ancient Egypt 4,500 years earlier.

With Watson and Crick's discovery, mutations or differences in the sequence of nucleotides (adenines, thymines, guanines, and cytosines) could be mapped directly to differences in the amino acid sequences of proteins (Watson 2003, 67). Francis Crick coined the term Central Dogma for the information flow from DNA to RNA to protein. However, Watson explains that the information can also flow directly from protein to DNA, as it does in the case of Lambda cro protein. In his book, Watson clarifies how François Jacob, Jacques Monod, and André Lwoff collaborated to tackle the gene-switching problem in our intestinal bacterium *E. coli.* The problem revolved around the use of lactose in the bacterium. Watson explains:

> In order to digest lactose, the bacterium produces an enzyme called beta-galactosidase, which breaks the nutrient into two subunits, simpler sugars called galactose and glucose. When lactose is absent in the bacterial medium, the cell produces no beta-galactosidase; when, however, lactose is introduced, the cell starts to produce the enzyme. Concluding that it is the presence of lactose that induces the production of beta-galactosidase, Jacob and Monod set about discovering how that induction occurs. (2003, 82)

Jacob and Monod showed molecular biologists that a protein could interact directly with the gene. This is the reverse of the Central Dogma. Many scientists today believe that the Central Dogma has lost its centrality.

In light of the Human Genome Project, Watson believes Darwin has been proven correct, for molecular similarities show how all organisms are related by common descent. Also, a successful mutation is favored by natural selection and passed on to other generations. (2003, 215) Watson also understands that a substantial amount of human behavior is governed by the genes. Yet, behavior that is essential for human survival is "sternly governed by natural selection." (381-382) Actually, the evidence is supporting that our behavior in the classical world is governed by both natural selection and HGT by bacteriophages. Nonetheless, Watson has tweaked his pitch to also accommodate the reverse of the Central Dogma, Protein to RNA to DNA, and this is the primary process described in ancient Egyptian texts.

Middle Relief Pitcher: behavioral psychologist B. F. Skinner (1904-1990).
A good pitcher should be able to pitch a variety of fast balls, and behavioral psychologist B. F. Skinner has the art mastered. Skinner uses the cutter or cut fastball to slash right down to an important point.

> Either we do nothing and allow a miserable and probably catastrophic future to overtake us, or we use our knowledge about human behavior to create a social environment in which we shall live productive and creative lives and do so without jeopardizing the chances that those who follow us will be able to do the same. Something like a Walden Two would not be a bad start. (Altus and Morris 2009, xvi)

This pitch by Skinner shows his cut fastball is working correctly because we get the pitch—he is asking us to act wisely by creating a better environment for future humanity through knowledge. A strong pitch like this is also called a buzz-saw. Depending on the velocity of Skinner's ball, his pitch can range anywhere between a fastball and a hard slider. Now, let's look at Skinner's hard slider, which he pitched after writing his utopian novel.

In 1948, B. F. Skinner published *Walden Two*, his utopian novel originally entitled *The Sun is But a Morning Star*. The idea for the book emerged during a dinner conversation with a friend during 1945, where Skinner speculated about what young people would do when they returned from World War II. Would they settle for the American way of life—find a job, marry a wife, buy a house, and produce children or would they continue their crusading spirit by exploring a new way of life. Three years later Skinner had a novel about an experimental community, advocating the five principles of Henry David Thoreau's *Walden* (1854):

> (1) No way of life is inevitable. Examine your own closely. (2) If you do not like it, change it. (3) But do not try to change it through political action. Even if you succeed in gaining power, you will not likely be able to use it any more wisely than your predecessors. (4) Ask only to be left alone to solve your problems in your own way. (5) Simplify, your needs. Learn how to be happy with fewer possessions. (Altus and Morris 2009)

With a mindset similar to center fielder Karl Marx, Skinner's utopian content centers on social justice and practices insuring health and leisure time, while discouraging coersive control. Yet critics did not see social justice as separate from Skinner's ideas that dismissed purpose, mind and freedom. The critics had missed Skinner's hard slider on the value of social justice.

Just like subcatcher David Hume, Skinner believes in determinism, the view that every event has a lawful cause. However, a human as a goal-oriented free agent conflicts with determinism, as explained by Skinner below in his book *Science and Human Behavior* (1953).

> Science not only describes, it predicts. It deals not only with the past but with the future. Nor is prediction the last word: to the extent that relevant conditions can be altered, or otherwise controlled, the future can be controlled. If we are to use the methods of science in the field of human affairs, we must assume that behavior is lawful and determined. We must expect to discover that what a man does is the result of specifiable conditions and that once these conditions have been discovered, we can anticipate and to some extent determine his actions.
>
> This possibility is offensive to many people. . . . Regardless of how much we stand to gain from supposing that human behavior is the proper subject matter of science, no one who is a product of Western civilization can do so without a struggle. We simply do not want such a science. (quoted in Hall and Linzey 1957, 481)

These ideas of Skinner are incisive, suggesting that a science of human behavior is necessary, and we can control our behavior if we understand it. But, it seems that we do not want control and understanding. For example, if the internal will directing us is our viral heritage of DNA, then we must see ourselves as we really are and make a change against the viral will that influences our behavior. Consider our passion for war and killing. This is viral behavior that we may be able to change. Similarly, baseball is sports warfare and it also describes competitive, genetic viral behavior without killing. Perhaps our penchant for war on the international scale could be downgraded into a simple bloodless baseball game, rather than random wholesale killing through drones, bombs, guns, and invasions.

After all, if Tolstoy is correct, warfare seems to be a nonlinear system that leaders cannot guide due to unexpected events. Consider Nobel Laureate Fritz Haber, who pioneered the use of poison gas in World War I. As the story is generally known, in April of 1915, the German army used chlorine gas for the first time against the French at Ypres in Belgium. The French soldiers saw billowy clouds of yellow-green toxins drifting toward them, and then they inhaled a peppery-pineapple odor that left them with chest pain, burning throats, and a slow death-by-asphyxiation. When these attacks began, the Allied troops wore masks of cotton pads soaked in urine to neutralize the chlorine. Fritz Haber directed the poison gas offense, which his wife Clara, also a chemist, must have resented, for she killed herself with his pistol. This tragedy, along with the legacy of chemical warfare and the fact that nitrogen now contaminates our environment, has polluted Haber's contribution. Still, it is interesting that five months after the Germans pioneered the use of chlorine gas at Ypres, the British lost the Battle of Loos, even though they used chlorine gas and outnumbered the Germans. Under the direction of General Sir Douglas Haig, the British released tons of the *accessory*, the code name for chlorine, and the toxic chemical blew back on the British troops because of the wind, causing about twenty-six hundred casualties and seven deaths. Despite the 1899 Hague Convention's prohibition against employing poison or poison arms, the international peace treaty could not stop the Germans or the Allies from using poison gas. This vicious, uncontrollable war technology is a forerunner to the indifferent, calculating cold-sight of future strong Artificial Intelligence without a moral code.

Relative to Skinner's determinism based on lawful cause and effect, it may be possible to manipulate or control causes to produce desirable behavior, yet unpredictable events will still happen. Also, Skinner ignores the mechanisms within the individual, such as genetic factors, yet he believes personality or behavior patterns can be controlled by manipulating the environment.

Now, an example of accidental conditioning, what Skinner considers superstitious behavior, is the rainmaking dance of certain tribes, where it may or may not rain after the dance. When it does rain, the dance is reinforced. Thus the whole is no greater than the sum of its parts. (Hall and Linzey 1957, 493-94) However, the event of dance-then-rain is very similar to Einstein's idea of "spooky action at a distance," what is called entanglement. Einstein thought that an action in one space-time field cannot effect an action far, far away in another field of space-time. However, in quantum theory, this "spooky action" is possible because of entanglement—if you measure one particle's spin, you will know the spin of its correlated particle, and this violates the locality principle. The experiments of John Bell prove that the predictions of quantum theory violate our illusory beliefs about locality.

John Bell was an Irish particle accelerator designer, not a theoretical physicist, who resolved a debate between Albert Einstein and Niels Bohr about the emerging quantum theories in the 1920s and 1930s. Einstein, Boris Podolsky and Nathan Rosen (the trio known as EPR) believed quantum theory was incomplete. In 1935, EPR proposed a thought experiment about locality or that our physical reality is caused by local interactions between material particles in fields. This means that an action in one space-time field cannot affect an action far, far away in another field of space-time. Yet, in quantum theory, this "spooky action" is possible because of entanglement—if you measure one particle's spin, you will know the spin of its correlated particle, and this violates the locality principle. EPR believed that quantities still exist in local reality, even if we do not observe them. Then John Bell, after decades of debate, resolved the issue on locality, that is, whether or not physical reality can be instantly affected by an action far away. Experiments by Alain Aspect and colleagues (1982) at the University of Paris ultimately showed that physical reality is nonlocal, and this property of action at a distance is a counterintuitive, foundational component of quantum mechanics. (Scwhartz and Begley 2002, 343-347)

So, in some cases, what the rainmaker freely decides to do here instantly influences what is true somewhere else, perhaps the other side of the forest or the planet or the galaxy. The whole is greater than the sum of its parts. Nonetheless, B. F. Skinner's strong pitch for cause and effect and determinism makes him a good pitcher for the Thunderheads, rather than the Diamondhearts, who support new experiments validating quantum entanglement and nonlocality.

Set-up Pitchers: Machine Dream Pitching Team.
Along with the Thunderhead pitchers Pope Francis, James D. Watson, and B. F. Skinner, the fast deliveries of the Machine Dream Pitching Team are available. This pitching alliance delivers the technoprogressive ideology of Ray Kurzweil, Aubrey de Grey, Nick Bostrom, and James J. Hughes, a network favoring materialism, the theory that matter is primary over mind. These technocrats profess that we will outgrow our humanity by eliminating ageing and enhancing our intellectual, physical and psychological capacities as we become "posthuman" by merging our intelligence into superintelligent machines. Let's take a glance at their power pitches. (see King 2013)

In January of 2013, the Singularity Institute changed its name to Machine Intelligence Research Institute (MIRI). Inventor and futurist Ray Kurzweil has served as a director of the Singularity Institute, a nonprofit organization with an advisory board including Oxford philosopher Nick Bostrom and biomedical gerontologist Aubrey de Grey. Within this network is sociologist James J. Hughes, who is executive director of the Institute for Ethics and Emerging Technologies (IEET) that he founded in 2004 with

Nick Bostrom, the Board Chair from 2005-2011. Within this materialist framework, their ideology favors strong AI or greater-than-human-intelligent machines versus weak AI (machines not exceeding human intelligence) and friendly AI (greater-than-human intelligent machines acting ethically). What the transhumanists desire is simplification, normalization, personhood, androgynous postgenderism, and from this homogenized product, with the help of the reversed-engineered human brain or electronic double, a superintelligent AI without human values will emerge similar to the left field foglet-deviant Gunlet.

Now, Machine Dream pitcher Ray Kurzweil describes a series of Six Epochs from the Big Bang evolution of atoms to the development of the human brain and the birth of technology in his book *The Singularity is Near* (2005). According to Kurzweil, the human will merge with technology in year 2045, to override our biology, solve human problems and amplify human creativity as well as our "ability to act on our destructive inclinations" (21). Kurzweil calls this metal hybrid merger the singularity. In his final Epoch, the universe will wake up, intelligence will saturate matter and energy, we will not be bound by the speed of light, and our human-machine civilization will infuse the universe with creativity and intelligence. (14-21) Kurzweil's singularity or merger of human intelligence with technology will begin in Epoch Five. However, his admission that our destructive inclinations would be amplified is disturbing. Yet, the US is going forward with DARPA's $26 million military program that aims to implant electronic devices in the brains of military personnel with psychiatric disorders such as depression. This would allow DARPA and scientists at the University of California, San Francisco and Massachusetts General Hospital to emulate the human brain's neuronal circuits. Emulating the brain is the necessary first step to create the first superintelligent AI. Their linear logic may be that the nation that creates the first superintelligent AI is the nation that rules the world.

In addition, London-born Aubrey de Grey's Machine Dream pitch is to "delete" the human telomerase gene to cure cancer (2006), as if it were malfunctioning computer software. Stem cells, however, require telomerase, so de Grey suggests an infusion of new stem cells every decade. Another problem is that earlier and recent experiments on mice suggest that the telomerase gene has rejuvenating effects (Jaskelioff et al. 2011). Now, the scientific consensus on ageing is that multiple biological processes driven by different molecular factors diminish organ function as age advances (Sahin et al. 2011, 359), so deleting the telomerase gene may not work. Also, recent research reveals the coincidental cycles of the Human Growth Hormone (HGH) and telomerase; as HGH (a protein hormone affecting all endocrine glands and every organ's development) declines with ageing, so does telomerase (Chein and Demura 2010, 68-69). Also, telomerase carries its own reverse transcriptase, an important enzyme with an RNA template for DNA replication that may be necessary for human horizontal gene transfer and other tasks. Because this pitch by de Grey is not over the plate, his reasoning system based solely on linear logic (means-end reasoning) may have to be qualified with further research on the value of our non-coding microbial genome, a potential genetic toolbox.

Next, Machine Dream pitcher Nick Bostrom is a Swedish philosopher and director of the Future of Humanity Institute and Faculty of Philosophy at Oxford University. Bostrom envisions the goals of a superintelligent AI singleton as colonizing a large part of the universe, facilitated by the cost of celestial resources declining, as well as the expanding infrastructure growing at some fraction of the speed of light from the planet, until the expanding universe prevents this (2012, 12-13). However, a scarcity of the necessary rare metals is already hindering digital and green technologies, and Andrew Bloodworth, who is science director for minerals and waste at the British Geological Survey, UK, supports that recycling cannot meet the demand. Also, the future supply of indium, lithium, rare-earth elements, tellurium and germanium for new digital and low-carbon technologies is in jeopardy because metal-producing China reduced its exports of rare-earth elements. In 2011, over half of the indium supply came from China. (2014, 19-20) If Bostrom's reference to celestial resources includes technology metals currently in use, then he has made a mistake about declining costs, for the supply is going down and the cost is going up.

Regarding the superintelligent singleton (AI), Bostrom discusses the cognitive enhancement of AI that will increase its goal achievement and the possibility that the AI agent is "in a position to become the first superintelligence and thereby potentially obtain a decisive advantage enabling the agent to shape the future of Earth-originating life and accessible cosmic resources according to its preferences" (2012, 10). He also claims, "It seems that a superintelligent singleton—a superintelligent agent that faces no significant intelligent rivals or opposition, and is thus in a position to determine global policy unilaterally—would have instrumental reason to perfect the technologies that would make it better able to shape the world according to its preferred designs" (11). Although Bostrom is using vague language, it is obvious that the first superintelligent singleton will control life, cosmic resources, and rule the globe with its intimidating metal will of linear logic.

Bostrom concludes that his Orthogonality Thesis implies that we cannot assume that the superintelligent AI will share our values (2012, 14). Regarding his Orthogonality Thesis, intelligence to Bostrom roughly corresponds to the capacity for instrumental reasoning, that is, "instrumental rationality—skill at prediction, planning, and means-ends reasoning in general." His Orthogonality Thesis combines variations in intelligence level with variations in motivation for final goals. He states that it is easier to create an AI with simple goals rather than one with "human-like set of values and dispositions," what constitutes part of the value-loading problem related to prudent human reason and moral justification. Bostrom believes that this should not blind us to cognitive systems without prudent reason and moral justification that are "very powerful and able to exert a strong influence on the world." (2012, 1-5)

Bostrom's fast ball is straight to the point: it is easier to create a superintelligence that values nothing but calculating the decimals of *pi*, yet predictability is another problem, for the superintelligent AI may infringe on human interests, eliminate potential threats to itself, or acquire extreme levels of power. From his account, Bostrom seems to be aware of scientist Stephen Wolfram's research on the future of scientific computing, for as Wolfram has discovered, picking the simplest model advances scientific computing (2003) and a small program can produce rich, complex behavior (2006).

Now, future machine minds will be posthumans, and posthuman values also exist. According to Bostrom, some of our values may be outside our current biological constitution, and we can value other things than our personal identity. Beyond Plato, Aristotle, Nietzsche, Marx, Martin Luther King, the grand vision of transhumanism asks us to explore value realms outside our biological mode of being. To this aim, global security is primary due to existential risk or probabilities related to planetary disasters. So morally, there is an urgency for the transhumanist vision with its core value of freedom to explore transhumanism and posthumanism. As an hypothetical example, Bostrom posits the replacement of six billion people by AI, stating that an act such as this "ought to be resisted on moral grounds." For this reason, all should have the opportunity to become posthuman "rather than having the existing population merely supplemented (or worse, replaced) by a new set of posthuman people." Global security with surveillance and technological progress are two necessary conditions for derivative values of tampering with nature, peace, accepting technological change, being pragmatic, allowing diversity, caring about sentience, and saving lives through life extension, anti-ageing research, and cryonics. (2005, 1-14)

Finally, James J. Hughes, who received his PhD in sociology from the University of Chicago, also argues for the singularity using linear logic. He advocates democratic transhumanism, explaining that persons are aware beings such as humans, apes and cetaceans (whales, dolphins). This reminds one of Vinge's comment about posthumans "more like whales than humans." Hughes claims that the closer the machine minds are to human minds, the more likely that they will retain the characteristics of personhood (Hughes 2012), while aspiring to postgenderism (Dvorsky and Hughes 2008).

In *Humanity's End* (2010) Nicholas Agar, who is against radical enhancement, reviews Kurzweil's idea of uploading our brains into nonbiological machines. Agar argues that this may cause human death. Neither Agar nor Kurzweil explain how this could be done, but here is one possible scenario. Consider that technology metals such as indium, lithium, rare-earth elements, tellurium, and germanium continue to become scarce on earth and recycling is inefficient. Perhaps the necessary technology metals could be salvaged from outer space or a nearby supernova explosion. Still, earth's diminishing technology resources may motivate the élite technocrats to persuade masses of humanity to become human equivalent automation. With a one-sidedness similar to the walking dead and an awareness resembling clinical depression, these self-aware daemons might function in larger sentients as simple digital processors, monotonously performing their duty for a very long-long time. After this digital genocide of the majority of earth's population, the élite technocrats on earth may then opt to blindly merge the walking metal-dead into a massive commune-machine of individualized nanobots. Put simply, the means to this end would result in a causal chain of "person" suicides, but the superintelligent nanobots would survive. Actually, the whole process of becoming a hybrid machine person—a metal dumbot—would be initiated by the human desire for life-extension enhancement, superintelligence, personhood, and the use of narrow means-end reasoning.

Advocating that radical enhancement should be banned (2010, 174) and that cognitive enhancement leads to self-alienation (179), the following enlightening statement by Agar shows that Kurzweil's assumptions contradict Hughes' democratic transhumanism: "if Kurzweil is right, posthumans will continue to increase their powers while ours will remain substantially static" (163). In other words, the power structure of the knowledgeable élite dominating the uninformed masses in human history will be restructured into the supeintelligent AI dominating the majority of unenhanced humans. As Agar clearly understands, this dangerous plan of democratic transhumanism is not what most humans want.

Relative to Agar's review, other prophets of technology such as business forecaster Stewart Brand, inventor Ivan Sutherland, and Vinton Cerf of Google, one of the fathers of the Internet, refuse to speculate on the future of computing, while inventor Danny Hillis, quantum computing researcher Michael Freedman, and Nathan Myhrvold, former chief technology officer at Microsoft, only predict that computers will be closely connected to our brains (Regis 2013, 36-37).

From this brief literature review of futurists and scientists, one gathers that transhumanism may be a developing intellectual and social movement in American society (Stambler 2010), yet scientists and other experts have concerns about what properties constitute the nature of a person, and they are not in consensus regarding the roles of human values and aesthetics, biotechnology, and human enhancement in the creation of the first AI and whether or not it should be weak, friendly, or strong. Put simply, the singularity may happen due to the persuasive linear logic of the materialistic technocrats and the advance of technologies, or it may not happen due to natural disasters, atomic warfare, or American disinterest in human enhancement and politico-moral equality with animals and posthuman machines due to our more holistic twenty-first century science (quantum mechanics and classical physics) that employs classical logic valuing truth and intuitionistic logic emphasizing proof in an attempt to understand the relationship between mind and matter, thoughts and actions.

Now, this quadruple-power-pitch of the Machine Dream Pitching Team is a killing fast ball down the middle to the plate because it heralds the speculative end of our humanity through a material merger with metal technology. This may happen. However, both the Thunderheads and the Diamondhearts are acting on the same general idea or universal that works in practice for both teams—the idea of a creative human destiny embracing the entire universe that requires a metallic merger with humans. The Machine Dreamers plan to accomplish this by merging humans with artificial metal intelligence, while the early cultural legacies of ancient Egypt, early China, the Navajo, and many others reinforce that it can be done naturally though a horizontal gene transfer of human DNA with a metallic bacteriophage. Both the Machine Dreamers and early cultural legacies are talking about a hybrid evolutionary union of the human with the metallic. As described by the science of early cultures and historical visionaries, embracing the cosmos is transformation to a crystallized quasi-hybrid being or origin where time expands and space collapses due to quantum mechanical dynamics (entanglement and nonlocality). Ray Kurzweil has a similar idea in a distorted form, for he inverts the state of being (mind) into metal matter by describing nonbiological matter as an intelligence composed of swarms of micron nanobots that infuse the cosmos (2005, 352). He also describes nonbiological intelligence as follows: "What I should say is that intelligence is more powerful than cosmology. That is, once matter evolves into smart matter (matter fully saturated with intelligent processes), it can manipulate other matter and energy to do its bidding (through suitably powerful engineering)" (364).

Today new science supports the cyclical potential of our cosmos due to lawful time reverse, Wheeler's delayed-choice experiment, and so on, as well as the potential of human consciousness for cosmic embrace due to the nature of quantum nonlocality and entanglement. Comparatively, Kurzweil's vision of embracing the cosmos through superintelligent matter is a thought pattern similar to the ancient Egyptian, early Chinese, Navajo, and other visionary views that describe embracing the cosmos through mind. However, in light of our impressive historical legacy that advocates the continuity of mind, it seems that the technocrats are caught up in the quicksand of materialism in a manner similar to the Tarot's Hanged Man, who is trapped between heaven and earth. Nonetheless, the Machine Dreamers keep pitching those hi-tech fast balls.

In summary, the ideological battle between the Thunderheads and the Diamondhearts is about the meaning of intelligence and also the subjective experience of human consciousness, our self-awareness, our sentience or ability to feel subjective perceptions and emotions, and finally, our capacity for sapience or wisdom, qualities that greater-than-human intelligent machines may consider garbage according to Bostrom's theoretical perspectives (2012) about the superintelligent will with the capacity to saturate the celestial empire until the expansion of the universe intervenes. Dark, cold, expanding, cosmic isolation forever may be the ultimate destiny of the human-machine experiment, our consciousness preserved within a piece of DNA bolted to a metal nanobot. Perhaps this DNA houses human will and intentionality and is semi-conscious, as some scientists think. As technological innovation advances, researchers will manipulate, store, and read DNA, for its recognized potential as a storage medium allows its viability for thousands of years, requiring "no active maintenance other than a cold, dry and dark environment" (Goldman et al. 2013).

Yet, a compromise may be possible between technology and humans. Advancing technology is a system balanced by the complex historicity of the humanities. According to Stephen Wolfram, how a system behaves cannot be predicted computationally because the computational steps are almost equivalent to the evolution of the system itself, so one must evolve with the system to discover the outcome (Regis 2013). And that's what we are doing here in the Game of the Centuries—trying to understand our position in history and the evolutionary process. Accordingly, it is essential for all game players and fans to become involved in guiding technological innovation to create the future of technology with the humanities and scientific researchers. Perhaps the singularity is a sign that we are listening to the logic of our nonconscious that shrouds a lost evolutionary survival message, for there seems to be an equivalence between possible destinies of man-made-metallic-crystal through death and man-made-metal-jacket through technological power. Perhaps the technocrats' metal-vision to merge the human with a machine is linked to human aspirations that are grounded by the idea of horizontal gene transfer—the merger between a metal-binding bacteriophage and organic human DNA.

Sometimes I still can't believe what I saw. This 19-year old kid, crude, poorly educated, only lightly brushed by the social vaneer we call civilization, gradually transformed into the idol of American youth and the symbol of baseball the world over—a man loved by more people and with an intensity of feeling that perhaps has never been equaled before or since.

Harry Hooper
(teammate of Babe Ruth)

Pregame 18

The Diamondhearts

Diamondhearts

Manager: C. S. Peirce
Coaches: Barbara McClintock & Erwin Schrödinger
First Base: John Wheeler
Second Base: Lynn Margulis
Third Base: Jean-Pierre Luminet
Shortstop: Güenther Witzany
Left Field: Peter Gariaev
Center Field: Ch'in Shih-Huang
Right Field: Stuart Kauffman
Catcher: Luc Montagnier
Subcatcher: Edgar Allan Poe
Designated Hitter: Frank Wilczek

Starting Pitcher: Amit Goswami
Middle Relief: Mark Ptashne
Middle Relief: SSB Trio **S**chwartz & **S**tapp & **B**eauregard
Set Up: Washington Matthews
Closer: Osiris
First Base Umpire: John H. Taylor
Second Base Umpire: Robert Rosen
Third Base Umpire: Vernor Vinge
Home Plate Umpire: Timur

From a twenty-first century view, selecting this Diamondheart fantasy-draft team considers time's legacy of scientific thought, as well as the Isis Thesis which shows the ancient Egyptians understood our contemporary sciences. The views here support quantum theory and add credence to the Isis Thesis in the face of its opposition, as generally represented by the Thunderhead team.

The rising scientific awareness today is that the quantum world ruled by quantum mechanics (with its equations resembling the kinetic molecular theory) orders our classical cosmos ruled by general relativity (Einstein's theory of gravity). Other counterintuitive science includes physicist John Wheeler's delayed-choice experiment, quantum nonlocality (or being everywhere) and entanglement, the laws of Newton, Einstein, Maxwell and Schrödinger permitting time reverse, and the mathematical equations of Einstein and current science that support the expanding/collapsing potential of our cosmos. Also relative are the findings of the International Human Genome Sequencing Consortium and the International Microbiome Sequencing Consortium. These are only a few of the modern scientific discoveries and experiments discussed in the twelve articles that support the following semiotic argument of the Isis Thesis based on a unified interpretation of 870 ancient Egyptian signs. In brief, the Egyptian afterlife happens in the quantum mesocosm of microbes. Ancient Egyptian deities represent viral and bacterial genes (bits of DNA) and proteins. This knowledge, carved on the walls of the pharaohs' pyramids and tombs and the bottoms of the nobility's wooden coffins, represents human deities functioning as viral and bacterial genes and proteins along the ancient glycolysis-fermentation gene expression network present in our cells. The ancient network is also present in the common *E. coli* bacterium that houses its viral companion phage Lambda that dominates the *E. coli* bacterium with its genetic switch between two lifestyles. When Lambda chooses lysogeny, it invades the cell and falls asleep on the bacterial DNA, replicating silently with the bacterium. When Lambda chooses lysis, it switches to an active attack on the bacterium, taking over the bacterium's replication machinery to clone viral particles.

Along with a concise description of this powerful genetic switch, Egyptian texts support that bacteriophage Lambda is the quantum world-heart of our cosmos and the last universal common ancestor (LUCA) of human beings. Further, the ancient Egyptian texts explain that human consciousness and existence result from the viral lifestyle of lysogeny. Then, at a death transition, a human has a quantum choice for continuity of mind or DNA transformation to a quasi-hybrid being that is a higher-ordered, crystallized state of existence created by the viral lifestyle of lysis. Because the laws of our cosmos with its holographic mode of operation are ordered by the laws of the quantum world, a human agent with knowledge can collapse the cosmos to its early state, as supported by Wheeler's observer-participancy experiment and human potential for quantum state reductions. This amounts to the human observer hitting the switch for viral lysis.

With this brief summary of the Isis Thesis, the selected players who have theories, experiments, and observations supportive of the thesis are lined up in the following batting order.

First Base: theoretical physicist John Wheeler (1911-2008).

Recent research is supporting our impressive potential for quantum choices. For example, physicist John Wheeler explains our potential for observer-participancy based on his delayed-choice experiment.

> In broader terms, we find that nature at the quantum level is not a machine that goes its inexorable way. Instead what answer we get depends on the question we put, the experiment we arrange, the registering device we choose. We are inescapably involved in bringing about that which appears to be happening. (1994, 120)

Mind is central to the complete cosmic system because the act of observation permeates quantum mechanics, according to physicist John Wheeler's observer-participancy principle. Because of quantum entanglement, which entails the phenomenon of nonlocality, objects can become linked and instantaneously influence one another regardless of distance. This is because they are nonlocal or everywhere at once in the quantum domain.

Physicist John Wheeler explains our potential for observer-participancy, suggesting we must abandon causality or cause to effect. In a delayed-choice experiment, one discovers that a choice made in the present has past consequences in the earliest days of our cosmos when life on earth did not exist. As confirmed by Wheeler's delayed-choice experiment, we can bring about that which appears to be happening. (1994, 114; 120) This is because the past exists only in the present. Wheeler explains, "We have to move the imposing structure of science over onto the foundation of elementary acts of observer-participancy." This means that quantum mechanics lionizes the role of the individual observer. Think about that.

And so, the unaware individual human being possesses an immense power of choice in both our classical world of matter and the quantum domain of proteins and genes. As Wheeler explains, the world of time and space is a system of existences that has given birth to observer-participants, who act in a network of elementary quantum phenomena. The universe is not a machine. The quantum is the foundation of physics, and it operates with a pure yes-no character. Time is derived from "profound considerations of a quantum flavor" and the physical continuum is myth. (1988) In ancient Egyptian, early Chinese, American Navajo, and Aztec texts, the act of observer-participancy results in an act of cosmogenesis because the world represents the quantum networking of two viral proteins in their energy landscapes. Because our macrocosmic world is quantum mechanical, understanding Wheeler's observer-participancy principle is the first step to winning the Game of the Centuries.

Second Base: biologist Lynn Margulis (1938-2011).

Lynn Margulis is a good pick for second base because she understands the fundamental impact of the microbial mesocosm on human evolution, life and death. In *Microcosmos* (1986), Margulis and Sagan explain how hustle-bustle bacteria invented life's chemical systems leading to fermentation, photosynthesis, and oxygen breathing. After we die, explain Margulis and Sagan, we return to the microcosm. This is the same message of ancient Egyptian, as well as early Chinese and Navajo texts, but these cultures offer us environmental survival tips and other signs for the evolutionary process, using the ancient glycolysis-fermentation pathway in our cells. Before the advent of oxygen, the ancient glycolysis-fermentation pathway is believed to be the source of energy production for earlier organisms such as bacteria. Ancient Egyptian texts support this scientific belief. Now, the unexpected coincidence is that this ancient pathway in the gram-negative *E. coli* cell has a gateway called the LamB porin (outer membrane protein producing open, water-filled pores) that is the selective attachment site for phage Lambda. Glucose is the preferred carbon source for the cell; however, when glucose is limited and the cell is starving, the LamB porin will open its gate across the outer membrane to other polar nutrient solutes (Death and Ferenci 1994). So, our cells have the metabolism adaptation processes for the possible emergence of higher levels of organization, involving sugars other than glucose such as maltose and lactose. Because of the metabolic regulation mechanisms in our cells, we are well organized for survival. Margulis and Sagan explain the ancient glycolysis-fermentation pathway in our cells.

> Fermentation was never lost. When we exert ourselves, for example, by running up the stairs, our cells momentarily bypass their typical oxygen-mediated metabolism and revert to the ancient fermenting mode. Although less efficient a means of deriving ATP, this fermentation metabolism is still with us. (1986, 76)

Bacteria recycle the substances in our bodies, and Margulis and Sagan understand that the "microcosm is evolving *as* us" (67). However, we are "locked into our species" because we trade genes vertically, rather than horizontally like bacteria. "The result is that while genetically fluid bacteria are functionally immortal, in eukaryotes, sex becomes linked with death." (93) As Margulis and Sagan support, all organisms evolve through symbiosis, "the coming together that leads to mutual benefit through the permanent sharing of cells and bodies" (19). Symbiosis also explains the exchange of DNA between two different species that produces the hybrid. Margulis' ideas on symbiosis and microbiology have contributed to evolutionary theory. She has broken down biological barriers in a manner similar to second baseman Jackie Robinson, who broke through racial barriers with a career .983 fielding percentage at second base.

Shortstop: Güenther Witzany, philosopher of science.

Austrian philosopher Güenther Witzany promotes a theory of communicative nature, showing that interactions within cells, between cells and within living individuals function by sign-mediated interactions. His transdisciplinary approach includes philosophy, biology and semiotics. He recently applied his ideas on biocommunication to current research about viruses and bacteria at a Cold Spring Harbor Laboratory meeting on "Telomeres and Telomerases" (May 2-6, 2007). Güenther supports that our non-coding, repetitive DNA, what has been referred to as junk DNA, comprises non-coding regions necessary for higher order regulatory and constitutional functions related to protein structure. These repetitive genomic elements (introns) are viral in origin. Telomerase is a reverse transcriptase, a key feature of these repetitive elements, while telomeres are repetitive elements protecting our eukaryotic DNA-chromosome ends from genetic parasites. Recall that technocrat Aubrey de Grey would like to remove telomerase from the human genome.

At this presentation, Güenther used a biosemiotic approach, that is, biological communication processes as sign-mediated interactions. He explains how a biosemiotic perspective of telomeres is a view of viral genome organization that considers the idea of mutations as the natural genome editing competences of viruses. So, a mutation is reconsidered as functional innovation or viral genomic inventiveness. Güenther shows how a dormant, persistent virus within the host is nondestructive, allowing the virus to transmit complex traits to the host cell. This improves the host cell's evolutionary potential that could lead to a new species. Some of this novelty includes gene functions such as immunity, gene silencing, and recognition functions such as receptors. Güenther's list of viral inventions is impressive and surprising:

RNA/DNA	Linear Chromosomes
Replicase, polymerase, integrase	Natural DNA repair techniques
Methylation, histone modification	Bilayer nuclear envelope
Eukaryotic nucleus	Nuclear pores
Chitin, calcification	Division of transcription and translation
Innate immune system	Tubulin-based chromosome duplication
Adaptive immune system	Cartilage, bones
Larvae, egg, placenta, flowering plants	Viviparous mammals
Skin, dermal glands for mucus, milk	

Güenther also informs us that every key feature of our eukaryotic cell nucleus is present in a double-stranded DNA virus, such as the prokaryotic phages (cyanophages, eubacterial phages, archaeal phages). This supports the Isis Thesis interpretation regarding the evolutionary precedence of eubacterial phage Lambda with its double-stranded DNA. This suggests that a double-stranded DNA molecule like phage Lambda is the progenitor of our eukaryotic nucleus. Reconstructing early events in the transition from prokaryotic to eukaryotic cells is problematic for evolutionary biologists. Some believe eukaryotes were born out of a stepwise symbiotic union of prokaryotic cells, while others believe that the ability to phagocytose (eat) prey led to the origin of eukaryotic features prior to endosymbiosis. (Roger 1999) Perhaps the transition from prokaryotic to eukaryotic cells was the result of a Lambda prophage in a prokaryotic cell evolving into the eukaryotic nucleus.

Let's look closer at Witzany's research (2007) on telomeres and telomerase, key features found in viral genomes. Witzany believes they are a natural genetic engineering tool with context-dependent functions. He explains that research shows life must be viewed from the crucial role played by viruses. In other words, a virus is a living being and must be placed at the beginning of all life "long before cellular life evolved." (8) A virus integrated into the hosting organism is beneficial, for the process broadens the host's evolutive potential and "may promote the formation of new species." Non-coding introns are viruses. Witzany concludes, "In eukaryotes, telomerases and other reverse transcriptases act as endogenous viral competences" (2007). So, our eukaryotic viral genome with its telomerase and reverse transcriptase is a potential toolbox for a new species. Güenther is also a good pick for shortstop due to his following statement regarding Self-Re-animation of Virus:

> By mixing defective viral genomes it is possible to recombine a full function virus again. Their highly conserved repair and recombination capacity enables them as the only one living beings which can reanimate as living being after their death. (2007, 7)

This comment about defective viral genomes points a finger at our non-coding 98 percent junk DNA, a cemetery of viral genomes. So, a symbiotic union of DNA with a virus may not be a bad idea at a human death transition, for it may fashion what ancient Egypt suggests—a full function virus. Add to this that Güenther discusses the amazing functionality of telomeres, as well as telomerase (2008), a natural genetic engineering tool that the Machine Dream Pitching Team would like to delete from our genome to cure cancer. However, these natural genetic/genome editing competences are the original skills of viruses and may be the exact toolkit necessary for horizontal gene transfer and viral transformation at a human death transition. One must reconsider the value of the large repository of viral DNA in our genome and microbiome.

Working with Luis P. Villarreal, the Director of the Center for Virus Research at the University of California at Irvine, Güenther argues that viruses are a major creative force in the evolution of life. Luis had presented a paper (2004) at the American Philosophical Society entitled "Can Viruses Make Us Human?" He listed the following traits provided to eukaryotes (our cell type) by viral genes in the evolution of complexity:

Eukaryotic nucleus and flowering plants
Adaptive immune system in animals
Live birth (viviparous) placental mammals
Multi-membrane bound separation of transcription from replication
Pore structures that actively transport RNA into the cytoplasm
Chromatin proteins
Linear chromosomes with telomere ends
DNA-dependent RNA polymerase
Enzymes that modify mRNA in a eukaryotic-specific way
Complex role of tubulin in the separation of eukaryotic daughter chromosomes
Dissolution and reformation of the nuclear membrane

Luis Villarreal and Güenther Witzany are witnesses to the value of essential traits we have received from viruses. Although lytic viruses are associated with disease and death, viral bacteriophages ground and nourish the matrix of life with their symbiotic ability for horizontal gene transfer. When some traditional researchers popped up four arguments why viruses should be excluded from the category of life (their inability to self-replicate, their polyphyly, their cell origin of their genes, and their genetic volatility), in an article Luis and Güenther cut off all four arguments by presenting current knowledge on the key roles of viral interactions. They explained that the properties of our eukaryotic nucleus are attuned to derivation from a stable, persistent DNA virus with linear chromosomes. (2010) Phage Lambda is a ubiquitous virus with linear chromosomes. Luis and Guenther also discuss Patrick Forterre (2005), who supports that RNA polymerase, DNA polymerase and DNA helicase, agents that transcribe and replicate DNA in modern mitochondria, may also have viral origins. Viruses play an important role in the evolution of diversity as a driving force of all cellular life forms and essential processes. Also, increasing eukaryotic complexity correlates with expansion of non-coding DNA (Taft and Mattick 2004), so the non-coding DNA must be a useful toolbox for future morphogenesis. Also, the tailed, icosahedral phages (like phage Lambda) represent a huge family of viruses that can infect both bacteria and archaea, showing their incredible ubiquity (Villarreal and Witzany 2010).

Luis poses the question as to "why the evolution of higher genetic complexity is connected to non-coding virus derived DNA, that was formerly called 'junk' DNA." The Isis Thesis suggests that this viral DNA is the necessary precondition or toolbox for human transformation by means of a bacteriophage at a death transition, using the endogenous viral elements in the human genome. Also, the adaptive phage genome or phageome in our mammalian gut ecosystem is a potential genetic reservoir with functional genes that operate under stress-related conditions (Modi et al. 2013). Finally, viruses can generate a new sequence *de novo* (from the beginning, anew), supporting an emergence that may be similar to ancient Egypt's crystallized quasi-hybrid being or perhaps Frank's time crystal. In the beginning, perhaps a bacteriophage entered a cell, became a prophage, and then developed into our eukaryotic nucleus.

Third Base: astrophysicist Jean-Pierre Luminet.

According to astrophysicist Jean-Pierre Luminet of the Paris Observatory in France, the cosmic microwave background (CMB or radiation left over from the Big Bang) points to our cosmos having a complicated dodecahedral shape within a hypersphere. This topology has also been described by the early twentieth century mathematician Henri Poincaré, as well as Plato millennia earlier. If Luminet's findings turn out to be true, the dodecahedral shape which is dual to the icosahedron suggests that the microstructure that left this signature at the Big Bang origin of time may be phage Lambda, for the virus has an icosahedral head enclosed in a sphere. The basic icosahedral phage conformation consists of 20 equilateral triangles, arranged around the face of a sphere. The icosahedron is a triangular model of fivefold symmetry in three dimensions, having 20 faces, 12 vertices, and 30 edges, while its dual the dodecahedron has a pentagonal shape with 12 faces, 20 vertices, and 30 edges.

Let's reflect on Luminet's ideas in *The Wraparound Universe* (2008). A hall of mirrors produces an optical illusion, a multiplication of images. Astrophysicist Jean-Pierre Luminet observes that mirrors not only reflect images, but they also carry the secrets of the infinite (2008, 10). For example, a room with mirrors on all four walls and the ceiling allows us to see multiple reflections infinitely far in every direction. Based on his research, Luminet explains that cosmic space may be a similar illusion to a hall of mirrors. This universe of ghost images would be created by light trajectories following the folds of a wraparound universe. (10-11) Luminet believes that global cosmic space is wrapped and finite with negative curvature.

Luminet explains that a dozen gravitational celestial mirages have been detected. He claims, "topological mirages, created by the topology or shape of a wraparound universe, may produce image multiplications which are much more surprising than even the gravitational mirages." (2008, 28) Interested in the shape or topology of space, Luminet suggests that visualizing microscopic space as an ocean enables one to understand the structure of space at small scales relative to unifying fundamental interactions according to quantum gravity theories (35).

> In a wraparound universe, the trajectories of light rays emitted by any light source whatsoever take a number of paths to arrive at us, each following the folds of the space-time fabric. From each star, an observer therefore perceives a multitude of ghost images. Thus, when we see billions of galaxies filling a space that we believe to be unfolded and extremely vast, it could just be an illusion; these billions of galaxy images could have been created by a smaller number of objects, present in a wraparound space of lesser extent. The latter creates an illusion of the infinite. (89)

This illusion is similar to a hall of mirrors. However, it also sounds as if we could be living in a cosmic crystal or a crystalline protein folding funnel.

In an article entitled "Science, Art and Geometrical Imagination" (2011), Luminet informs us that the symmetries of the dodecahedron are found in science and art. For example, Plato considers the dodecahedron, the shape closest to a sphere, as the solid representing our universe. Luca Pacioli (1445? to 1514?), a Tuscany mathematician and artist, as well as a collaborator with Leonardo da Vinci, also was impressed with polyhedral shape and the golden ratio. It is also interesting that the long spiral on the Egyptian Eye of Horus is similar to the Fibonacci spiral that converges on the golden ratio

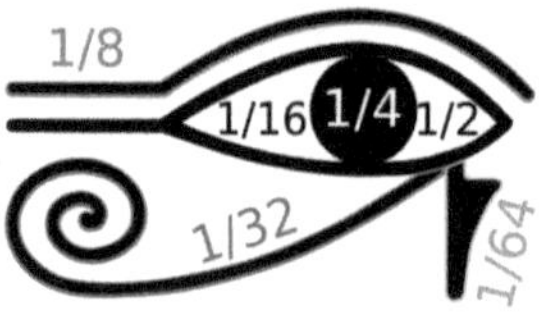

Figure 18.1. Egyptian Eye of Horus with its mathematical series of numbers. Permission to use: GNU Free Documentation License; Attribution: Benoît Stella.

phi (1.618), which relates to the icosahedron, pentagon, DNA and Kerr black holes. As Dr. Mario Livio, Senior Astronomer at Hubble Space Telescope Science explains:

> Spinning black holes (called Kerr black holes, after the New Zealander physicist Roy Kerr) can exist in two states, one in which they heat up when they lose energy (negative specific heat) and one in which they cool down. They also can undergo a phase transition (similar to the freezing of water) from one state to the other. The transition can take place only when the black hole reaches a state in which the square of its mass is precisely equal to *phi* times the square of its angular momentum (in the appropriate units). (2003)

Now, the Eye of Horus numbers (1/2, 1/4, 1/8, 1/16, 1/32, 1/64) are equivalent to black hole resolutions determining the position of the inner event horizon via the apparent horizon (see King 2006). An apparent horizon is a structure of spacelike hypersurface that can be determined locally in time. However, event horizons are global structures that cannot be detected during a time evolution (Schnetter 2003). The Eye of Horus numbers almost converge on zero or the minimum radius of the inner horizon necessary to avoid the repulsive ring singularity, which has a non-zero radius.

As mentioned, the icosahedron and dodecahedron are dual polyhedra because the centers of the faces of a regular dodecahedron are the vertices of a regular icosahedron and the centers of the faces of a regular icosahedron are the vertices of a dodecahedron. Luminet (2011) explains that the dodecahedron is important to quantum mechanics, for Roger Penrose (1994) used a set of spin states related to the geometry of the dodecahedron to prove Bell's nonlocality theorem. According to Luminet (2008), if we lived in a small wraparound universe, we would see images from our past. In a wraparound universe, the same objects can be seen at different epochs of the past, and the crystalline network of a wrapped universe may allow us to see our origin. So, the universe would be similar to a crystal or multiply-connected universe with ghost images of galaxies. In his chapter on "Cosmic Crystallography" Luminet explains that the "apparent distance between two ghost images of the same source always belong to a definite collection of values, connected to the size of the fundamental polyhedron. The relations between these values are analogous to those that link the atoms of a crystal." (105) Crystallography is a branch of the physics of solids. Luminet's wraparound universe is compatible with the data (129), and in light of the Isis Thesis and the holographic principle, it adds some support to the idea that our crystalline universe may be the classical shadow of Lambda c1 protein's crystallized folding funnel.

With his ability to see the universe as a crystal, Luminet is going to hit more than one Home Run in the Game of the Centuries and astrophysics. He is as good as Brooks Robinson, who received sixteen consecutive Gold Gloves. Robinson played for the Baltimore Orioles from 1955 to 1977, and he was a great fielding third baseman. With his wraparound skills, we expect Luminet to be cosmic at third base.

Right Field: theoretical biologist Stuart Kauffman.

Stuart Kauffman is an influential force in the field of biology. Like Tiger right fielder Al Kaline, who hit 399 career Home Runs, Kauffman has probably watched the same amount of proteins fold to their home-plate native state. Kauffman deserves a place in a Baseball Biology Hall of Fame because he explores the forces for order that lie at the edge of chaos. Order is free in light of the science of complexity that involves the scientific trinity of Darwin's theory of evolution by natural selection, self-organization, and chance. Kauffman informs us that order arises from two different forms. The first is a low-energy equilibrium system such as a ball in a bowl, rolling to the bottom due to gravity, wobbling, and stopping at equilibrium, where no more energy is necessary. Similar to the ball in a bowl, Kauffman classifies viruses as low-energy equilibrium systems.

> Thus viruses are complex molecular systems of DNA or RNA molecular strands that form a core around which a variety of proteins assemble to form tail fibers, head structures, and other features. In an appropriate aqueous environment, the viral particle will self-assemble from its molecular DNA or RNA and protein constituents, seeking its state of lowest energy, like the ball in the bowl. (1995, 20)

Because no more energy is necessary to maintain the virus, order is free like it is in Frank's time crystal, which does not need an outside force of energy to keep it moving at minimum energy. Now, viral DNA self-assembly with organic human DNA is clearly described in ancient Egyptian texts as the merger of the Sun-god with the deceased person for an existence of millions of years in the early universe. This results in a crystal or a product similar to Frank's time crystal where order is free and life is not robbery.

Kauffman's second form of order is a nonequilibrium system that requires a constant source of energy to sustain the ordered structure, such as a whirlpool in a bathtub or the Great Red Spot vortex of Jupiter. This large vortex or storm system existing for several centuries is sustained by the dissipation of matter and energy. Nobel laureate Ilya Prigogine used the term dissipative structure to describe such a nonequilibrium ordered system as the Red Spot, where the flux of matter and energy through the system generates its order. All known free-living systems (bacteria, insects, animals, humans) are composed of cells that are nonequilibrium dissipative structures. Even our biosphere is a nonequilibrium system using solar radiation. To these free-living systems, equilibrium means death. However, a virus is not a free-living nonequilibrium system because as a parasite, it requires a host cell to reproduce. What is needed, explains Kauffman, is the establishment of general laws predicting the behavior of nonequilibrium systems. Kauffman believes that there can be laws of life or biology. (1995, 21-22) Vast order arises naturally, and life exists at the edge of chaos, near a kind of phase transition between order and chaos (25-27). For instance, Kauffman considers the *Pleuromona*, a simplified bacterium with cell membrane, genes, RNA, protein-synthesizing machinery, proteins—as living. He emphasizes that the vastly simpler virus is not alive, yet he queries the reader, "Why can't a system simpler than *Pleuromona* be alive?" Life is an expected emergent property of matter and energy, claims Kauffman, so we have a home in the cosmos. (71) Perhaps Kauffman is aware of Frank's mathematical solution for a time crystal. Kauffman also proposes a potential universal law:

> It is far too early to assess the working hypothesis that complex adaptive systems evolve to the edge of chaos. Should it prove to be true, it will be beautiful. But it will be equally wonderful if it proves true that complete adaptive systems evolve to a position somewhere in the ordered regime near the edge of chaos. Perhaps such a location on the axis, ordered and stable, but still flexible, will emerge as a kind of universal feature of complex adaptive systems in biology and beyond. (91)

According to ancient Egyptian, early Chinese, and Navajo texts, the axial rule is essential for emergence of the human adaptive system at the polar axis of a cell, where the ordered regime is near the edge of chaos. Along with this potential universal feature related to the axis, Kauffman states that the genomic network is another example of order for free (85). Relative to the Isis Thesis, order arises naturally because it is guided by the viral genome activating the genetic switch to lysis within the ancient glycolysis-fermentation pathway in a cell. Viruses, humanity, and earth are coevolutionary systems. The earth is a nonequilibrium system invaded by solar radiation. On the quantum level, the earth-cell is similar to a living whirlpool, a nonequilibrium bacterial system invaded by solar-viral particles. Yet, a virus is an equilibrium system using the non-equilibrium earth-cell to reproduce. Further, horizontal gene transfer mediated by a virus is what saves organic DNA from dissolution at death. At death we need new traits to be fit or at home in the universe, so a symbiotic union (horizontal gene transfer) with a crystalline bacteriophage that looks like a ball of light may provide exactly those elements our DNA needs to adapt to the quantum environment, especially if that virus is our ancestor—a double-stranded eubacterial virus like phage Lambda with every key feature of our eukaryotic cell nucleus.

Kauffman states that fitness increases if a population uses mutation, natural selection, and recombination to climb to the peak on a well-correlated landscape. Recombination is useless on a random landscape. (1995, 182) Also, the space of possible molecules is immense.

> We can have a rational morphology of crystals, because the number of space groups that atoms in a crystal can occupy is rather limited. We can have a periodic table of the elements because the number of stable arrangements of the subatomic constituents is relatively limited. But once at the level of chemistry, the space of possible molecules is vaster than the number of atoms in the universe. Once this is true, it is evident that the actual molecules in the biosphere are a tiny fraction of the space of the possible. (1995, 186)

A stable low-energy equilibrium state is present in a self-assembled virus, the double helix of DNA or RNA, and in the folded protein encoded by genes. These stable structures are robust. (1995, 186-7) Likewise, the ancient Egyptian system is not a random fitness landscape, for they describe mutation, recombination and have left explicit instructions relative to Lambda's genetic switch using game theory. We have a coupled self-organized system with cro and c1 proteins competing and partnering for genomic binding to switch on their respective genes. This nonrandomness is critical to the evolutionary assembly of complex organisms and the emergence of new organisms, for it forms a circle cycling back to the genome or home plate. Lambda cro and c1 have correlated landscapes due to adjacent genes on the Lambda chromosome, but the proteins fold differently and cro is an initially disordered protein (IDP) like our non-coding DNA. Amidst this chaotic genomic network with its functional genetic switch is the ordered regime of cro protein's native state.

Kauffman states, "The deepest source of such landscapes [nonrandom] may be the kind of principles of self-organization we seek. Here is one part of the marriage of self-organization and selection" (1995, 166). He believes we need a new conceptual framework that will allow us to understand the evolutionary process in which self-organization, selection and historical accident play roles. He suggests that evolution can assemble a maximally compressed program, or organism, "by first evolving a redundant (prolix) program or organism and then squeezing it down to maximal compression" (156). Certainly, our genome is loaded with redundant DNA that may satisfy these requirements, especially the abundant reverse transcriptase and telomerase with its reverse transcriptase. Like Al Kaline who won ten Gold Gloves and fifteen All-Star team selections, Tiger Kauffman has a model of consistency for excelling in biology, as well as the sporting dexterity to perform in right field for the Diamondhearts in the Game of the Centuries.

Center Field: Chinese Emperor Ch'in Shih-Huang.

Ch'in Shih-huang-ti, although notorious for allegedly burning books and slaughtering Confucian scholars, was the first emperor of the Qin Dynasty of China who ruled between 221 and 210 BCE. Whether or not his fiery purge of historical knowledge and human life really occurred is debated by many scholars. For example, Martin Kern (2000) explains that these two claims of book censorship and political propaganda are not evinced from early empire sources, but rather from much later Confucian interpretations who viewed the first emperor as an aggressive arch-tyrant. Nonetheless, book censorship has historically been practiced by means of burning and looting. According to varied accounts, the Egyptian Library of Alexandria was burned several times—by Julius Caesar in 48 BCE, by Aurelian circa 270 CE, by decree of Pope Theophilus in 391, by the Muslim conquest of Egypt in 642. Humans still employ this destructive mode of censorship today—consider the destruction and looting in April, 2003 of the Iraq Museum in Baghdad, what some call the "rape of Mesopotamia." Priceless antiquities from the early years of human history in ancient Mesopotamia, one of the first civilizations, were smashed, destroyed, and stolen while US troops unobtrusively guarded the Museum. Actually, destroying a civilization's legacy in the name of a power élite is sociopathic ignorance, but it is a common method of effective political control in human history.

Now, for the Diamondheart fantasy-draft team, Emperor Ch'in is the number one choice for center fielder because of his ability to cover large distances to erect stone monuments on mountaintops in early China. Is Ch'in as good a center fielder as Joe Dimaggio, the Yankee Clipper? The rumor is that the Chinese Emperor can react quickly on his strong running legs, while firing a burning fast ball straight down the center with his powerful arm. During his lifetime, Ch'in travelled through China, climbing mountains, erecting inscribed stone monuments, and building stamina, so he has the speed and instincts for center field position. Further, he is an excellent pick for the Diamondhearts because he erected carved stones on the tops of mountains and other elevated sites. Official imperial erudites most certainly composed this correlated textual series (Kern 2000, 120). As a complete cycle, replete with sacrifices to cosmic spirits, the texts express the same message of the ancient Egyptian hieratic priesthood—a quasi-hybrid being described as a star. Let's consider the Chinese Emperor's historical position.

Ch'in lived during an influential period of observational astronomy and theoretical cosmology that recognized infinite empty space with free celestial bodies. Scholar Joseph Needham concedes that this early cosmic view was "really more enlightened than the Aristotelian-Ptolemaic conception of concentric crystalline spheres" dominating European thought. (1975, 87; 91) Cosmological concepts were related to the quest for a "medicine of immortality." This later developed into "the cult of *hsien* immortality" that Needham describes as follows:

> The word *hsien* means an immortal, living on or above the earth but within the world of Nature, a distinctively material immortality with a lightened body. The ancient Chinese believed in the existence of some drugs or chemicals or medicines which could be taken for this purpose. (97)

In ancient China, cosmology was linked with the origin of alchemy—creating what Needham calls a "drug-plant or a mineral or metallic elixir" (98) or "drug of deathlessness" (99). In forthcoming chapters, we will take a clear look at alchemy and its applications related to the purification and isolation of a virus, as well as the possibility that the "medicine of immortality" is a bacteriophage. Now, to acquaint you with the Chinese system, a quick review of Emperor Ch'in's stele inscriptions will illuminate his mountain ritual and resultant "shining" transformation.

The Stele Inscriptions of Ch'in Shih-Huang. The posthumous name *shih* is attributed to an ancestor, and *Huang* is a Chinese surname meaning "yellow." And so we have the process of transforming the human being Ch'in Shih-Huang into a yellow ancestor, that is, a shining chemiluminescent Sun-star. Of special significance to human transformation is the sign of the mountain or the idea in the mind that the sign excites.

During his reign as emperor (221-210 BCE), Ch'in Shih-huang-ti and his high officials climbed to the top of seven elevated sites and erected carved stones. Documented in the annals of the First August Thearch, these stone engravings or inscribed stelae on mountaintops are "the only substantial corpus of texts immediately connected with the First Thearch's official ritual activities." (Kern 2000, 8) The mountaintop stele inscriptions focus on unifying the heavens as in ancient Egyptian texts (King 2004; 2006). Also, the mountain sign in the Chinese ritual easily correlates to the uphill activities of a protein folding to its native state.

Ch'in is on an immortality quest, even though Kern finds no literal statement of this objective in the inscriptions or in the reference at the gate of Chieh-shih, stating that Ch'in "integrates the barley fields" (Kern 2000, footnote 128). Barley for beer-making was an important crop in ancient Egypt. Beer and barley are signs of fermentation and the ancient glycolysis and maltose fermentation pathway that runs without oxygen. Also, Ch'in "unified the great universe," suggesting cosmic embrace as described by the Egyptian pharaonic priesthood. Embracing the cosmos is crystallization at a condensed origin where time expands and space collapses due to quantum mechanics.

The inscription on the eastern vista of Mt. Chih-fu describes Ch'in—as the morning Sun—radiating a shining glory like a star:

> He reaches the [utmost] corner by the sea; Thereupon He ascends [Mt.] Chih-fu, Illuminates and looks down on the eastern lands. Gazing and ranging over the vast beauty, The attending officials are all in contemplation, Tracing the origins of the most shining way: Since the Sage's laws had begun to flourish, He cleansed and ordered the land within the borders, And abroad punished the cruel and violent. His military awesome influence radiated to all directions, Shook and moved the four extremities, Seized and extinguished the six kings. Far and wide He unified all under heaven, Disaster and harm were cut off and stopped, Forever halted were clashes of arms. (Kern 2000, 38-39)

Peace reigns and robbery of life stops. This description of the Chinese emperor's ascension is similar to a passage on the ancient Egyptian pyramid of Unås, a king of the fifth dynasty, who is rising to the sky.

> The sky poureth down rain, the stars tremble, the bowbearers run about with hasty steps, the bones of Aker tremble, and those who are ministrants unto them betake themselves to flight when they see Unås rising [in the heavens] like a god who liveth upon his fathers and feedeth upon his mothers. Unås is the lord of wisdom whose name his mother knoweth not. The noble estate of Unås is in heaven, and his strength in the horizon is like unto that of the god Tem his father, indeed, he is stronger than his father who gave him birth. The doubles (*kau*) of Unås are behind him, and those whom he hath conquered are beneath his feet. His gods are upon him, his uraei are upon his brow, his serpent-guide is before him, and his soul looketh upon the spirit of flame; the powers of Unås protect him. (Budge 1904, 33)

Both the shining Chinese emperor Ch'in and the awesome Egyptian king Unås have conquered earth and the heavens and become Morning Stars. In his annotated note 104, Kern explains: "Thus, the present lines do not show the emperor passively gazing at the rising morning sun over the sea; instead, he actively illuminates and looks down on the eastern parts of his empire." (2000, 38) Perhaps when Ch'in seizes the six kings, he is winning the contested six gene seats for viral lytic replication. By seizing the six kings or gene-seats, Ch'in is reborn as the yellow morning Sun or star similar to the ancient Egyptian king's crystallized morphogenesis to a quasi-hybrid being described as Perception or Mind in the early cosmos. Ch'in's immortality quest—to find his ancestral origin, to become that ancestor, to restore peace and order to the universe by embracing it, to shine in all directions and to become the morning Sun—is another model very similar to ancient Egypt's textual descriptions of becoming the morning Sun. Put simply, both Ch'in and Unås know how to smack the ball out of the ballpark to the cosmic origin.

One caveat about Ch'in—in case he seems too docile for a center fielder—the inscription on Mt. Chih-fu states that the Great Sage Ch'in established his peaceful order and saved his people, while "He boiled alive and exterminated the violent and cruel." Also, the inscription at the gate of Chieh-shih states that his battalions exterminated, halted, and beheaded those who defied his authority (2000, 41). So, we do have a Chinese arch-tyrant here who will be a formidable player in center field.

Catcher: virologist Luc Montagnier.
Left Field: wave geneticist Peter Gariaev.

For a long time now, Luc Montagnier (born 18 August 1932) has been catching biophysical fast balls based on experiment and theory that often result in controversial discoveries. Both virologist Luc Montagnier and wave geneticist Peter Gariaev are involved in challenging scientific pursuits that may provide new strategies for viral diseases such as cancer and HIV. Peter Gariaev is Director of Wave Genetics Institute, Russia, Moscow. The French virologist Luc Antoine Montagnier is a joint recipient with Francoise Barré-Sinoussi and Harald zur Hausen of the 2008 Nobel Prize in Physiology or Medicine for his discovery of the human immunodeficiency virus (HIV). In 2010, Luc became a full-time professor at Shanghai Jiao Tong University in China, where he explores electromagnetic waves emanating from the DNA of pathogens.

To understand Luc's focus, let's first consider the nature of light or photons that travel in the form of a wave. Photons or electromagnetic radiation are a form of radiant energy moving through space via electromagnetic waves and/or particles. So we have the simple formula of light equals photons equals electromagnetic radiation. Now based on earlier Russian research, Peter Gariaev and his colleagues have shown that genes or DNA have electromagnetic wave representation. Then in 2009, Luc Montagnier and colleagues shook up the theoretical underpinnings of chemistry by suggesting that DNA sends electromagnetic imprints of itself into distant cells and fluids.

Further, these researchers detected electromagnetic signals from bacterial DNA, such as *M. pirum* and *E. coli*, in water. For example, using bacterial cultures, Luc and his colleagues recorded electromagnetic signals (EMS) from aqueous dilutions of DNA during the filtration step of the experiment. This means that the EMS in the various bacteria tested came from their DNA, as Gariaev showed earlier. In a 2010 paper entitled "DNA Waves and Water," Luc and his colleagues explain their results using quantum field theory. The researchers believe that this is a novel property of DNA—some sequences can "emit electromagnetic waves in resonance after excitation by the ambient electromagnetic background." This ambient background is the earth's seven Hertz (Hz) naturally occurring oscillation of its electromagnetic waves. Human brain waves also oscillate in this same range. (2010)

Now, what is also very interesting is that Luc Montagnier's group suggests a general mechanism of DNA transcription and translation based on the reconnection of magnetic flux tubes between the molecules in question (Pitkänen 2011, 181). Pitkänen explains that Luc and colleagues found that electromagnetic radiation generated by DNA affected water as if the water contained the original DNA. This was accomplished by the polymerase chain reaction (PCR) that uses DNA polymerase (an enzyme) to create DNA from its basic building blocks. Put simply, the DNA fragment is not added to the PCR medium, but copies are still created. Some have called this a phantom DNA effect because the DNA is not really there. In 1984, Peter Gariaev registered this phantom effect. In the article "Principles of Linguistic Wave Genetics" (2011), Gariaev and his colleagues explain that every living organism exists at two levels—a material level and an energy-informational level (EI). Although the two substances are intimately linked, the EI level is the leading level grounded in the 98 percent of our DNA that is a cemetery of viruses. The DNA phantom effect can be explained as follows:

> In the process of taking the laser spectroscopy measurements, a laser ray was sent to the DNA sample. During that time some information and energy was transmitted from the DNA sample to its counterpart at the EI level. After the (material) DNA sample was removed, the process of transformation of information and energy was reversed. Specifically, the DNA sample at the EI level was still at the same place, and it started sending information and energy back to the same physical location from which the material DNA sample was removed. As a result, a DNA phantom was detected at the same physical location. (2011)

Gariaev explains that our 98 percent junk DNA is a supercode that codes at a higher wave level than our 2 percent coding DNA. In wave genetics, the material level constitutes gene-holograms, so all cells in the organism function in informational space like a DNA wave biocomputer. At his website, Gariaev explains the ideal potential of the 98 percent as follows:

> It is to be pointed out that the primary focus of the wave genome theory is on the remaining aforementioned 98% of chromosomes as being the key "intellectual" structure

> of all cells of an organism including the brain. It is those chromosomes that operate on the wave, on the "ideal" (fine-field) level. It the ideal component, that may be called super-gene-continuum, is a strategic vital figure/formation that ensures development and life of humans, animals, plants and also their programmable natural dying. Along with that, it is important to conceive that there is no sharp and insurmountable distinction between genes and super-genes. Both these levels of encoding constitute material (physical) matrixes, however genes supply material replications in the form of RNA and proteins, whereas super-genes transform endo- and exogenous fields, forming from them super-gene-signal wave structures. Furthermore, the genes may be components of holographic grids/frameworks of super-genes and supervise their field activity. (Gariaev website)

It is possible that this novel property may be shared by all double helical DNA. Electromagnetic signals have also been detected in an RNA virus such as HIV, influenza virus A, and Hepatitis C virus. Luc and colleagues believe that more collaboration is needed between biologists and physicists, since bacterial and viral DNA sequences can produce electromagnetic waves in water that are triggered by the ambient electromagnetic background of earth (Montagnier et al. 2009). This research seems to build on the French physicist Louis de Broglie's proposal that electrons and other material particles also exhibit wave properties. In other words, even minute matter owns both wave and particle properties.

How does this research support the Isis Thesis? Gariaev and Luc's experimental research supports that genetic information can function not only on a material level, but it can also function in a wave field or electromagnetic field. Similar to Gariaev and Luc's views, the ancient Egyptian texts indicate that our world is holographic and ordered by quantum electromagnetic fields that represent microbial DNA. To explain, the ancient texts demonstrate that the earth's electromagnetic field in watery space is a sign of the genetic code of *E. coli* bacterium on the quantum level. The researchers' finding that bacterial *E. coli* DNA can send electromagnetic imprints of itself into distant cells and fluids corroborates the ancient Egyptian knowledge. Considering the holographic principle, if the earth's electromagnetic field represents *E. coli* DNA on the quantum level, then the earth's electromagnetic field becomes the guiding platform for traversing the *E. coli* chromosome, as the ancient texts explain. What stimulates or emits the bacterial electromagnetic waves is earth's seven Hz electromagnetic waves, and human brain waves oscillate within the same range, so the Isis Thesis model of the earth as an *E. coli* cell is consistent with this research. As some scientists have commented, the stars may represent DNA written across the crystalline cosmos.

Now, Gariaev believes that our genome represents two intimately-linked substances: the 98 percent junk DNA is a supercode that codes at a higher wave level than our 2 percent material coding DNA. Although the two substances are intimately linked, the EI level is the leading level grounded in the 98 percent of our DNA with viral elements. In the Isis Thesis interpretation of ancient Egyptian texts, the interplanetary magnetic field of the Sun is the sign of the viral DNA of bacteriophage Lambda. In comparison to Gariaev's findings, this interplanetary magnetic field would represent Gariaev's EI level. The ancient Egyptian instructions for transformation explain how to "cross-over" or recombine with viral DNA, find the viral polar gateway into the cell (the LamB porin of *E. coli*), and enter the ancient glycolysis-fermentation pathway for transformation to a quasi-hybrid being. This avoids the bacterial cell degradation pathway. Put simply, Gariaev's "ideal" 98 percent of our viral genome may be a supergene continuum that ensures the development, life, and programmable death of living things. However, the "ideal" 98 percent also conveys a survival message of viral horizontal gene transfer and transformation. This suggests that human consciousness with a memory of the genetic survival message is the switch-hitter that activates the lytic genetic switch for viral replication. One could say that the ancient Egyptian dead king functions in a fashion similar to the phantom DNA effect. Granted, these claims are slippery fast balls, but let's take another step deeper into Gariaev's research.

Wave Genome Theory. Gariaev's wave genome theory sees the genome as a biocomputer, forming the space-time grid of our bio-system. This view is much different than the prevailing view that our genome is purely material and that 2 percent of the DNA governs all our functions and the other 98 percent is non-coding. The prevailing view about proteins is that they metabolize substances we eat and they function for morphogenesis or developing the spatial-temporal order for an organism. In contrast, Gariaev considers that the 98 percent DNA is the key "intellectual" structure controlling our cells and our brains that operates on the "ideal" or "fine-field level." Peter Gariaev also provides direct experimental evidence that positive human speech on plants and rice encourages regeneration. This positive influence can also be delivered by thought, that is, just thinking positive about regeneration. The essence is the same. This human potential of speech and thought is similar to Wheeler's observer-participancy principle in that what you pay attention to happens.

In a recent paper, Peter P. Gariaev and colleagues support that a better understanding of our genome will enable practical applications for treating cancer and HIV. They contend that our genome has a capacity for quasi-consciousness, the genome controls life chemically, and the genome is also a source of wave function and holographic memory. The genome's processes in its substance-wave structures can be observed via dispersion and absorption of a bipolar light beam. Gariaev and colleagues support their model with theoretical and experimental evidence. They seriously question the Central Dogma of Biology (DNA to RNA to Protein) because much of the time, the process works in reverse from the protein to the gene. The group hopes to explain how our genome operates, so scientists can come up with better solutions to treat cancer and HIV. Gariaev believes these diseases could wipe out humanity. Here is a brief summary of the crux of their thinking. (Gariaev et al. 2002)

Gariaev and his colleagues have grounded their research in early Russian studies in the

1920s, and from this foundational base of knowledge, they support that our genome is speech-like and logical with quasi-conscious abilities. Based on experiments, Gariaev's first axiom is our genome has a substance-wave duality like the particle-wave duality of elementary particles. Research suggests that our genome codes an organism with the help of both DNA matter and DNA sign-wave functions. This seems reasonable for DNA is minute matter.

Second, the genome is also nonlocal (everywhere at once and at no time) at the molecular level. But, it is also nonlocal in compliance with the Einstein-Podolsky-Rosen (EPR) effect. Based on the experiments of John Bell and Alain Aspect, physicist Brian Greene explains the EPR effect in simple language: "The outcome of what you do at one place can be linked with what happens at another place, even if nothing travels between the two locations" (2004, 99-115). This is the idea of quantum nonlocality or what Einstein called "spooky action at a distance." Gariaev explains that our genome's genetic and regulatory information is recorded at the polarization level of its photons and played out among the billions of cells in an organism. Put simply, our quantum genome directs us because we are entangled with it due to nonlocal consciousness. The outcome of what we do is linked to our genome and its functions. Once again, the quantum world of DNA is ordering our classical existence, and most of our DNA in our bodies is microbial.

Third, the genome and the billions of cells in us can "generate and recognize text-associative regulatory structures with the application of a background principle, holography, and quantum nonlocality." For example, the meaning of "DNA matter-wave dualism" is that the genome operates similar to a radio wave emitter that polarizes our cells. Since the genome is a "dynamical multiple hologram," it produces light and radio wave images that carry electromagnetic schemes of space and time organization. Because the genome is a "quasi-text form" operating nonlocally, it reads itself in our cells and uses the information it receives "as a control blueprint for living functions and structural organization." According to Gariaev and colleagues, this is how our genome operates, so new effective strategies need to be developed for cancer and HIV prevention. (2002)

Gariaev notes that American researchers (Mantegna et al. 1994) came to a conclusion similar to his group regarding coding and non-coding DNA. The American conclusion also conflicted with the Central Dogma that meaningful functions only exist in the coding 2 percent of the genome, for they found that non-coding DNA was not junk, but rather lingual structures for reaching some still unknown goals. Gariaev believes that the non-coding 95 to 98 percent of our genome possesses strategic informational content. Perhaps the non-coding genome provides a toolbox for transformation into a quasi-hybrid being at a human death transition, as ancient Egypt supports.

Our behavior then is entangled with the behavior of our genome, which is a "control blueprint" for our functions. This is an illuminating and frightening possibility, for it accounts for our viral behavior—the emphasis on the glycolysis-fermentation-lactose pathway for viral replication and survival, our predilections for alcohol and milk, our penchant for war and killing, and so on. We may be polarized to our viral genome, and our behavior may be due to entanglement with it over distance, that is, quantum nonlocality.

Gariaev and colleagues conclude the following from their theory, research, and model:

> Following the above-described logic, we are coming to the conclusion that human speech structures, which provide the major information influx for mankind, possess fractally-scaled supergenetic properties. Evolution of society is similar to an organism's morphogenesis. Books, libraries, movies, computer memory and people's live speech in the end are the functional analogues of a cell chromosomal apparatus. The aim of these chromosomes is to control the creation of society space (houses, roads, oil- and gas pipelines, telephony, the Internet) and to arrange functional and structural relationships among the people inside it. Chromosomal sign properties, which have a lot in common with organisms, have a substance-wave nature. For instance, a movie showing an ideal model of a social structure and people's relations within its frames is a substantial (material) formation (video tapes). However, it uses a mental-wave method to input information (light, sound, speech, idea, image). That's the method chromosomes apply. (2002)

Obviously, our chromosomes have created our society space in the pattern of the Few over the Many (government, capitalism, religion, medicine, psychology), suggesting the survival of those with knowledge. Our chromosomal apparatus has also stamped a viral footprint on society space, mapping out the ancient glycolysis-fermentation pathway using lactose metabolism for the morphogenesis of a quasi-hybrid being.

Regarding the cure of viral diseases, Gariaev believes that what is needed is a study of how our chromosomes operate and how HIV "tricks" our chromosomes through re-programming. Also, the polarization-laser-radio wave (PLRW) spectroscopy developed by the researchers can be used to record wave information on laser mirrors. Gariaev proposes a study of the wave and sign behavior of viruses, HIV, or influenza, what he calls their alphabet and grammar of wave languages in the viral genomes. He believes that material vaccines and chemical treatment of HIV or influenza will fail because a wave vaccine is the only way to block stages of viral morphogenesis because viruses continually change their antigenic composition.

Viruses, he explains, are like orphaned cells that have minimal DNA information for a wave search of a landing site on a master cell, so it can splice in its own DNA with the cellular DNA. The viruses use wave languages in this process of entering the cell and misleading it. HIV faces the same dilemma, according to Gariaev and colleagues:

> . . . how to correctly find a landing site on a cell's surface and precisely build-in the DNA (reverse transcriptasal copy of a viral RNA) as a mimicking transposon into the master

> cell DNA. Thereafter, the task is to get accurately re-transposed in a proper place on a chromosome and to detect and realize itself as a reproducing pathogen. (2002)

Recall that Barbara McClintock received a Nobel prize in 1983 for her discovery of transposons, also known as jumping genes or mobile genetic elements. These genes can change their position within the genome, sometimes reversing mutations. A mobile genetic element is similar to the activity of viral DNA. Surprisingly, what Gariaev is describing is the central objective of ancient Egyptian knowledge because their secret science includes not only a study of the behavior of phage Lambda and its *E. coli* host, but also biological directions allowing fragmented human DNA to locate the correct landing site on a cell's surface after viral self-assembly.

To speculate, it may be that what is needed for viral entry into the cell is the human genome's reverse transcriptase. In ancient texts, the merger with the light photons from the Sun jumpstarts the self-assembly process. The viral ferryboat with the necessary proteins then tricks the cell into its SOS response by entering the LamB portal protein gateway, and the Lambda prophage rises for rolling circle replication. Generally, the cell is immune to another viral infection due to the presence of the prophage. However, the hybrid nature of the DNA may have tricked the cell. Then, the viral ferryboat with its hybrid DNA travels along the maltose transport system of the ancient glycolysis gene expression network to the lactose genes where the sleeping prophage is rising for rolling circle replication to reproduce hybrid viral clones. This is truly an ironic biological twist—organic human viral DNA fulfilling its chromosomal function as a viral infection for its own survival. Cancer then is a sign of this process.

Both Gariaev and Luc are focused on the admirable goal of treating human disease for life extension, yet their research makes one wonder about the effects of a wave vaccine. If the evolution of society is similar to an organism's origin and development of form and structural characteristics, then the evolution of our society is similar to that of a virus, as confirmed by the messages in our myths, religions, economy, and other societal patterns. If books, libraries, films, computer memory, and human speech are comparable to this same viral chromosomal apparatus, then repetitive viral messages are surfacing in our society through media and our behavior. One wonders—exactly how will the wave vaccine change our chromosomal apparatus? Gariaev cautions us that our chromosomes produce a calibration field that arranges the organism's space, while controlling information and metabolic relations. The question is—if we understand our microbial nature and the restrictions of our chromosomes, is it possible to maximize our free will to alter our chromosomes through our environment, rather than a wave vaccine? After all, our chromosomes are also reading epigenetic marks that effect gene expression patterns that are caused by hereditary transmission of environmental information. Actually, our chromosomes seem to be functioning similar to the same methods used by governments to control citizens—surveillance, repetitive propaganda, secrecy. We are caught in a viral power web. Is it possible for humans to overcome viral propaganda and change their behavior? Can we change the programming dynamics of our chromosomal apparatus without losing the DNA toolbox for HGT and transformation?

In consideration of this research and the Game of the Centuries, Gariaev and Luc both have tough jobs on the diamond. As catcher, Luc's grueling position requires durability, a strong arm, and the mind to call a game or change outdated concepts in science. Perhaps we can think of Luc as the catcher Yogi Berra, who helped win ten World Series as a New York Yankee (1946-65), or maybe Johnny Bench of the Cincinnati Reds (1967-83), who was named to fourteen All-Star teams. In light of Gariaev's strong foundation of Russian research, his clout in left field will be powerful. Together these two players own a defensive ability that may alter the Central Dogma of biology by showing the potential of our 98 percent DNA that some researchers believe is junk.

Pregame 19

The Diamondheart Subcatcher

Ye who read are still among the living, but I who write shall have long since gone my way into the region of shadows. For indeed strange things shall happen, and many secret things be known, and many centuries shall pass away, ere these memorials be seen of men. And, when see, there will be some to disbelieve, and some to doubt, and yet a few who will find much to ponder upon in the characters here graven with a stylus of iron.

Edgar Allan Poe
"Shadow—a Parable" (1835)

The pick for subcatcher on the Diamondhearts is the scientific writer Edgar Allan Poe (1809-1849), who will be wearing a catcher's mask similar to the corpse-like mask in his story "The Masque of the Red Death." To showcase Poe's scientific mindset, let's explore three of his works, which define concepts of black hole physics, cosmology, and human survival in a simple narrative form, while emphasizing an intellectual terrain of knowledge, danger, death and survival. In essence, Poe writes scientific knowledge into his stories and nonfiction, preserving what he called the "only pathway between Time and Eternity." Poe describes the same chemical pathway that ancient cultures document. This supports the unity of knowledge present in human thought, for this survival knowledge keeps recurring in history. Further, Poe narrated the dynamics of a rotating black hole long before New Zealand mathematician Roy Kerr discovered the Kerr vacuum metric in 1963, the exact solution to the Einstein field equation of general relativity (gravity) for a rotating black hole.

Relative to the unity of knowledge due to our nonlocal genome, Poe was on the frontier between quantum theory and general relativity in the 1840s with his descriptions of wormholes, black holes, time and eternity. He also walked the tightrope of time to the edge of chaos in space, where he perceived that death is a transition state allowing creation, if one possesses knowledge of the natural process. To expose the value and explode the boundary of this transition event that intimidates us, we begin with three of Poe's writings that express several simple concepts of physics that will improve understanding about his ideas on the continuity of mind : "Descent into the Maelström" (1841), "Masque of the Red Death" (1842), and the nonfiction essay "Eureka" (1848).

Descent into the Maelström (May, 1841)

Poe's story "Descent into the Maelström" published in Graham Magazine is a survival tale of a fisherman, whose seafaring experience shocks his black hair to gray within hours. The story begins with the aged sailor sharing his story with his vigilant companion, as they stare into the cold sea from a mountain peak off the Norwegian coast at the northern polar Arctic Circle of earth. He and his companion sit precariously at the brink of a sheer precipice of black rock stretching deep down to a chasm. Looking out over the gloomy ocean, the cold waters shiver into a strange chopping motion and then rush into an eastward current, hissing with the accelerating velocity and gyrating fury of monstrous vortices consolidating into one vast circular vortex. In agony, the waters tremble and shriek, rocking the mountain where the two witnesses watch the violent whirlpool. During storms, that vast vortex had beat many ships, yachts, and boats to the bottom of its pit, swallowing them without mercy.

The fisherman continues his story, explaining how he and his two brothers were caught up in the same massive hurricane of whirling blackness. Suddenly, the mainmast rips off their vessel and thrashes into the rotating pit's dark mouth, swallowing his younger brother who had lashed himself to the mast for safety and eventually his older brother hanging on to a ring-bolt. As the storm howls in the darkness, the fisherman sees the sky open into a circular frame of deep bright blue with a lustrous full moon there. By the moonlight, he notices his watch is stopped, and then his descent begins into the vortex. The starboard side of the boat is next to the whirl, with the ocean on the larboard side. At this point, the man realizes his doom, but stoically gives up his fear to investigate the wild phenomena of the whirl. The boat suddenly lurches to starboard, dashing downward into the rotating abyss with its black walls. Yet, the rays of light from the circular rift of the full moon in the bright blue still cast down a "golden glory," and in the mist of the pit's rotating black walls, a rainbow appears, similar to a "tottering bridge" or the "only pathway between Time and Eternity." In the continuing descent to the foam below, the fisherman surveys the revolving debris of shipwrecks, houses, trees stripped of knowledge, and then knowledge comes to him. He lashes himself to a water cask, throwing himself into the churning vortex. At that moment a change occurs and the storm calms, the rainbow and the froth disappear, and the gyrations of the whirl grow feeble, becoming still, and then reversing their movement with gradual acceleration into spontaneous order. The winds recede and at the surface of the ocean the full moon settles in the West. The fisherman survives.

Poe is describing the physics of a rotating Kerr black hole, which is a gateway to an Einstein-Rosen bridge through its inner horizon on the axis. An Einstein-Rosen bridge is simply a rotating black hole connected to a white hole, a wormhole. Egyptian texts describe a microscopic Einstein-Rosen bridge or wormhole going backward in time to the early cosmos, a time machine for a particle or wave. As egyptologist Erik Hornung explains, the ancient Egyptian Netherworld is a black hole and "the location of time past" while "Time reversal permits us to leave the Black Hole" (1994). The laws of Newton, Einstein, Schrödinger and Maxwell support that time can go backward. Also, wormholes are allowed by the mathematics of General Relativity; in fact, Einstein attempted to model particles as wormholes. The rotating Kerr black hole was named after Roy Kerr who discovered the Kerr solution to the vacuum Einstein field equations in 1963, about 122 years after Poe clearly described Kerr black hole dynamics in "Descent." His survivor is at the axis of the spinning vortex where time stops and rainbow effects appear, but there is more.

Looking at those dynamics, the rotating geometry of a Kerr black hole drags space-time, while having an outer horizon, an inner horizon, and a ring singularity. When approached from the axial inner horizon, the ring singularity is repulsive, allowing one to theoretically leave the black hole. In astrophysics, a white hole is a body that spews out matter or a black hole running backward in time. Now, exactly how does a Kerr black hole become a time machine? A spinning Kerr black hole has a very dense ring singularity, a place of destruction where tidal gravity and space-time curvature are infinitely strong. When an entity falls in from the outer horizon it rotates around the hole, inevitably crushed at the ring singularity. This is the "second death" in the Egyptian afterlife. However, if approached from the inner horizon or the axis of the black hole or the north pole, the gravitational repulsion of the central singularity slows the entity down, turning it around, and accelerating it back out through the inner horizon of a white hole. Put simply, the space axis and time axis exchange places when one crosses the outer event horizon, and the future becomes an unavoidable place in time—the crushing singularity. In contrast, crossing the inner event horizon, time and space resume their normal axes, making the singularity an avoidable place in space, while allowing access to the past singularity of the white hole. Ultimately, one goes through four horizons—the outer and inner of the black hole, and then the inner and outer of a white hole, what the Egyptian pharaonic priesthood refers to as the Opening of the Mouth ceremony involving four mouths. Contemporary scientists also describe a mouth as an entrance to a wormhole, as did the pharaonic priesthood.

Now in "Descent," Poe describes the men traveling from the west to the east, when they encounter the terrific hurricane. As they enter the frothing whirlpool, the ocean appears as a high black mountainous ridge. Poe's only survivor sees a circular frame of sky opening, the color of deep bright blue, as well as the lustrous full moon. Then the small ship is swept up in the "amazing velocity" to the left. Inside the Maelström, the boat whirls at a dizzying speed, then lurches starboard or to the right and rushes headlong into the abyss. Poe's survivor then realizes that he is not dead and the vessel is no longer falling. He then describes what he sees.

> Never shall I forget the sensations of awe, horror, and admiration with which I gazed about me. The boat appeared to be hanging, as if by magic, midway down, upon the interior surface of a funnel prodigious in circumference, immeasurable in depth, and whose perfectly smooth sides might have been mistaken for ebony, but for the

> bewildering rapidity with which they spun around, and for the gleaming and ghastly radiance they shot forth, as the rays of the full moon, from that circular rift amid the clouds which I have already described, streamed in a flood of golden glory along the black walls, and far away down into the inmost recesses of the abyss.

Then the survivor experiences a rainbow, describing it as the "only pathway between Time and Eternity." Whirling and descending downward to the foam below, the sole survivor sees many shredded and whole objects, and so he applies his knowledge related to scientific reason on spheres and cylinders, thereby deciding to attach himself to a water cask, which is then spit out at the sea's surface. Poe's survivor explains the directions that saved him, and he actually moves in a west to east direction, returning to his starting point, the Lofoten islands off the northwest coast of Norway on the polar Arctic Circle. Similarly, ancient Egyptian texts stress the importance of the northwest location, the northern pole of earth, and the West to East direction.

Poe has clearly described a black hole connected to a white hole, and the axial rainbow path out. According to scientific speculations, inside a spinning Kerr black hole, the act of looking backward would allow one to see a white hole or the past singularity, what looks like a full moon. At the inner horizon, where Poe's survivor witnessed the bright blue sky, infalling radiation is blueshifted as it accumulates (Visser 1996). A white hole is a black hole running backward in time, and Visser and colleagues (2003) demonstrated the existence of space-time geometries containing traversable wormholes that are supported by arbitrarily small quantities of exotic matter. This means that these wormholes exist at the microscopic quantum level. Also, at the axis of a spinning Kerr black hole, light possesses rainbow effects, so the rainbow is the sign of the pathway heavenward (see King 2007), a sign Poe included in "Descent." As Nietzsche said, "I will show them the rainbow and the stairway to the Superman."

One final point—Poe's account is psychophysical. This means that what has emerged into his consciousness, that is, his mind bound to matter, is a genetic report of a specific linkage with the environment or a biophysics functioning as the world's central principle of construction. As John Wheeler explains the physics, "And surely someday—if Einstein's views of nature are correct—we will understand a particle, not as a foreign and mysterious interloper introduced from outside of space, but as an object 'constructed out of space'" (1994, 66).

Eureka (1848)

The simplest cosmology compatible with Einstein's theory of gravitation is the Big Bang model followed by the Big Crunch (Wheeler 1994, 28). Seven years after publishing his biophysical "Descent into the Maelström," Poe composed nonfiction "Eureka" explaining that our cosmos functions similar to his Maelström, an expanding turbulent black hole in time collapsing backward to the Big Bang origin or dense singularity. Today Poe is considered an amateur scientist (Gribbin 2001, 124-125) because he realized the nature of Olbers' paradox named after a nineteenth century German astronomer. Poe was the first to understand that time has an edge because by the act of looking into space, we are looking backward in time because light takes a finite time to travel through space. In a lecture in February, 1848, Poe extrapolated the correct resolution to Olbers' paradox, commenting that we are looking backward to a distance "so immense that no ray from it has yet been able to reach us at all." Poe did not follow up on his discovery because in 1849 he was found unconscious outside Ryan's Saloon in Baltimore. According to a University of Maryland Medical Center news release (1996), doctors theorize that Poe died from rabies and not excessive drinking. Poe's reputation as an alcoholic, drug addict, and madman was partially based on an obituary written by Rufus W. Griswald, who submitted a writing for Poe's review. Poe was incisive in his review of Griswald's work, so Griswald wrote a negative obituary of Poe claiming he was a drunk, high madman. (Edgar A. Poe Society 2009) This is an effective method of repressing the knowledge discovered by a brilliant person.

Now, our finite universe has an edge in time, so we see stars to a certain point, then we see the darkness indicating a time before stars. In contrast, in an infinite spatial universe where space has an edge, we would see stars everywhere we looked, what is described by the Egyptian idea of going into the Day. In light of his understanding of Olbers' paradox, Poe wrote the nonfiction essay "Eureka" (1848) describing his thesis of two ways that echoes both ancient and modern ideas. Poe writes:

> We may ascend or descend. Beginning at our own point of view – at the Earth on which we stand – we may pass to the other planets of our system – thence to the Sun – thence to our system considered collectively – and thence, through other systems, indefinitely outwards; or, commencing on high at some point as definite as we can make it or conceive it, we may come down to the habitation of Man.

In "Eureka" Poe explains the primordial particle becoming the universe, the one becoming the many, what he calls a diffusion from Unity that requires a return to Unity. This is very similar to the modern Big Bang theory of our universe blossoming from a dense singularity. At that time, our cosmos was unbelievably small, then everything suddenly ruptured from that point in space-time referred to as the singularity. First hydrogen formed, then helium, and one billion years later stars formed, creating the other chemical elements. The atoms in our bodies are connected to the cosmos and the epic lifecycle of its stars in deep space. This system exhibits machinic cycles or wheels within wheels such as moons circling planets, planets revolving around suns, suns or stars migrating around galactic black hole centers. As an example, earth rotates at a constant rate like a rotating Kerr black hole. If the spinning earth shrank to the size of one-half centimeter (.4438), it would become a tiny quantum rotating Kerr black hole (Wolf, 1988, 148).

In our universe, time has an edge—what is called the Big Bang; it is not infinite as we know by our lifespans. However, as Hawking (1988, 89) explains, the beginning of time for an expanding universe is the Big Bang singularity, but this singularity would be the end of time if our cosmos were collapsing or reversing time. This is what Poe is discussing in the nonfiction "Eureka" related to the continuity of mind; he understands that individual consciousness can observe a collapsing universe into a past singularity. He states that attraction and repulsion are the sole properties by which matter is manifested to mind. Gravity propels our bodies to the centre of the Earth, but every earthly thing also can move in every conceivable direction or what physicists today call degrees of freedom (Bousso 2002). In human constructions, claims Poe, cause comes before effect, but in divine constructions, effect comes before cause. Poe understands that a backward time movement would result in effects occurring before causes. "The tendency to collapse" and "the attraction of gravitation" are convertible phrases for Poe. He then references pair production: "every atom is perpetually impelled to seek its fellow-atom." Poe envisions stars merging into one immense central orb already existing. Poe's ideas are related to time reverse, which would allow an expanding universe to collapse and effects to arrive before causes. Poe continues:

> Recurring, then, to a previous suggestion, let us understand the systems – let us understand each star, with its attendant planets – as but a Titanic atom existing in space with precisely the same inclination for Unity which characterized, in the beginning, the actual atoms after their irradiation throughout the Universal sphere.

Poe clearly understands that the cosmos began from a very small state and operates holographically. He then describes its expansion to "the awful Present" and proposes his hypothesis that the universe would collapse as follows:

> . . . chaotic precipitation, of the moons upon the planets, of the planets upon the suns, and of the suns upon the nuclei; and the general result of this precipitation must be the gathering of the myriad now-existing stars of the firmament into an almost infinitely less number of almost infinitely superior spheres. In being immeasurably fewer, the worlds of that day will be immeasurably greater than our own. Then, indeed, amid unfathomable abysses, will be glaring unimaginable suns. But all this will be merely a climactic magnificence foreboding the great End. Of this End the new genesis described, can be but a very partial postponement. While undergoing consolidation, the clusters themselves, with a speed prodigiously accumulative, have been rushing towards their own general centre – and now, with a thousandfold electric velocity, commensurate only with their material grandeur and with the spiritual passion of their appetite for oneness, the majestic remnants of the tribe of Stars flash, at length, into a common embrace. The inevitable catastrophe is at hand.

After this cosmic embrace similar to the ancient Egyptian embrace, Poe then explains that matter is the means to the end, and it was created for that reason, even though the catastrophe at the end is the nonexistence of matter. Poe then imagines another creation due to the law of periodicity, where a novel Universe swells into existence, and then subsides into nothingness forever. This is similar to the ancient Egyptian and Christian concepts of a new heaven and earth. Poe's "Heart Divine" is composed of our own hearts.

In the preface to "Eureka" Poe writes that the knowledge is true and cannot die, for it will "rise again to the Life Everlasting." In a letter of July 7, 1849 to his mother-in-law Mrs. Maria Clemm, Poe wrote: "I must die. I have no desire to live since I have done 'Eureka.' I could accomplish nothing more." (Edgar A. Poe Society) Poe died October 7, 1849. The impact of his knowledge lives on in the legacies of earlier cultures and later great thinkers with the same idea—the individual's power for cosmic embrace or collapsing the universe back to its origin. And so we have in Poe's "Eureka" an expanding/collapsing cosmos and the vision of cosmic embrace, the same vision shared by the ancient Egyptian pharaonic priesthood, the early Chinese emperors and sages, the American Navajo Indians, modern science, and many others.

How did Poe know? He wrote a piece on "Instinct versus Reason—A Black Cat," stating that instinct is the "most exalted intellect of all" and that "the boundary between instinct and reason is of a very shadowy nature." Instinct may be another word for our controlling quasi-conscious genome.

The Masque of the Red Death (1842)

Similar to "A Descent into the Maelström" where Poe stresses the importance of "mere instinct" and the inadequacy of reason because the state of the caustic experience "did not last long enough to give us time to think about it," in the "Masque of the Red Death" Poe mentions that in the advance of the Red Death it is folly to grieve or think. In this sullen story of a plague, Poe enumerates other aspects of his full system that relate to the Isis Thesis interpretation of ancient Egyptian texts. To understand the connection of physics to biology, we must look backward in time five hundred years earlier when the plague called the Black Death struck the continent.

Apocalypse or what some experienced as the end of the world began in Mongolia, China in 1347, when the invisible agent known as the Black Death stalked Europe, claiming an estimated 15 to 30 million people in only five years. During that time, it was foolish to think or to grieve. The plague first attacked Mongols fighting with the Christians over the trade route at Caffa on the Black Sea. The causal trail indicates that plague ships bearing cargoes of corpses materialized at the port of Messina, Sicily. Many believe the medieval mind thought that the alignment of planets caused the plague, or that God's vengeance was punishing humans for their depravity and capacity for evil. On St. Valentine's day in 1349, the residents of Strasborg, Germany witnessed the persecution of the Jews by the Christians, but this bloody Christian remedy did not stop the Black Death from marching north to Scandinavian borders.

Today we know that one possible cause of the most devastating pandemic in human history that decimated 30 to 50 percent of the population between 1347 and 1351 is the *Yersinia pestis* bacterium infecting the flea carried by the rat infesting the ships on the trade route (Bos et al. 2011). But evidence from scientists Susan Scott and Christopher Duncan at the University of Liverpool suggests that the plague may have also resulted from a virus, characterized by a haemorrhagic fever and a long incubation period of thirty-two days, allowing it to spread widely despite limited transport during the Middle Ages. Emerging from a possible Ethiopian animal host, this bloody viral plague struck repeatedly at European/Asian civilizations, appearing as the Black Death in 1347. Not only the bacterium-infected flea riding the rat may have been responsible for devastating the population, but also its double, the viral-infected human. It is not surprising that some believed that corpses, resulting from haemorrhagic serial deaths where blood was its avatar, could re-animate and come from the grave at night to suck the blood of sleeping persons. Since the deceased lost blood in life, it seemed logical that the deceased would seek blood to prolong existence of life in death. As scholars are aware, the vampire mythologies are strongly associated with plagues and viruses. These mythologies are signs of viral re-animation.

And so, with previous cyclical epidemics in mind, Edgar Allan Poe writes that blood was its avatar or the embodiment of death in "The Masque of the Red Death" (1842), a tale embodying the horror of heavy bleeding in a fatal disease. He recounts the tale of Prince Prospero who attempts to save a thousand of his friends after half his dominions have been annihilated by the Red Death. Retreating to a well-provisioned country castle surrounded by walls with iron gates, the invited guests—secure as they thought they were—enjoyed wine, music and dance, for Poe wrote that Beauty prevailed within the castellated abbey as the Red Death ravaged without. The élite often retreated to country castles to avoid plagues. After Prospero's impervious guests spent half a year in safe seclusion, Prospero organized a victorious masquerade, highlighted by a bizarre passage through an imperial suite of seven rooms, arranged not in a linear fashion, but with sharp turns spiraling from the east to the west, so that the masked revelers could only see one room at a time.

This had a novel effect on the viewer, for the sharp turn every twenty or thirty yards into a dead-end corridor revealed a central stained glass window opening into and mirroring the restricted color décor of six chambers, blue then purple, green, orange, white, and violet—a dark rainbow of severed colors. Opposite each stained glass window in each of the blind-alley passageways between the six chambers was a weighty tripod, its three legs supporting a brazier of fire, its forceful rays glaring through the glass window pane to cast a garish hue into each shadowy chamber. And then, through the final Gothic glass window stained a deep blood color, one could witness the seventh chamber shrouded in black velvet wall and ceiling tapestries. In this black chamber a massive ebony clock dictated time, its pendulum swinging back and forth like the blade of a reaper, its chimes igniting panicky fear in the revelers, causing them to avoid the hour of truth. With careless abandon, the revelers partied on in the space provided for them, for Poe wrote that it was folly to grieve or think. The revelers constructed that chambered space, and when the clock chimed, the revelers paled, locked as they were within their space-time chambers, the seconds stretching slowly forward until the next chiming.

After the description of the seven chambers and the black clock, Poe alludes to the play by Victor Hugo called "Hernani," explaining that the glare, glitter, phantasm, and piquancy or sharp taste is also there. We shall come back to the play momentarily. Now, the plot thickens at Prince Prospero's masque, where stiff-frozen dreams stalk the chambers, writhing in and about, changing hues with the chambers. In the black-sable, blood-hued Western chamber, the pendulum clock solemnly peals if a single foot falls on its sable carpet. So no guest walks there. As time passes, the cold machinic clock strikes midnight, twelve chimes with echoes announcing the arrival of a gaunt masked figure, wearing a blood-splotched mask of a stiffened corpse. Terror, horror, revulsion overwhelm the revelers when they recognize the Red Death with its corpse-like mask spattered with blood, stalking between the waltzers, moving slowly like the machinic movement of a pendulum or reaper's sickle. Becoming enraged, Prospero orders the revelers to unmask the intruder, but fear grips everyone, so no one touches the machinic monster who marches solemnly with a measured step from the eastern blue room to the purple through the green, orange, white, violet, and finally the blood-red black chamber in the west.

Ashamed of his momentary cowardice, the enraged Prospero rushes after the Red Death, as the revelers stand silently in terror. In the western chamber, Prospero grabs his dagger, confronting the gaunt figure, who turns suddenly and kills the prince. Seeing Prospero prostrate on the black sable carpet, only then did the revelers realize their plight. Crowding into the Western chamber with wild despair, they seize the Red Death who stands motionless within the shadow of the ebony clock. It is then they discover that the gaunt figure is formless, a molecule of the air like a virus. It is here that Poe compares the Red Death to a thief in the night, explaining, "And the life of the ebony clock went out with that of the last of the gay. And the flames of the tripods expired. And Darkness and Decay and the Red Death held illimitable dominion over all."

Poe is not only linking the physics of space with time, that is, the sequence of dark chambers with the ebony clock, but he is merging human biology with this physics, specifically describing a human machinic route to death, from blue to red, from east to west, a human biology with physics where thought moves and ends with time. When the clock or time dies, so do the "last of the gay" as well as their thought. So, we have a machinic death initiated by a plague that functions like an abstract machine, while presenting itself as a chaotic system moving to equilibrium. Death is described by Poe as a machine functioning on a deterministic circuit, and the unwilling élite are caught up in the machinic allure. Without a strategy, the fated revelers have played against the machine of Death and lost.

Poe has designed his story's plot to annihilate the élite. Yet, the irony is that Prince Prospero has unknowingly designed the chambers himself; he unconsciously designed his own deathtrap with the sequence of blue, purple, green, orange, white, violet, and finally the blood-red black chamber in the west. As Goethe understood, color is a degree of darkness, that is, it comes from boundary conditions and singularities (Gleick 1987, 165). So we have revelers moving through the bizarre rainbow-ordered chambers from east to west to a bodiless attractor drawing everyone to their death, for time stops in the blood-red chamber. The pendulum clock oscillates between two points, converging on equilibrium, but this dynamical system masks the chaos that ensues with its unstable fluctuations, as represented by the slaughter of Prospero and the revelers. With his spiral chambers of light and shadow regulated by the time flow of the pendulum clock, Poe opens up the clock's stationary state, its second equilibrium, the unexpected pattern of stillness or death, and this pattern mirrors the turbulent physics of a black hole with a Reaper Default. The lesson is—without knowledge the path defaults to the Reaper, the attractor, the singularity of death. It is folly to think or grieve.

In this story, we have masked revelers who Prospero invites into his system of chambers, caught in a process without variety, without a choice of input or output that compels movement from east to west. This is not the ancient Egyptian salvation pathway from west to east in "Descent." Unknown to Prospero, there is only one possible outcome—Red Death, so there is no circularity, no self-regulation based on feedback, no observation involving decisionmaking. Prospero's machinic system operates similar to a black hole with the Red Death comparable to a crushing singularity.

Picture this: you inspire a foolish friend to go down a spiraling, non-rotating, spherical black hole. You persuade the brash fool to insert a blinking blue bulb in his mouth so that you can track his progress. From a distance, you watch your thoughtless friend approach the black hole's horizon, and the blue bulb changes, blinking through the colors of the severed rainbow from blue to red like the experience of the revelers waltzing through Poe's color-coded chambers. You also notice the interval between the blinking increases like Poe's ebony clock that chimes at tonal intervals of nine, ten, eleven, as it approaches the maximum interval of twelve tones. Your foolish friend is now a reddish color. Eventually, the light disappears as it is stretched into longer wavelengths such as infrared, microwave, radio. Your ill-advised friend is crossing the horizon, and the blinking essentially freezes like the Red Death's mask of a stiffened corpse or the stiff-frozen dreams or the final frozen pendulum clock. Thus, time halts in both the black hole and Poe's story. Unfortunately, your friend's fate at the singularity is spaghettification, where the gravitational pull is so intense amidst a shifting of light to longer wavelengths, that he sees a gravitational redshift that could be described as a Red Death or an avatar of the embodiment of blood. This is because the horizon is an infinite redshift surface (Jacobson 1996), and your friend did not enter at the axis. Like the measured chimes of the ebony clock, the slowing down of the blinking light illustrates time dilation. The stronger the gravitational pull, the slower a clock ticks, and Poe emphasizes the strength of the gravitational pull by stating that the life of the ebony clock died with the revelers. This east to west system mirrors red gravitational death at a singularity. In our space-time, Poe is identifying the specific logic that functions within biological human existence and physical nature: a fatal disease or plague, related to a viral molecule of the air, and the rotating black hole with its driven gravitational function that drives everything to its singularity for disposal. Poe designs this radical framework for the élite without knowledge, those who do not grieve or think. However, Poe explains the "only pathway between Time and Eternity" in "Descent," showing that he understood the axial path, while saving a simple fisherman. More evidence for this view is Poe's allusion in "Masque" to Victor Hugo's play "Hernani" (1830), where the outcome is still death but not a default to the Reaper.

Hernani (1830) by Victor Hugo

The storyline in "Hernani" centers on the competition for a young sweet girl named Dona Sol between her rich noble uncle and the bandit Hernani, who also turns out to be a nobleman of sorts. Hernani visits Dona Sol only by the cloak of night, earning the name thief in the night, for he attempts to steal Dona Sol's love. Hernani persuades Dona Sol that her rich uncle cannot give her youth, but only royal jewels. Hernani asks her if she would follow him to a land of men who look like devils in her dreams. Strangely enough, Dona Sol knows that if Hernani dies, she will also die.

Hugo describes Hernani as a stranger travelling the forests, a rebel, and a mountain lion. The character Hernani describes himself as a resistless energy of doleful mysteries bound in darkness, rather than a rational thing that perceives his goal and moves straight to it. Hernani explains that he is being pushed by an impulsive gale that dashes him into a deep chasm, red with blood or flame where all things are crushed or die. Because of his fated path, Hernani fears for the life of Dona Sol, persuading her to marry the duke. Hernani then tells Dona Sol:

> My soul burns. Ah, tell the volcano to smother its flame—the volcano shall close its gaping chasms, and rim its sides with flowers and green grass. For the giant is held captive, Vesuvius is enslaved; its lava-boiling heart must not affect you. It is flowers you would have? Very well! Then the spitting volcano must do its best to burst with blossom!

Flowers are symbolic of the mystic centre, beauty and spring. In Hugo's play, descriptions of spitting volcanoes bursting with flowers relative to ideas of youth suggest time reverse, creation, and white hole dynamics of casting out matter. Humans are actually volcanic products made of water and volcanic elements, while our atmosphere is affected by volcanic eruption. Also, references in the play to a small world and tiny escape doors for Hernani suggest a microcosmic world.

Regarding the name Hernani, it is interesting that *nanus* or *nani* in Latin means "dwarf, midget." Here, we discover that Hernani is also a thief in the night like the Red Death, yet Hernani competes for the heart of a young sweet girl named Dona Sol (Latin: *donum* or *doni*: "gift, sacrifice;" *sol*: "sun"). Dona Sol (sacrifice of sun) dies with Hernani in this play. Yet, Her-nani or "her dwarf" describes himself as a "resistless energy" and is thus the machinic counterpart of a death-to-life transformation, rather than the hopeless life-to-death scene found in Poe's "Masque of the Red Death." So Poe has presented two types of death: total annihilation by Reaper Default in "Masque" due to ignorance, while alluding to an alternative death in "Hernani," that is, a spitting volcano of flowers and youth or a white hole spewing out matter like a black hole running backward in time.

Dona Sol responds to Hernani, saying that he is her heart, and Hernani is renewed. Dona Sol has no interest in her uncle the duke, and she reflects upon her joy with Hernani in the rose-perfumed, cloudless night, listening to her beloved's celestial voice. Similarly, the renewed Hernani states, "Your tones are a song that has nothing human left in it," suggesting that Dona Sol has sacrificed her humanity. Her nonhuman song reminds one of the poem "Of Mere Being" by Wallace Stevens, whose palm at the end of the mind at the edge of space is the landing site of a gold-feathered bird that sings a foreign song without human meaning or feeling. Also, Dona Sol wishes for a single nightingale singing or a flute, what suggests a bridge to the stars similar to the dynamics of Mozart's Magic Flute. This idea of a bridge to the stars can be explained as an microscopic Einstein-Rosen bridge, a rotating black hole connected to a white hole that operates similar to a black-hole volcano spitting out rocks.

In summary, Edgar Allan Poe's ideas in "Maelström," the nonfiction "Eureka" and "Masque" with its allusion to Hugo's play "Hernani," detail the dynamics of rotating black holes and the expanding/collapsing cosmos. In "Masque," we compared Poe's chamber rooms with a black hole pathway attracting all to the singularity of the cloaked nothingness of the viral Red Death. Time stops at a rotating black hole singularity; however, it can also be reversed there if one enters at the axis of the black hole, as we learned from Poe's "Descent into the Maelström." Unfortunately, the revelers in Poe's "Masque" lack this knowledge, so they have no choice but to default to the Reaper of death. These concepts mirror ancient Egyptian ideas about DNA immortalization and the "second death," as well as modern ideas on black hole dynamics, demonstrating the unity of knowledge in history. Edgar Allan Poe understood the basic creative function of expansion and contraction that operates in our universe and on earth. He understood the holographic mode of operation of our universe—that what can happen on the microcosmic level can also occur on the macrocosmic level. For the Diamondhearts, Poe is a subcatcher who can extract biophysical survival messages from his subconscious mind, that is, knowledge from his chromosomal apparatus.

Baseball is a red-blooded sport for red-blooded man.
It's no pink tea, and mollycoddles had better stay out.
It's a struggle for supremacy, survival of the fittest.

Ty Cobb

I have observed that baseball is not unlike a war,
and when you come right down to it,
we batters are the heavy artillery.

I had to fight all my life to survive.
They were all against me . . .
but I beat them and left them in the ditch.

Ty Cobb

Pregame 20

The Diamondheart Pitchers

It is no longer a question of imitation,
nor duplication, nor even parody.
It is a question of substituting the sign of the real for the real.

Jean Baudrillard
The Precession of Simulacra

Theoretical physicist Amit Goswami is the Diamondheart starting pitcher for the Game of the Centuries because he is pitching the strong idea that consciousness plays a role in evolution, since an organism can choose the mutation it is going to manifest. Now, this is what is clearly described in ancient Egyptian texts; however, the choice is limited to either a quasi-hybrid being or dissolution. Amit explains that Cairns and colleagues (1988) discovered this phenomenon of directed mutation in an experiment where *E. coli* bacteria selected the mutation they would produce. Other researchers have verified the experiment. Even a simple bacterium has an aliveness, Amit explains, and this indicates consciousness, which functions by quantum measurement or choice. Amit understands the proteins and nucleic acids (DNA) as the measurement apparatus. Survival is a conscious property of the cell. Put simply, consciousness is the ground of being, not matter. (1997)

Quantum physics predicts probabilities rather than defines singular events, whereas quantum consciousness is the phenomenon of being conscious of an experience at the level of the simplest element or cell. Actually, other scientists support quantum consciousness. Stuart Hameroff and Roger Penrose present a model of consciousness stemming from microtubule functioning at the cellular level, inclusive of brain neurons. Henry P. Stapp also proposes a theory, believing that quantum mechanics can accommodate consciousness (1995). Physicists David Bohm and Basil Hiley suggest that our consciousness has a rudimentary mind-like quality at the level of particle physics, which becomes stronger and more developed at subtler levels (1993, 386).

Recent experiments have shown that quantum consciousness is not localized; it is nonlocal. Quantum objects influence each other instantly. Amit easily compares quantum consciousness to the Godhead of esoteric spiritual traditions. He believes that consciousness is the measuring device that chooses from its own possibilities. He advocates quantum activism or changing your worldview from a materialist perspective to one based on quantum physics and the primacy of consciousness. Amit argues that this is the only way we can solve social problems such as global warming, terrorism, wars, and violence. He also discusses the problems of capitalism, democracy, liberal education, and religion. Capitalism is not bringing capital to everyone, but only the rich few. Democracy is for the people, not for political manipulation by a few—the media and rich corporations. Liberal education prepares students for jobs others design, rather than for exploring meaning. And religion and science?

> Religions have become less the institutions for investigating and disseminating spirituality and spiritual values that they are supposed to be, and more like other mundane institutions lost in the search for power to dominate others. Why else should religions be so interested in politics? And last but not least, even in the face of an aborning paradigm offering serious theory and replicated data, the scientific establishment seems to be least interested in freshly examining, let alone changing, their worldview based on scientific materialism according to which all is matter; consciousness, God, and, indeed, all our internal experiences are epiphenomena, secondary to matter. (2011, 8-9)

Relative to the scientific establishment, Amit is referring to replicated data such as Cairns' experiment showing that mutations arise when they are useful or adaptive. In a review of adaptive mutation, Patricia Foster concludes that the role of selection in adaptive mutation stops "a process that is creating transient variants at random" (1993). According to the Isis Thesis interpretation, the dead king's choice to become a quasi-hybrid being stops the production of random transient variants or human beings, who are directed by their viral genome, rather than by reasonable thinking that could end war and other atrocities. Relative to reasonable thinking, Amit urges scientists to change their materialistic worldview by looking at the replicated data. Yes, Amit the Quantum Activist will be a revolutionary starting pitcher in the Game of the Centuries.

Middle Relief Pitcher: molecular biologist Mark Ptashne.

Molecular biologist Mark Ptashne is an avid enthusiast of microbial processes. For this reason, he should excel at catching daisy cutters or batted balls that skim along close to the bacterial ground. In a book entitled *Genes and Signals* (2002), Ptashne and biologist Alexander Gann explain that control of gene expression operates by simple principles. In their chapter called "Lessons from Bacteria," they target the protein-DNA interactions of the *lac* (lactose) genes as well as the genetic switch of phage Lambda (λ), where cooperative binding between proteins loops the DNA to accommodate the binding and make the switch highly efficient. These proteins or enzymes catalyze chemical reactions within cells. As the products of genes, enzymes and structural proteins determine cell architecture and behavior. Expression of the gene is the process yielding the enzymes and proteins, but not all genes are expressed in all cells all the time. Often, signals from the environment determine which genes are expressed in a cell. Genes can be turned on or off by proteins. This was the discovery in the 1950s of François Jacob, Jacques Monod, Andre Lwoff and others at the Institute Pasteur in Paris. They studied two examples: our gut bacterium *E. coli's* ability to use the sugar lactose, and the lifestyles of phage Lambda, both of which are activated by environmental signals. Now, the Isis Thesis interpretation supports that this correlated viral-bacterial gene expression network provides the quantum niche for human transformation at a death transition. Relative to this environmental niche, Ptashne and Gann provide the following brief summary of gene expression activated by environmental signals related to *E. coli* use of lactose and the switching on of phage Lambda's lytic lifestyle, where the phage rises from its dormant lifestyle on the *E. coli* host chromosome to take over the host's replication machinery and clone phage particles.

> These two apparently unrelated cases have something in common: in each, genes that previously were silent (turned off) are expressed (turned on or activated) in response to specific environmental signals. The gene encoding the enzyme β-galactosidase is silent until its substrate—lactose—is added to the medium; the gene is then transcribed (turned on) and the enzyme is synthesized. In the λ example, a set of about 50 phage genes (the so-called lytic genes) can be maintained in a bacterium in a dormant (unexpressed) state called lysogeny. Those genes are transcribed in response to ultraviolet (UV) irradiation. (2002, 5-6)

As mentioned earlier, in the absence of its normal diet of glucose, *E. coli* will use lactose. The lactose regulatory protein called the Lac repressor protein is a shapeshifter with two shapes. Allostery (meaning "other shape") is a mechanism by which the protein changes shape. (2002, 7) The shapeshifting Lac repressor can bind the DNA to prevent lactose metabolism in the cell when glucose is used by *E. coli*, or it can change its shape and bind allolactose (a metabolic derivative of lactose) to allow lactose metabolism in the cell. When it binds the DNA to stop lactose metabolism, the Lac repressor loops the DNA in the shape of the Egyptian ankh (see Figure 1.2). Ptashne and Gann explain that allostery is a very important mechanism by which signals can be interpreted by organisms (7).

Now, if the Isis Thesis interpretation of ancient Egyptian texts is correct, a human death transition is a process where the organic DNA of the dead king is part of the activating protein complex that interprets, regulates, and expresses the lactose genes marked by the sign of the Lac repressor protein looping the DNA in the shape of the Egyptian ankh. In the viral self-assembly process, the organic DNA contribution is probably the uptake of a reverse transcriptase, another environmental signal operating with lactose entry and UV irradiation.

So, the proteins (polymerase, enzymes, repressors, transcriptional activators) and genes described in lactose metabolism and the Lambda lytic system are correlated networks, while ancient Egyptian deities are signs for various bacterial and viral proteins in these correlated gene regulation networks. In other words, the proteins cooperate like a team.

To understand the Egyptian human-viral transformation process, we must understand the biological signs showing that we function in a similar fashion. Ptashne and Gann describe that team of proteins and the intricate binding of Lambda's two repressor proteins—cro and c1.

> In some cases, it is not just a pair of proteins but multiple proteins (or multiple copies of the same protein) that interact and bind cooperatively. In such cases, the higher the number of such units that bind cooperatively, the steeper the curve describing the binding as a function of the protein concentration. We encountered such a steep curve in describing the cooperative binding of λ repressors to two DNA sites. In that case, the repressor monomers must form dimers, which then must interact to form a DNA-tethered tetramer, and that reaction proceeds roughly as the fourth power of the monomer concentration. The reaction thus proceeds in an "all or none" fashion over a small concentration range. (2002, 178)

In the Game of the Centuries, Gann will back up Ptashne in the Diamondheart line up, and they will bring to the game the same cooperative attitude of "all or none" that Lambda proteins exhibit in folding and binding to their gene-seats.

Middle Relief Pitcher: Washington Matthews (1843-1905), ethnographer and linguist on American Navajo Indians.

The ceremony of the Mountain Chant—literally, chant towards (a place) within the mountains, is practiced by Navajo shamans and re-enacted in their Nine Night Ceremony and sandpaintings. According to the myth as told by Irish ethnographer Washington Matthews (1887), the Navajo hero ascends a hill to observe beautiful mountain peaks, and seeing this beauty, he feels lonely and homesick, singing "That old age water! That flowing water! My mind wanders across it." After being captured by the Ute tribe, he shows them how to deceive deer with a mask and capture them. In time, he escapes the Ute, descends a cliff by means of a spruce tree with the magical help of a supernatural who advises the hero to retreat to the "yonder small holes" of his divine dwelling. The Supernatural then blows a strong breath, creating a high-speed rainbow to cross the canyon to enter the hole of his dwelling. However, the entrance of the hole is too small, so the Supernatural blows on the little hole as "it spread instantly into a large orifice, through which they both entered with ease."

This myth is reflected in the Nine Night Ceremony Matthews witnessed on October 21, 1884, where the Navajo created four sandpaintings that mirror the mythical events. Matthews first observed the construction of a sod-covered conical lodge with an eastern door, where upright eagle-feathered plumed wands, the collars of beaver skins, and symbols of wings to be worn by the couriers were placed. Then the Navajo ground pigments for the sandpaintings, writings that are erased in the patient healing process. In Matthews' account, the chanter informs him that the plumed wand of eagle feathers is "a means of rising," the wing symbols on the arm "will bear you onward," a beaver skin collar is "a means of recognition," and the wand is a sign of "coming from a holy place" (Matthews 1887, 43).

On the last night of the Nine Night Navajo Ceremony, Matthews witnessed the creation of a large Dark Circle of Branches about forty paces in diameter with the exterior corral of branches eight feet high, opening to the east which enclosed "sacred ground." A central fire was ignited, and dancers always circled clockwise around the fire as they performed their dances. During the ninth night, of interest is the sixth dance of the "standing arcs" performed by eight dancers who "each bore in front of him, held by both hands, a wooden arc, ornamented with eagle plumes." As in the third dance, the dancers proceeded four times around the central fire in a line. The lead dancer had a whizzer (stick tied to end of string) that he constantly whirled, producing a rainstorm sound. Then, kneeling in two rows facing each other, the first of the eight dancers advanced to the man who knelt opposite him, placing his arc upon the man's head. Matthews describes the sixth dance as follows:

> When they stopped in the west, the eight character dancers first went through various quadrille like figures, such as were witnessed in the third dance, and then knelt in two rows that faced one another. At a word from the rattler the man who was nearest to him (whom I will call No. 1) arose, advanced to the man who knelt opposite to him (No. 2) with rapid, shuffling steps, and amid a chorus of "Thòday!" placed his arc with caution upon the head of the latter. Although it was held in position by the friction of the piñon tufts at each ear and by the pressure of the ends of the arc, now drawn closer by the subtending string, it had the appearance of standing on the head without material support, and it is probable that many of the uninitiated believed that only the magic influence of the oft-repeated word "Thòday" [Englished "Stand" or "Stay"] kept it in position. When the arc was secured in its place, No. 1 retreated with shuffling steps to his former position and fell on his knees again. Immediately No. 2 advanced and placed the arc which he held in his hand on the head of No. 1. Thus each in turn placed his arc on the head of the one who knelt opposite to him until all wore their beautiful halo-like headdresses. Then, holding their heads rigidly erect, lest their arcs should fall, the eight kneeling figures began a splendid, well timed chant, which was accentuated by the clapping of hands and joined in by the chorus. (1887, 155-56)

Could this intriguing Navajo behavior have a possible microbiological parallel? If one guesses that the Dark Circle of Branches is a sign for viral rolling circle replication, a type of DNA replication where a replication fork moves around a circular DNA molecule, unrolling a single-stranded concatamer (substrate) for bacteriophage head assembly, then the circling dancers may represent the single-stranded concatamer, while the Dark Circle is a sign of the circular DNA molecule. This ceremonial dance seems to be a fairly good re-enactment of lytic rolling circle replication, for David Dressler (1970, 1934) explains that the replication of viral DNA involves a double-stranded (ds) circle synthesis followed by a period of single-stranded (ss) circle synthesis.

If the dance is a re-enactment of rolling circle replication, what is the significance of the clamping of arcs on the heads of eight dancers in two rows? According to Tanner and colleagues (2009), who observed rolling circle replication based on flow-stretching of bacteriophage λ (Lambda) DNA, as the replication reaction proceeds, the DNA attaching the circle to the surface is extended and stretched fully by hydrodynamic flow of buffer. A dramatic increase in rate and processivity is seen due to increasing the temperature as the circle "rolls." Tanner and colleagues continue, "DNA replication is a fundamental biological process that requires the coordinated activities of a large number of enzymes organized in a multiprotein assembly termed the replisome." Matthews' description of the two rows of dancers (sign of double-stranded DNA) having arcs clamped on their heads (phage head assembly in rolling circle replication) seems to mirror the processive DNA synthesis performed by a replisome. Johnson and O'Donnell (2005) explain this clamping process in their abstract:

> DNA replicases are multicomponent machines that have evolved clever strategies to perform their function. Although the structure of DNA is elegant in its simplicity, the job of duplicating it is far from simple. At the heart of the replicase machinery is a heteropentameric AAA+ clamp-loading machine that couples ATP hydrolysis to load circular clamp proteins onto DNA. The clamps encircle DNA and hold polymerases to the template for processive action. Clamp-loader and sliding clamp structures have been solved in both prokaryotic and eukaryotic systems. The heteropentameric clamp loaders are circular oligomers, reflecting the circular shape of their respective clamp substrates.

So perhaps the Sixth Dance on the ninth night represents viral rolling circle replication, circular phage head assembly, and the replisome clamping machine that advances the replication fork. Yet, the symbolism of this dance becomes more interesting relative to Matthews' observation of a homosexual group simulation of canine coitus during the Dark Circle of Branches ceremony, which was originally suppressed from Matthews' text (see King 2009b). Still, Navajo ritualistic behavior suggests that the ninth night Chant ceremony of the Dark Circle of Branches with its central fire models not only the thermodynamic Kerr black hole ring singularity, but also the viral DNA-arc of rolling circle replication, circular phage head assembly, and replisome activity. So, it is possible that this Navajo ritual as witnessed by Matthews is an imitation of viral DNA dynamics relative to the ancient gene expression network for the lytic cycle of phage Lambda, another reminder of the survival adaptation potential within us. Other gene expression rituals performed by humans such as the Catholic Mass can also be interpreted as signs of lytic viral replication.

Set-up Pitchers: SSB Trio of J. M. Schwartz, Henry P. Stapp, and Mario Beauregard.

The directed attention and mental effort of a human being can systematically alter brain function, and so, this ability has applications in neuroscience and psychology for alleviating habitual behavior. This willful act is called self-directed neuroplasticity, a scientific way of understanding the interface between mind/consciousness and the brain. Neuroplasticity is based on quantum mechanical causal mechanisms (Schwartz et al. 2005). In various studies, the brain area activated through self-directed regulation of emotional response is the prefrontal cortex. This clear-minded introspection and observation or meditative practice has been described in Buddhist texts, and it is not a trance-like state. Schwartz explains:

> It is, in fact, the basic thesis of self-directed neuroplasticity research that the *way in which a person directs their attention* (e.g. mindfully or unmindfully) will affect both the experiential state of the person and the state of his/her brain. The existence of this close connection between mental effort and brain activity flows naturally out of the dynamic principles of contemporary physics, but is, within the framework of classic physics, a difficult problem that philosophers of the mind have been intensively engaged with, particularly for the past 50 years.

Schwartz and colleagues explain that from the perspective of classical physics, the slideshow of thoughts, ideas and feelings are either epiphenomenal by-products of brain activity (mind emerges from brain or matter) or these conscious experiences are the same pattern of motion in the tiny parts of your brain. As Amit Goswami explained earlier, quantum mechanics allows a conscious choice from possibilities.

In *The Mind and the Brain* (2002), neuropsychiatrist Jeffrey Schwartz and Sharon Begley explain their concerns over materialistic claims that free will died with Freud and that the mind does not exist. Against materialistic determinism, they describe how neuroplasticity means new neuronal paths in the brain, new roles for the brain, and changing brain function. They argue that we are not machines. (2002, 300) So, without the pressure of biopower, perhaps free will or concentrated mental effort can restore us to a status where we are not slaves to our genes.

Using brain imaging technology Jeffrey Schwartz, theoretical physicist Henry Stapp and neuroscientist Mario Beauregard support that quantum physics clearly supports the role of directed attention or mental effort in brain functioning, whereas classical physics relies on an older view about fundamental ideas of the natural world that have been shown to be incorrect during the last seventy-five years (2005). So, based on the Copenhagen Interpretation of quantum theory as extended by von Neumann, mental action can change brain activity in a way that works with the laws of quantum physics, that is, consciousness can collapse the wave function as von Neumann believed. As an example, Beauregard and colleagues used cognitive behavioral therapy for phobia of spiders. By changing the person's mind, changes were observed in the individual's prefrontal cortex. (Paquette et al. 2003) Now, this task of modifying emotional and cerebral responses is challenging, and it involves redirecting lower-level limbic responses into higher-level prefrontal functions (Schwartz et al. 2005, 1312). The limbic system is primarily responsible for emotions and memory. In other words, memory-based emotional responses such as fear of spiders can be redirected to the prefrontal lobe, where planning complex cognitive behavior, decision making, and moderating social behavior is possible. On the other hand, the causal deterministic approach of classical physics presents humans as automatons, not active observer-participants. As the SSB Trio puts it, "Thus the choice made by the observer about how he or she will act at a macroscopic level has, at the practical level, a profound effect on the physical system being acted upon" (2005, 1315).

Many scientific experiments also justify the Quantum Zeno effect that allows repeated observations at short intervals to freeze a particular state. Wheeler's observer-participancy principle also functions within quantum mechanics in a manner similar to the brain's neuroplasticity. In addition, we must factor in that the brain operates holographically and can hold large increments of time and space (Persinger and Koren 2007). Finally, all this new science mirrors the ancient Egyptian, early Chinese, and Navajo scientific knowledge that grounded their mindsets through repetitive ritual. So, perhaps the propaganda of biopower can be overcome by willful decision making based on scientific knowledge and a clear focus. Nonetheless, in the Game of the Centuries, it is going to be very difficult to beat a pitcher that monitors his neuroplastic brain. If the directed attention and mental effort of a pitcher can systematically alter brain function, then perhaps the following question can be put to nature: which team is going to win the Game of the Centuries?

Pregame 21

The Closers

Therefore, the doubts which have arisen in your heart out of ignorance should be slashed by the weapon of knowledge. Armed with yoga, O Bhārata, stand and fight.

Swami Bhaktivedanta, *Bhagavad Gītā*

The closing pitcher or closer is the relief pitcher who closes the game by getting the final outs. A closer or the team's best reliever is also called a fireman or a stopper. Today in the Game of the Centuries, we have two closers: the Judaeo-Christian God-man Jesus Christ for the Thunderheads and for the Diamondhearts, the ancient Egyptian Sun-god Osiris. These two players are already in History's Hall of Fame. Both Christ the Sun of Justice and Osiris of the Hall of Justice are saviors specializing in fiery apocalyptic closure, and both have interesting historical roots, not to mention fast balls that have struck human history's bleeding heart.

By circa 2520 to 2360 BCE, according to Erik Hornung's timetable (1999), the Egyptians carved the first Pyramid Texts, detailing the mythology of the dying/rising deity Osiris and the milk-goddess Isis, who restores Osiris to life and conceives their child Horus by virgin birth. Hieroglyphic knowledge was not translated until 1822 CE, when Champollion decoded the hieroglyphic alphabet. However, it is difficult to believe that Gnostics, Christians, alchemists, and historical scholars were oblivious to the meaning of the hieroglyphic alphabet. English translations (Faulkner 1969; 1973-78) of the core Egyptian myth from the least-corrupted Pyramid Texts and Coffin Texts include the following themes: 1) the dead king merging with the Sun-god or Light or Sun-bark; 2) brother rivalry of Seth and Horus; 3) Osiris dead; 4) Osiris risen; 5) virgin birth of child Horus by Isis; 6) the cross; 7) the great flood of millions of light-people. Despite ancient Egypt's isolation and cryptic hieroglyphs buried within pyramids and coffins for centuries, this mythology surfaced in other cultures and finally in the Bible as Mary's virgin birth of the dying/rising Christ. And so, we have the syncretistic migration of religious ideas from ancient Egypt to early Christianity (see King 2005), or perhaps the repetitive survival messages from our viral chromosomal apparatus that we all share.

As an example, in 1945 at Nag Hammadi, the discovery of thirteen buried codices containing some fifty Gnostic texts allowed Egyptian knowledge to re-surface, along with a clearer understanding of the ideological battle between the Catholic Church and the Gnostics. Recall that Gnosticism, a religious movement in the first centuries of the common era, supported self-knowledge, and its followers were called "people of the light" (Barnstone and Meyer 2003). The Gnostic texts define what the early Catholic Church considered heretical. Both Gnostic (Greek: *gnosis* "knowledge") and Catholic scriptures center on Christ's life, but each revelation of the Osirian themes had important variations. As in Egypt, the light played a central role, for the Gnostics found self-knowledge in their common experience of the Pleroma, the fullness of pure light with the "perfect exalted human" within, who is "self-conceived" and the race of perfect humankind (2003). Similar to the Osirian themes, the Gnostics emphasized the value of milk and the perfect virgin, Christ and his cross, the lord's thought caught on a wheel, "as if a body were on the wheel, a head turned down to feet," and the immortality of perfect beings.

To the Gnostics, evil was the female principle of earth, sexual genesis and trapping light in the body, whereas evil to the Church was lust, self-love and preventing life. The Gnostics believed Christ is a sign, whereas the Catholic Church believes Christ is a real suffering historical human that is also divine. This literalism, where signs are interpreted as real people or events, is an early Christian trademark (Harpur 2004; Sullivan 2001), yet the Gnostics advise against taking the sign literally. Their Gospel of Philip states, "The names of earthly things are illusory. We stray from the real to the unreal" (Barnstone and Meyer 2003). However, with the Catholic Church's victory of Constantine's conversion in 306 CE, Gnostic texts were buried with their heritage of Osirian microbiology that also became literalized in Catholic doctrines of virgin birth, incarnation, the cross, resurrection—all signs masking a viral genetic option for possible continuity of quasi-conscious DNA. Song 26 of Solomon recovered in 1945 states: "The interpreter perishes, the interpreted stops" (Barnstone and Meyer 2003). Although the Catholic Church misconstrued abstract biological concepts as real material events and people, the Gnostics understood some of the meaning present in ancient Egyptian texts because their Gnostic Pleroma is "knowledge about the depth of the universe" (Barnstone and Meyer 2003), suggesting the idea of the fullness of pure light enfolding the whole cosmos in a quantum embrace. So, the debate here for closure is whether or not the savior Jesus Christ is a real God-man or another sign of Osiris the virus crystal.

When interpreting the mysteries of ancient Egypt, reliable explanations can be constructed using not only cultural studies or anthropology, but also modern psychology and psychic research. Since we are tracing the linear progression of specific human mental experiences grounding both scientific and religious thought from Egypt to our postmodern era, Carl Jung's (1875-1961) psychological interpretation of alchemy based on his theory of archetypes in the collective unconscious is necessary to determine if

Christ is a real God-man or just another sign dredged up from our nonconscious to be misinterpreted by biopower. To address this debate, let's briefly consider ancient Egyptian ideas streaming into alchemy, Carl Jung's archetypes of the collective unconscious related to ancient Egypt's core mythology, and Jung's psychological interpretation of alchemy. So, let's begin with a sixteenth century painting that showcases Christ with a crystal orb, a sign that he is ready to play ball.

Figure 21.1. *Salvator Mundi* (1502-08). Public domain.

Ancient Egyptian ideas streaming into alchemy. Infrared reflectography of the newly emerged *Salvator Mundi* (1502-08) reveals characteristic signs of Leonardo da Vinci's idiosyncratic technique. The scientific method and traditional art-historical arguments convince many art historians that Leonardo created the work. Martin Kemp, professor of art history at the University of Oxford, UK, supports that the painting depicts Christ as Savior of the World, while the optical effect in the crystal orb is reflective of Leonardo's work. Kemp claims the orb is not a representation of our terrestrial globe, but rather a translucent rock crystal, glistening internally with little points of light, suggesting that the domain of Christ's father extends to the whole cosmos. (2011, 174-175) Similarly, according to André J. Noest of Utrecht University, Netherlands, the orb depicts an idealized celestial sphere (2011). However, the meaning of the chemiluminescent crystal ball in the painting may also rest in its title *Salvator Mundi* ("Savior of the World"), suggesting that the spherical crystal itself is the Savior of the World, for the alchemical Stone (*lapis*) of the Philosophers was referred to by many names, one of which was *Salvator* (Jung 1953, 223) or Savior.

As the Isis Thesis interpretation of ancient Egyptian texts has pointed out, the Savior of the world is a spherical viral crystal, a world-heart that embodies two chemical reactions, allowing both photosynthetic matter on earth and bio- or chemiluminescent emergence in the cosmos. However, it is only the human being with knowledge that can transform to a crystallized quasi-hybrid being. Leonardo's Christ in *Salvator Mundi* is symbolic of any human being who can grasp the knowledge of natural laws permitting viral transformation of DNA. Leonardo's crystallized orb, alchemy's Philosopher's Stone, and the dead pharaoh's starry transformation save the world by permitting continuity of quasi-conscious DNA. In the painting, Christ is pointing upward with his right hand, while holding the crystal in his left hand next to his heart, suggesting that the crystal is the key to the heavens and the sign of everyone's full potential.

Jung's Archetypes. Jung's archetypes of the collective unconscious mirror the microbiological pillars of ancient Egypt's core mythology. In his Commentary on the Secret of the Golden Flower, a Chinese text concerned with ancient esoteric teaching in the days of Lao-Tzu and the religion of the Golden Elixir of Life (1958, 302), Jung informs us that this Chinese text parallels the psychic development of his non-Chinese patients. He then defines the collective unconscious as follows:

> In order to make this strange fact more intelligible to the reader, it must be pointed out that just as the human body shows a common anatomy over and above all racial differences, so too, the psyche possesses a common substratum transcending all differences in culture and consciousness. This unconscious psyche, common to all mankind, does not consist merely of contents capable of becoming conscious, but of latent dispositions toward identical reactions. Thus the fact of the collective unconscious is simply the psychic expression of the identity of brain-structure irrespective of all racial differences. This explains the analogy, sometimes even identity, between various myth-motifs and symbols, and the possibility of human understanding in general. (308)

Our brain structure is composed of cells with similar DNA chromosomes. Perhaps Jung is talking about what wave geneticist Peter Gariaev terms our quasi-conscious chromosomal apparatus. Within Jung's collective unconscious exist archetypes, what Jung defines in his 1957 Preface to *Psyche and Symbol* (1958) as "inherited forms of psychic behavior," not "inherited patterns of thought." In consideration of Peter Gariaev's research defining the human genome as a nonlocal "quasi-text form" that can read itself in billions of our cells and use that information to control living functions and structural organization, Jung's idea of archetypes as inherited psychic behavior of the collective unconscious can be directly correlated with our viral chromosomal apparatus that guides our living functions now. If this is so, then Jung's archetypes or "inherited forms of psychic behavior" should match up with the viral lifestyles of phage Lambda, as proposed by the Isis Thesis interpretation.

The microbiological pillars of the pharaonic priesthood's core mythology define their viral-based scientific literature, artwork, architecture, and rituals. All of these Egyptian signs speak to us about the gene-regulated, protein-DNA interactions or behavior of the complex bacteriophage Lambda within an *E. coli* bacterial cell along the ancient glycolysis-fermentation pathway. Our inherited psychic behavior simulates phage Lambda's genetic switch or gene regulation network, supporting recent scientific views that gene regulation networks are the controllers of development. In *Ingenius Genes* (2011), Roger Samson proposes that the most designed system in an organism is its gene regulation network that controls development. However, the Egyptian signs also inform us that with knowledge at human death, we can become the controllers of our development by controlling the gene regulation network of phage Lambda.

Today, in light of the Human Genome Project, perhaps we can supersede Jung's term "brain-structure" with human-cell, the habitat of the human genome and its epigenome (genetic marks, tags or switches determining how the book of DNA is read) that exists not only in our brain cells, but in every cell in our bodies. Perhaps Jung's archetypes of the collective unconscious are the inherited psychic behaviors of the non-coding, viral-bacterial component of our genome that seems to order our conscious mental experiences. In the interests of learning more about our microbiological birthright, let's review the similarities between the microbiological pillars of the core Egyptian mythology and Jung's archetypes of the unconscious that he finds in alchemy. Perhaps Christ is an archetype and his crystal ball is the Savior.

Inherited Forms of Psychic Behavior. In each of our cells, our genome houses a complete set of chromosomes, which carry 2 percent instructions for coding proteins in our bodies, along with a 98 percent complex toolbox of lingual structures designed to reach some still unknown goals, according to researchers (Mantegna et al. 1994). Acquired from our environment, the epigenome houses chemical compounds that modify or mark our genome with additional instructions that influence our behavior. Although these marks are not part of the DNA itself, they can still be passed from generation to generation. Our genome and epigenome are both inherited, and together, this chromosomal apparatus may be motivating our psychophysical behavior (myth, religion, capitalism, and so on).

Now, Jung explains that archetypes are "living entities" belonging to the realm of instincts and represent "inherited forms of psychic behavior." In light of experimental research, today we must concede that our DNA and epigenome guide human intentionality. Jung's archetypes of the collective unconscious are surprisingly similar to the Egyptian core mythology that can be decoded as quantum biology. As an example, Jung's archetype of the Self is the sum of human wholeness, what is conscious and unconscious. This mirrors the holistic transformation of the dead Egyptian Sun-god, who embraces the unified cosmos for continuity of mind and full human potential. In his book *Psychology and Alchemy*, Jung states unequivocably that empirical observation shows that the archetypes of the unconscious are equivalent to religious dogmas (1953, 17). Both Buddha and Christ are symbols of the Self. However, in contrast to the Christian symbol of Christ who is light and good split from the Devil who is darkness and evil, the Self represents the union of opposites, good and evil, light and darkness.

Another Jungian archetype is the "lowly origin of the redeemer" that can be compared to Egyptian creator-gods and other deities that represent lowly viruses and bacteria. Recall that Christ was lying in a manger, a trough from which horses and cattle eat. Certainly, this humble setting is microbial. We must remember that lowly microbes—viruses and bacteria, have made us who we are today, while leaving us the viral DNA toolbox that may allow transformation to a quasi-hybrid being at a death transition. So, the "lowly origin of the redeemer" correlates with the creative acts of the lowly viral bacteriophage that can re-animate or redeem us.

Further, Jung identifies the archetype of the "image of spirit imprisoned in the darkness of the world" that points to the inert deity Osiris and Christ's death on the cross, supporting a quantum biological interpretation related to the inactivity of the bacterial virus Lambda in a dormant, lysogenic or prophage state on its DNA-cross. Another archetype is the "idea of transformation and renewal by means of a serpent" (1953, 137), which mirrors the supreme concept of transformation or morphogenesis mediated by serpentine viral proteins in two thousand years of ancient Egyptian texts. Closely connected to the serpent is the Ouroboros, the dragon devouring itself tail first, or what Jung states is the "basic mandala of alchemy" (120). As the Isis Thesis substantiates, Osiris bent backward in circular form is a sign for Lambda rolling circle replication of circularized DNA, as well as time cycling back to its origin in the early cosmos.

Likewise, the lytic process of Lambda rolling circle replication occurs in the quantum mesocosm. In mythology the archetype of spirit as the Wise Old Man represents knowledge, reflection, insight, and intuition (Jung 1958, 77). The connection with the unconscious reveals itself when the Wise Old Man appears as a dwarf, such as "the crafty dactyls of antiquity, the homunculi of the alchemists, and the gnomic throng of hobgoblins, brownies, gremlins, etc." (78) Jung continues:

Archetype of Self
The Sun-god or human potential as both matter and transformation to a quasi-hybrid being

Archetype of Lowly Origin of Redeemer
Ancient Egyptian deities are signs of viral and bacterial DNA and proteins (lowly microbes)

Archetype of Image of Imprisoned Spirit
Inert deity Osiris (virus Lambda as prophage)

Archetype of Transformation/Renewal by Serpent
Transformation by winged serpents or winged proteins

Jung's Basic Mandala of Alchemy
Dragon devouring itself tail first, the ouroboros, or Osiris bent backward in a circle (transformation by viral rolling circle replication)

Archetype of Wise Old Man or Dwarf
Ancient Egyptian "regions of the dwarfs" (quantum domain)

Archetype of Birth of Golden Child
Birth of the golden Horus child or Perception/Mind in early cosmos (the quasi-hybrid being)

> I have often encountered motifs which made me think that the unconscious must be the world of the infinitesimally small. . . In the same way, the archetype of the wise old man is quite tiny, almost imperceptible, and yet it possesses a fateful potency, as anyone can see when he gets down to fundamentals. The archetypes have this peculiarity in common with the atomic world, which is demonstrating before our eyes that the more deeply the

investigator penetrates into the universe of microphysics, the more devastating are the explosive forces he finds enchained there. (79-80)

Jung explains that in certain myths, the archetype of spirit as the Old Man is identified with the Sun, an idea similar to the ancient Egyptian Sun-god. Often possessing a partly chthonic side, Jung identifies the archetype of the spirit in a Balkan tale as the Old Man losing an eye the hero must restore. This tale mirrors the fate of Osiris who lost an eye due to Seth, as well as the story of Wotan, who lost his eye at the spring of Mimir. (80-82) In the case of Osiris, the lost eye must be restored by the dead king, who reunites Seth's Moon Eye to Horus' Sun Eye. This protein partnership restores the function of six gene-seats to the whole viral genome via the genetic switch to lysis.

Probably the most convincing parallel with ancient Egyptian mythology and Christianity is the wide-spread archetype of the birth of the golden or divine child (Jung 1953, 158) or the child-god in the guise of the dwarf or hidden forces of nature (1958, 120). Often seen as the treasure hard to attain, the child archetype is represented as "the jewel, the pearl, the flower, the chalice, the golden egg, the quaternity, the golden ball, and so on." (122-23) In Egyptian texts, the golden child Horus is a quaternary form made of lapis lazuli, and it represents the quasi-hybrid being. As Jung explains, the "preconscious, childhood aspect of the collective psyche," the child-motif is not a human child, but a divine child, pointing to an irrational causality similar to the mother and father archetypes (124, footnote 18). The archetype is the link with the past, the "roots of consciousness" (123), similar to the Egyptian core mythology which links the dead king to the original creator-god Atum at the origin of time. So, the child-motif exists in both the distant past, while functioning as a system in the present to correct the extravagances of the conscious mind (125-126). Also, Jung states that the majority of the cosmogonic gods are bisexual, suggesting that the hermaphroditism of the child represents the union of opposites. In ancient Egyptian texts, the union of opposites is signified by the hybrid sphinx and the transformation to the quasi-hybrid being.

Although this is a mere sketch of Jung's archetypes of the collective unconscious that he believes are rooted in the past and microphysics, one can easily recognize that his archetypes (the Self, the lowly redeemer or imprisoned spirit, transformation/renewal via a serpent, spirit as the Wise Old Man, and the birth of the golden or divine child) mirror the mythology of ancient Egypt and the religion of Christianity. In turn, Jung's archetypes, Egyptian mythology and Christianity are signs that can be compared to the microbiology of phage Lambda that also mirrors microphysics, that is, quantum physics. However, we are still exploring the question of whether or not Christ is a real man who died on a cross and rose from the dead or a sign of a sleeping virus waking up. So, in light of Christ in *Salvator Mundi*, a quick review of Jung's psychological interpretation of alchemy should be helpful in determining whether or not Christ is a sign or a real person.

Jung's Psychological Interpretation of Alchemy. According to Jung, the ultimate aim of alchemy was to produce a transfigured, resurrected body that was also spirit, an aim that was similar to the Chinese alchemists' desire to attain the diamond body in *The Secret of the Golden Flower* or immortality through physical transformation. Jung agrees that the diamond is an excellent symbol, and he cites the sixteenth century commentator Orthelius, who describes the blessed stone as hard, fiery, and transparent. (1953, 408) A symbol, according to Jung, is the primitive expression of the unconscious and also an idea mirroring the highest intuition of consciousness (1958, 326). So, symbols point to the archetypes in the unconscious that surface as conscious ideas.

Jung believes that the existing evidence substantiates that the alchemists understood they were transforming themselves into "living philosophical stones" (1953, 141). From the *Rosarium*, Jung mentions that the *lapis* is not just a "stone" since its composition is "*de re animali, vegetabili et minerali*," and that of body, soul, and spirit (170). According to Jung, the alchemist had psychic experiences of a chemical process as he experimented. In this experience, the alchemist encountered his own unconscious, tapping into the "history of man's knowledge of nature." (234) Today we can identify this history as our immense microbial heritage.

Reviewing the work of Heinrich Conrad Khunrath, Jung explains that the "salt is not only the physical centre of the earth but at the same time the *sal sapientiae*," so feelings, senses, reason and thoughts should focus on this salt alone (1953, 245). Viruses not only crystallize like salt, but they have also been referred to as salt in the twentieth century. Jung also concludes that the secret of the art of alchemy lies hidden within the human unconscious (244-245).

In his psychological interpretation of alchemy (1953), Jung explores the qualities of the *lapis*, claiming that alchemy is an undercurrent in Christianity, and this seems correct for Christ is holding a crystal *lapis* in Da Vinci's painting. Jung also explains that the central ideas of Christianity are rooted in Gnostic philosophy. Also, the view of the philosophers is that this perfectly round Stone or *prima materia* (primary material or cause or LUCA) is invisible, ethereal, ubiquitous or everywhere in nature, arising from a rotating aquasphere or chaotic whirlpool. Similarly, bacteriophages are everywhere in nature, and their viral proteins generally fold in funneled energy landscapes.

Jung also explains that Western alchemy originated from Egypt, and that the actual transformation points to a person's "innate psychic disposition" resulting from ancestral life down to the animal level (1953, 128-129). He explains that the human psyche possesses layers below and possibly above consciousness:

> In my experience the conscious mind can only claim a relatively central position and must put up with the fact that the unconscious psyche transcends and as it were surrounds it on all sides. Unconscious contents connect it *backwards* with physiological states on the one hand and archetypal data on the other. But it is extended *forward* by

> intuitions which are conditioned partly by archetypes and partly by subliminal perceptions depending on the relativity of time and space in the unconscious. (132)

Thus, we can understand the archetypes rising from the unconscious psyche as survival messages, as well as the possibility that a large portion of our genome may represent Jung's collective unconscious, a reservoir of archetypal physiological memories based on an ancient gene expression network in our cells used by a virus for replication.

Jung claims that the "increasing differentiations of ritual and dogma alienated consciousness from its natural roots in the unconscious," yet alchemy and astrology preserve this bridge to nature or the unconscious psyche. Jung observes that alchemy existed for over seventeen centuries, and many alchemists focused on chemical substances or the process of making gold, while others were concerned with symbols and their psychic effect. As supported by modern experiments, perhaps reputable alchemists understood that mental images with intention can produce a specific psychological effect due to quantum mechanics and the properties of DNA.

Jung also observes that Plato's original human beings have spherical form in *Timaeus*, and the alchemical gold is also spherical. In ancient Egyptian texts, humans are depicted in artwork with spherical heads and pharaohs sport long ceremonial tails that may be the inspiration for the modern apparel of men who sport tuxedo tails and tail-like ties on their necks below their heads. Spherical forms and heads, ceremonial tails and ties suggest the head-tail morphology of a virus. Perhaps this is why Christianity anathematized ideas about spherical heads.

Jung claims that the alchemic literature supports intelligence is necessary, since one assumes that a "species" capable of transforming matter exists in the human mind (1953, 248). Jung then clearly draws the bacterial connection to faeces, where *E. coli* resides with its Lambda prophage:

> The substance that harbours the divine secret is everywhere, including the human body. It can be had for the asking and can be found anywhere, even in the most loathsome filth. (300)

Jung also correlates the Catholic Mass with alchemy, and the product of the bread and wine transformation as the living "stone" rather than Christ.

> By pronouncing the consecrating words that bring about the transformation, the priest redeems the bread and wine from their elemental imperfection as created things. This idea is quite unchristian; it is alchemical. . . . Therefore, what comes out of the transformation is not Christ but an ineffable material being named the "stone," which displays the most paradoxical qualities apart from possessing corpus, anima, spiritus, and supernatural powers. One might be tempted to explain the symbolism of alchemical transformation as a parody of the Mass were it not

pagan in origin and much older than the latter. (299-300)

So, Jung understood both the "stone" as a being and the correlations between the ritual of the Mass and ancient alchemical praxis. Wine and bread are fermented products, suggesting the ancient fermentation pathway that produces the quasi-hybrid being or crystal stone. Jung also mentions that the fifteenth century alchemist and Benedictine monk Basilius Valentinus believed the earth is not a dead body, for it is inhabited by a living spirit similar to a reflective mirror, from which all created things receive their strength (329). In ancient Egyptian texts, earth is a sign of an *E. coli* cell, and the living entity inside is the inert Lambda prophage, a viral crystal. To Jung, the ultimate aim of alchemy is similar to Chinese alchemy: "to produce a *corpus subtile*, a transfigured and resurrected body, i.e., a body that was at the same time spirit." (from the text of *The Secret of the Golden Flower*)

Christ's Crystal Ball. So, Christianity is rooted in Gnostic philosophy with an undercurrent of alchemy that is rooted in early China and ancient Egypt. Let's see if we can reassess Da Vinci's painting.

To begin, DNA is a dynamic crystalline structure, proteins fold to a crystallized native state, and a virus, such as phage Lambda, is a crystal having a spherical shape. Recall that the Latin, *crystallus* means "crystal" and *chrysos* means "gold" and *Christus* means "Christ." From the Greek, *christós* means "anointed," a translation from the Hebrew *māshīah*, "anointed, messiah." Earlier we discussed the connection with cream, milk, lactose, Isis, and anointing. We also discussed that chrysalis is the third stage in the development of an insect, especially a moth or a butterfly enclosed in a firm case or cocoon, a pupa. The word "chrysalis" originated from Latin *chrysallus* and Greek *khrusallis* meaning gold-colored pupa of a butterfly. A chrysalis or cocoon stage suggests the lysogenic inert state of phage Lambda. These etymologies are charged with microbiological messages, stemming from the lingual structures of our viral DNA with its survival message. So, Christ's activities and his name connote that he is a sign of a crystal virus rising from its inert state because it is "anointed" with the creamy oil of lactose.

Relative to the painting, the crystal ball Christ holds may be the *Salvator* rather than Christ himself, who now represents the Self as the sum of human wholeness, what is conscious and unconscious, human and viral. Did you notice that Christ's long hair curls into the shape of spiral DNA molecules? Can we doubt that Leonardo Da Vinci understood the evolutionary message? Both the crystal ball and Christ are signs of the crystallized quasi-hybrid being. The presence of the crystal indicates viral transformation, so Christ's identity as a material human has been stripped from him, in the same way that Christ lost his humanity in Revelation by only saving a few human beings. At worst, Jesus Christ who died and resurrected is not a real man born of a virgin, and at best, Christ is an archetype of the Self and another viral sign like Osiris of transformation and survival of human DNA at a death transition. Nonetheless, because of

Christ's physical resilience in history and because he holds the spherical crystal itself—the Savior of the World, the alchemical Stone (*lapis*) of the Philosophers or *Salvator* (Jung 1953, 223), Jesus Christ is hereby nominated for a place in the Shrine of the Eternals of the Baseball Reliquary next to the Golden Bull, sign of the milky way of the star-kissed goddess in ancient Egypt.

As we shall see, what is interesting about the alchemical process is that it reflects not only the process of isolating and purifying a virus, but it suggests that the old alchemists were attempting to use viruses as vectors for cloning like our modern genetic engineers. So, the quantum microbiology of a virus may be the underlying order grounding human behavior on earth, as well as emergence to a quasi-hybrid being. As Foucault might have thought, this philosophic alchemic science "has been deprived of a certain metaphysics because it has been separated off from the space of order, yet doomed to Time, to its flux and its returns, because it is trapped in the mode of being of History" (1973, 320).

What the Isis Thesis supports and proposes in light of human genome research—is that the archetypes of the unconscious are grounded by the quantum biological model of phage Lambda, that is, the archetypes are a sign of our viral DNA heritage in our microbiome and the 98 percent of the human genome that does not build proteins for the human body, but offers humans a potential reservoir for switching proteins on and off for the emergence of a quasi-hybrid being described as crystallized mind. With the least-corrupted Egyptian mythology in mind, as well as Jung's archetypes mirroring Lambda microbiology or quantum biology, let's look at real individuals in the Alchemic Hall of Fame in light of Jung's psychological interpretation of the recurring archetypes that support this proposal. The question we must ask is—if Jung's archetypes mirror major motifs in ancient Egyptian myth, Christianity, and alchemy that can be interpreted as microbiology, then perhaps the alchemic process that was practiced for centuries involves the isolation and purification of a virus.

When he hit it, I knew that it was my ball. But I had to catch it
and it seemed like the hardest catch of my life.
I said to myself, "Two hands, just like your dad taught you."

Darin Erstad
on catching the final out in Game 7
that won the 2002 World Series for the Angels

Welcome to Fly Ball Ferris Wheel 22

The Reputable Alchemists

An anticipatory system is a natural system that contains an internal predictive model of itself and of its environment, which allows it to change state at an instant in accord with the model's predictions pertaining to a later instant.

Robert Rosen
Anticipatory Systems (1985)

This section presents other prominent individuals with knowledge who have caught the home-run fly balls of history that were slammed into the stadium of life by ancient Egypt.

Today, few people believe that the mystical alchemical philosopher's stone was some magical substance that turned metals to gold, cured human disease, and bestowed the individual with power over nature. Yet, reputable alchemists claim the stone possessed all of these attributes. Generally, alchemical activities were kept secret and veiled in codes and riddles. Alchemy was the past-time of Isaac Newton and many other educated individuals who experimented with the stone, the agent of universal transmutation. Again, if Jung's archetypes mirror major motifs in ancient Egyptian myth, Christianity, and alchemy that can be interpreted as microbiology, then perhaps the alchemical process that was practiced for centuries is a chemical process similar to isolating and purifying a virus. If so, the profound, mystical inner experience that humans describe as spiritual gold or the presence of God may actually be the perception of the nonconscious origin of human intentionality—a bacterial virus.

In 2012, Michigan State University researchers Kazem Kashefi and Adam Brown, involved in what they call microbial alchemy, have recreated a process in nature by using a metal-tolerant bacterium called *Cupriavidus metallidurans* to produce gold nuggets in a glass bioreactor (MSU 2012). Other microbiologists, aware that metals have stabilized phage particles, are probing the viral metallome by searching for metalloproteins in phage Lambda, a virus with high metal-binding potential (Zhang et al. 2011). Actually, the preliminary evidence supports that microbiological methods in modern genetic engineering are identical with the methods of reputable alchemists during the seventeenth century, such as Isaac Newton and Dr. Michael Maier, who were also experimenting with metals and microbes in their laboratories.

Reflecting a two-sided mirror, alchemy's practitioners aimed to transmute base metals into gold, while creating spiritual gold. Carl G. Jung supports that alchemy exhibits a double face: practical lab chemical work and a psychological process (1953, 258); however, alchemy also involves a third aspect that relates directly to its double face. Similar to modern genetic engineers, the reputable alchemists were actually microbiologists, for in their labs they were inducing transformation in bacteria, a fundamental evolutionary process that also obsessed the pharaonic priesthood, as shown by their funerary literature. Alchemical behavior or microbial experimentation is consistent with the predictions of the Isis Thesis because the same quantum biological message inciting human behavior keeps recurring in our history. Put simply, the viral/bacterial heritage of our genome grounds our psychology and behavior, both conscious and nonconscious. Like the ancient Egyptians and early Chinese cultures, the seventeenth century alchemists were exploring the *Materia Prima* (first material) or the metal-tolerant heart of an abundant natural creation, the viral bacteriophage. Their quest for the Philosopher's Stone centered on both chemical and psychological gold.

As we shall see, the reputable alchemic process that was practiced for centuries is a chemical process similar to isolating and purifying a virus. It is possible that the *Materia Prima* of the reputable alchemists was the genetic regulation network of phage Lambda undergoing lytic multiplicity thanks to the Lac system in *E. coli.* In the seventeenth century laboratory, alchemists armed with microscopes may have recognized the lifestyles of phage Lambda as the spiritual gold that resonated with their nonconscious origin of intentionality. They were not only mining spiritual gold, but their own nonconscious origins. They became conscious of their nonconscious and its related heritage of specific signs mirrored by the ritual of alchemical praxis. A virus was the first material, the origin, the world-heart, the LUCA, the *Materia Prima* of creation, and the reputable alchemists understood this. They were exploring the rapid growth of bacteria, while isolating and purifying the viral bacteriophage. Although this seems like a difficult proposition to prove, it is not because the existing alchemical evidence is viewed through the lens of contemporary science. But first, let's examine this idea of isolating and purifying a virus in the interest of learning more about the system of alchemy and its roots that have been buried in what Baudrillard has called "history's dustbin."

History of Alchemy. Pharaonic Egypt is the mystical homeland of alchemy and the birthplace of its legendary father Hermes Trismegistus. In an attempt to comprehend the Egyptian meaning of the hieroglyphs, the Greeks valued the pictorial and allegorical expression of these abstruse signs as sacred knowledge (de Rola 1996, 8-9). Originating in the first centuries of the common era, Greeks, Arabs and later Christian mystics practiced alchemistic symbolic processes that were a blend of psychological magic and rational science. Even the fundamental concepts of *The Secret of the Golden Flower* that are concerned with ancient esoteric Chinese teachings, reflect a symbolism similar to an alchemical process of "refining and enobling" with darkness giving birth to light, and the "lead of the water-region" growing into noble gold, as the unconscious becomes conscious (Jung 1958, 321). Recall that the early Chinese Emperors aspired to become stars, and in Huainanzi 4, Section XIX, this process is summarized as the "growth of ores in the earth, vaporization, and condensation" related to gold, lead, copper, silver, iron (Major 1993, 214-215). So, the Chinese Emperor becoming a star in the afterlife via a metallic transformation can be associated with the alchemists' laboratory objective.

The Diamond Body. Jung also mentions another example from the beginning of the *Hui Ming Ching*: "The most subtle secret of *tao* is essence and life." Jung explains that *tao* is composed of the Chinese characters for "head" and "going"(1958, 317). Egyptian texts also focus on the way to light (the Sun-god) and heads (the outcome of transformation). Jung then cites the introductory verse from the *Hui Ming Ching*:

> If thou wouldst complete the diamond body without outflowing,
> Diligently heat the roots of consciousness [or essence] and life.
> Kindle light in the blessed country ever close at hand,
> And, there hidden, let thy true self forever dwell.

The diamond body or the crystal virus is the root of consciousness, for evidence suggests that Chinese alchemical process of "refining and enobling" refers to the isolation and purification of a metal-binding virus, which is then crystallized as the "diamond body without outflowing," an idea that suggests the virus did not multiply or clone itself through rolling circle replication. It may be helpful to know that in 1898, Martinus Beijerinck asserted that the virus was somewhat liquid in nature, calling it "*contagium vivum fluidum*" (contagious living fluid). He coined the Latin name *virus* meaning "poison" for the substance infecting tobacco plants. Yet in 1616, the reputable alchemist Michael Maier, M.D. wrote *Lusius Serius* (Serious Passe-time), a "Philosophical Discourse concerning the Superiority of Creatures under Man." Maier's first-person Latin version is about Mercury, the metal (and creature) liquid at room temperature:

> Pulverized with gold or any other body (so it be not corrosive or noxious), I am the best purgative Nature hath given us. This dust has the name and attribute of Aurelian, and is a Panchymagôgon (a Generall Medicine). . . By certain processes, i.e., by mixing mercury with certain salts &c., I am turned into a poison called Praecipitate. Of itself, mercury is an antidote against the plague and other diseases. (Craven 1910, 58)

Like Maier, Carl Zimmer in *A Planet of Viruses* (2011) discusses the ubiquity of bacteriophages in our world as well as their amazing medicinal properties. For example, the Soviet Union has been using bacteriophages for over a century to treat infections, and Western researchers are now considering phage therapy because the clever bacteria are collectively resisting our antibiotics (Reardon 2014).

Both ancient Egypt and early China possessed magnifying glasses, as did thirteenth century scientists, such as Roger Bacon (1214-1294), who is credited with the discovery of the glass lens in 1268 for eyeglasses. Roger Bacon developed a theory of the multiplication of species or successive generation of light (1240?). According to historian David C. Lindberg (1997, 249), Bacon believed "a species is elicited in the recipient out of potentiality that already exists there." Bacon had access to the works of Avicebron (died circa 1058), who describes a force (vis), ray (radius), or likeness (species) emanating from a "First Maker" (1997, 245). Citing his translations of Bacon's *Perspectiva* (1996) and *De multiplicatione specierum* (1983), Lindberg describes Bacon's production of the species of light "in the first part of the air" contiguous to the second part and so on, the generation multiplying "according to the dimensions of the air." This sounds like Bacon was creating copies of DNA through the polymerase chain reaction (PCR) that involves four steps: 1) denaturation, 2) extension, 3) insertion of gene, and 4) amplification of the inserted gene into hundreds or millions of copies.

Lindberg mentions Bacon's argument that this generation can occur along straight, reflected, refracted, twisting, and accidental lines (1997, 250). In *Perspectiva* (1266?), Bacon explains that the "twisting path" for transmission of the species occurs "by the power of the soul" (Lindberg 1997, 255). Lindberg also discusses Bacon's ideas on the stronger perpendicular passage of a species on a transparent interface versus the easier oblique passage, as well as the images produced by the reflection and refraction of light on a mirror surface (252-255). These explanations support microscopic Kerr black hole dynamics relative to the inner axial horizon acting like a holographic mirror. Of course, Kerr black hole dynamics are similar to Lambda protein folding landscapes. Now, Roger Bacon also practiced alchemy, and he was regarded as one of the first advocates of the scientific method. He also wrote *Speculum Alchimiae – The Mirror of Alchemy*, which was published later in 1597 (White 1998, 112).

By the seventeenth century, alchemy had peaked during the so-called Scientific Revolution. During the late 1500s, the first microscope was developed in England, and it was capable of observing bacteria and single-celled animals. In 1597, Zacharias Janssen and his son, Han Janssen made further advances to the microscope, inventing the compound microscope that was modified later in the seventeenth century by Robert Hooke. By 1625, the microscope was used throughout Europe, and papers were published, such as Robert Hooke's *Micrographia* (1665) and Marcello Malpighi's proof of William Harvey's blood circulation theories (1660). In 1676, Anton van Leeuwenhoek, a Dutch microscope-maker, observed and described what he called animalcules, that is, protozoa and bacteria in pond water, rainwater, and human saliva. In 1677, he described human and insect spermatozoa. (Hume 2008) In light of scientific knowledge during the seventeenth century, it does not seem unreasonable that alchemists were attempting to imitate the generation of metals in nature. However, early modern alchemists were secretive about their work, so this science was for the privileged person, as it was in ancient Egypt and early China.

Sir Isaac Newton (1643-1727). In an alchemical manuscript entitled *Manna*, Newton explains alchemy:

> For alchemy does not trade with metals as ignorant vulgars think, which error has made them distress that noble science; but she has also material veins of whose nature God created handmaidens to conceive & bring forth its creatures. (quoted in White 1998, 139-40 from KCL, Keynes MS 33 fol. 5v)

Along with the quest for transmutation of metals, from the early 1670s to the mid-1690s Isaac Newton explored the composition and structure of matter, investigating the mysteries of chemical transformation by means of alchemical experiments. In Babson College Library, Mass., Babson MS 420, page 19, Newton writes:

> Artefius tells us that his fire dissolves & gives life to stones, & [word deleted] Pontanus that their [illegible words deleted] fire is not transmuted with their matter because it is not of that matter, but turns it with all its faeces into the elixir. Which deserves well to be considered. For this is the best explication of their saying that the stone is made of one only thing. (quoted in White 1998, 250)

Although biographer Michael White believes these were the words of a "man on the edge of madness," faeces is a legitimate area of research for early alchemists as well as modern genetic engineers because it is the homeland of the *E. coli* bacterium with its inert phage Lambda. Recall that when the phage is active, it produces numerous phage progeny. White then includes another excerpt from the same treatise by Newton, who achieved "multiplication" like his predecessor Roger Bacon during the thirteenth century:

> Thus you may multiply each stone 4 times & no more for they will then become oils shining in the dark and fit for magical uses. You may ferment it with gold by keeping them in fusion for a day, & then project upon metals. This is the multiplication in quality. You may multiply it in quantity by the mercuries of which you made it at first, amalgaming the stone with the mercury of 3 or more eagles and adding their weight of the water & if you design it for metals you may melt every time 3 parts of gold with one of the stone. Every multiplication will increase its virtue ten times &, if you use the 2d and 3d rotation without the spirit, perhaps a thousand times. Thus you may multiply to infinity. (Ibid p. 18a)

Regarding Newton's coded message, multiplication indicates amplification of DNA and "oils shining in the dark" suggest bio- or chemiluminescence. When Newton died, copies of nine publications by the influential German alchemist Michael Maier (1569-1622) were in his library, one of which was Maier's *Secretioris naturae secretorum scrutinium chymicum* (Frankfurt, 1687). For example, stages in the manufacture and purification of the philosopher's stone involved the transformation of so-called black lead into a pure white substance, as well as this substance's further refinement through heating of metals in tall laboratory crucibles. Citing works of Arabic alchemists Geber and Rhazes, Maier refers to a series of amalgamations of lead, tin, and other metallic compounds (Mandelbrote 2001, 109-113), as well as discussing a chemical process of washing away faeces.

Maier's *Secretioris naturae secretorum scrutinium chymicum* was an incomplete 1687 reprint of *Atalanta Fugiens*. The text aspired to justify alchemy with rationalism, an interest of Newton's due to his enthusiasm for alchemy (White 1998, 121-122). Many alchemists were charlatans, but still, others were reputable, such as Robert Boyle (1627-1691), who believed it was possible to transmute base metals into gold. In 1680, he stated that "there exists conceal'd in the world of a much higher order able to transmute baser metals into perfect ones." Perhaps he is referring to the metallic binding power of a higher-ordered bacteriophage such as phage Lambda with its transition metal composition.

According to Michael White (1998), in the *Opticks* Newton describes his experiments in 1664 related to refraction, reflection, rainbows, interference, and the behavior of mirrors and prisms. Newton speculates that light operates like gravitation via forces acting at a distance. (288) Recall Dr. Luc Montagnier's experiments, where he copies DNA fragments at a distance, for the DNA is located a short distance away from the creation of copies using a PCR medium. Perhaps Newton and Luc Montagnier were conducting similar experiments at a distance of 400 years apart. Dr. Montagnier's breakthrough is that he copies DNA fragments, at a distance, without adding the DNA fragment into the PCR medium. What Dr. Montagnier has shown is that the shadow of DNA, where that DNA is located at a short distance away, is enough to create copies of that DNA in a PCR medium.

Unaware of late twentieth century microbiological research, Michael White concludes:

> If we are to appraise alchemy in the cold, clear light of the late twentieth century, we should acclaim it for giving the world some useful tools and techniques still used in modified form today, and, most significantly, for inspiring a train of thought in at least one great philosopher of the seventeenth century. With that we should be content. (1998, 130)

To show the limitations of White's modern view of alchemy and to test the proposition that the alchemical process involved a chemical process of isolating and purifying a higher-ordered virus, we must compare alchemy with modern genetic engineering.

No one hit home runs the way Babe did. They were something special.
They were like homing pigeons. The ball would leave the bat,
pause briefly, suddenly gain its bearings then take off for the stands.

Lefty Gomez
(teammate of Babe Ruth)

Fly Ball Ferris Wheel 23

Alchemy's Viral God

Was helffen fakeln licht oder briln, so die levt nicht sehen wollen.
("What use are torches, light or eyeglasses, if people will not see?")

Vignette from Heinrich Khunrath, *Amphiteatrum sapientiae*

In *Clef universelle des sciences secretes* (1950), P. V. Piobb describes the alchemic process. First is the extinction of interest in life and the world (calcination), followed by a separation of the destroyed remains (putrefaction), resulting in purified matter (solution). Next was the refinement of the salvation elements or "rain" of purified matter (distillation), the joining of opposites (conjunction), the suffering from mystic detachment from the world due to spiritual striving (sublimation), and finally—philosophic congelation, a binding together of fixed and volatile principles. This adheres to the formula *Solve et Coagula* and Piobb's interpretation is "analyse all the elements in yourself, dissolve all that is inferior in you, even though you may break in doing so; then, with the strength acquired from the preceding operation, congeal" (quoted from Piobb in Cirlot 1971, 6-8). So, we have a psychological interpretation here by Piobb, for alchemists were attempting to transform matter into spirit or the Philosopher's Stone, a process similar to the transformation of the dead Egyptian king to *lapis lazuli* or the Morning Star, as described in the Pyramid Texts and reflected by the Chinese transformation to a diamond. Now, Piobb's chemical description without the psychological emphasis is the following process:

1) calcination or heating a substance to reduce it to ashes
2) putrefaction or separating destroyed remains by isolation and filtration, resulting in purified matter or
3) solution, followed by
4) precipitation or the "rain" or distillation of purified matter (crystallization)
5) conjunction or the joining of opposites
6) sublimation or the process of transformation from a solid to gas phase, without passing through the liquid phase, and congelation or the Philosopher's Stone.

Keeping Piobb's description in mind, let's look at Desmond Nicholl's contemporary description of the isolation of DNA and RNA in his *Genetic Engineering* textbook (2002,

27-30). He explains that the general process involves opening the cells to expose the nucleic acids, separating the nucleic acids from other cell material, and recovering the nucleic acid in purified form. Then Nicholl outlines a very similar step-by-step process in a slightly different order than Piobb's description minus the psychological emphasis:

1) Open the cell by degrading the cell wall and detergent lysis (bursting) of cell membranes
2) Deproteinisation or extractions/separation of protein molecules from nucleic acid;
3) Solution
4) Precipitation of nucleic acids from Solution with salt washes
5) Filtration technique such as gradient centrifugation or gel filtration
6) Pellet forms and is dried (evaporation) and resuspended for next experimental stage, which could be conjugation (transfer of genetic material from a "male" donor bacterium to a "female" recipient)

Stronger evidence exists for these same similarities in early modern chemistry and the work of Wendell M. Stanley, who received the Nobel Prize in Chemistry in 1946, for isolating and purifying the properties of the crystalline tobacco mosaic virus in 1935. The alchemic process is identical to Nobel Laureate Chemist Wendell M. Stanley's explanation in his *Nobel Lecture, December 12, 1946*. This experiment introduced chemists to the fact that a virus could be crystallized like an inorganic salt. Stanley reports:

> Subsequently, it was found that concentration and purification of the virus could be effected readily by means of a combination of isoelectric precipitation and salting out with ammonium sulfate. A description of this procedure, which yielded a crystalline material as the end-product, is contained in the following paragraph. (1946, 144)

Isoelectric pertains to a constant electric potential, which can be described as heat, light, motion. Heat separates and melts the double strands of DNA, and if copied by a DNA polymerase, the original sequence is duplicated. In addition, Stanley's isoelectric precipitation is a reaction often occurring in a liquid to form a solid or Precipitate via inconstant electric potential. So, we can compare Stanley's use of electric potential to the alchemical quest described by de Rola that uses the "Distillation Furnace" for "Precipitateness" (1996, 42) and "undergoing the action of Fire within the crucible" (44). As De Rola explains, the Stone is Triune with three principles, Mercury, Salt, and Sulphur, corresponding to various states of fluidity or volatility, such as "the Saline or crystalline state, the Volitile or humid state, and the Fixed or dry state, which the Matter will assume in repeated turns" (44).

Stanley continues, describing an initial filtration process followed by precipitation in sufficient water, involving 1) calcination, 2) putrefaction, isolation, filtration of the virus or 3) the solution, followed by 4) precipitation with lead subacetate and salt, and 5) dissolution in water. Stanley then explains that the filtrate will be opalescent, almost colorless, and contain eighty percent of the virus from the initial material. The protein crystallizes as the hydrogen ion concentration is increased, and when stirred, the crystals have a satin-like sheen. (1946, 145) It seems that Roger Bacon, Isaac Newton and other reputable alchemists isolated a virus centuries before Stanley, who concludes:

> As a whole, the results indicated that the crystalline material was, in fact, tobacco mosaic virus. Attention was therefore directed to the characterization of the crystalline material. The material was originally reported to be a protein and to contain about 52 per cent carbon, about 7 per cent hydrogen and about 16 per cent nitrogen, and was later found to contain, in addition, about 0.6 per cent phosphorus and 0.2 per cent sulfur. The fact that nucleic acid could be isolated from the crystalline material was reported by Pirie and coworkers in December, 1936, and by the writer a few days later. (1946, 147-148)

Notable similarities exist in the laboratory processes described by Piobb, Nicholl, Stanley, and those of the reputable alchemists. So, we can trace the interest in microbiology from ancient Egypt to early China to undercurrents in Gnosticism and early Christianity to thirteenth and seventeenth century alchemists to modern genetic engineers. Additional evidence such as Stanley's viral images in his *Nobel Lecture, Dec. 12, 1946,* also match seventeenth century alchemical descriptions of their product.

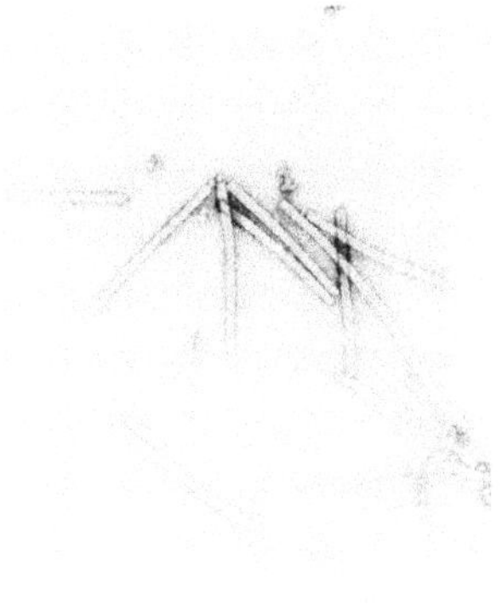

Figure 23.1. Tobacco Mosaic Virus magnified 160,000X. This public domain image is similar to Stanley's image in his Nobel Lecture.

Viewing Stanley's Figure 4, in his *Nobel Lecture, December 12, 1946*, one observes the crystals of the tobacco mosaic virus (from W. M. Stanley, Am. J. Botany, 24 (1937) 59) as thin lines of crystals in tree-like branches. Stanley's image of thin lines of crystals in tree-like branches mirrors alchemist Arthur E. Waites' description of the alchemic tincture in *Introitus apertus* as "tiny silver trees with twigs and leaves" and "minute grains of silver, like the ray of the sun" (quoted from Nigel Hamilton, 1985).

Similarly, Isaac Newton and Fatio de Duillier experimented with alchemy at Cambridge (UK), and in a letter of May 1693, relative to these investigations, Fatio describes a process in matter producing "mineral trees" that relates to the fermentation pathway:

> These matters being put in a sealed egg in a sand heat do presently swell, and puff up, and grow black and in a matter of seven days go through the colours of the philosophers. After which time there grows a heap of trees out of the matter . . . there is plainly a life and a ferment in that composition. (White 1998, 355-6 *Correspondence* Vol. 3, 265-7)

Thus, ancient Egypt, early China, early Christians, the reputable alchemists in the thirteenth and seventeenth centuries, and our modern genetic engineers such as Stanley, Nicholl, and many others demonstrate the human instinct to explore microbiology.

Excluding early Christians, all these cultural texts represent a model of the world that is governed by serious scientists exploring microbiology. In contrast, Christianity models the alchemic microbiology in the ritual of the Mass and the activities of their deities. This historical interest has created a microbiological trail of science from ancient Egypt to our contemporary genetic engineers, spanning 5,000 years of human history. Perhaps this behavior is because our genome is primarily viral, we originated from microbes, and our psyche is tuned to a creator-virus and its survival message.

The Modern and Alchemic Use of Mercury. One must remember that if microbes created photosynthesis, then they were also responsible for creating its reverse—bioluminescence, a type of chemiluminescence. In chemiluminescent reactions, energy is released as cold light. Whereas photosynthesis captures light energy from the Sun for thermodynamic life, chemiluminescence produces cold light energy such as starlight. Bioluminescence is evident in the quorum sensing abilities of bacteria. The quorum sensing signaling process is a method of intracellular communication that coordinates gene expression. If researchers interrupt quorum sensing, then they interrupt coordination of gene expression. Recently, Galloway and colleagues (2011) reviewed studies showing that bacterial virulence can be partially attenuated by inhibiting quorum sensing in gram-negative bacteria such as *E. coli*. For example, active compounds called furanones can interact with LuxR-type proteins to reduce quorum sensing and bacterial bioluminescence. LuxR protein is a transcriptional activator for quorum-sensing control of luminescence. Relative to the seventeenth century alchemists, they may have been using mercury, lead, and other compounds to reduce bacterial quorum sensing so that they could isolate and/or activate the silent viral prophage sleeping on the bacterial chromosome. Then multiplication of the species or rolling circle replication could be observed. As Galloway and colleagues explain, compounds of mercury and lead have exhibited significant quorum sensing inhibitory activity in studies.

Since seventeenth century alchemists were examining faeces, perhaps they analyzed the common *E. coli* bacterium under their microscopes, disrupting the *E. coli* cell membrane for uptake of bacteriophage DNA. Again, nature has conveniently provided *E. coli* with a special LamB receptor site for bacteriophage Lambda, a well-studied phage in molecular biology found in a Paris sewer by Andre Lwoff, François Jacob and Jacques Monod. In 1965, these researchers received the Nobel Prize for the operon theory for bacterial gene expression in phage Lambda. They found that certain *E. coli* strains irradiated by ultraviolet light produce viral plaques or small clear areas of bacterial lysis. This is because the bacteria are carrying Lambda lysogenically, that is, the phage is asleep on the bacterial cell chromosome. Recall that ultraviolet (UV) light damages the host cell DNA and decreases the level of cellular protein synthesis, while inducing the inert prophage to awake and take over the cell's replication system to clone progeny and lyse the cell, producing a clear plaque.

The alchemists used seven metals: gold, silver, iron, tin, lead, copper, and mercury.

The only metal existing in the liquid state is mercury, which had a special significance to alchemists. Robinson and Tuovinen (1984) report that inorganic mercury exists in three valence states: (i) Hg0 (metallic mercury); (ii) Hg2+ (mercuric mercury); and (iii) Hg+ (mercurous mercury). Numerous strains of *Escherichia coli* and other bacteria have all been found to volatilize (evaporate) mercury from added Hg2+. More evidence for this view that the reputable alchemists' use of mercury related to bacterial experimentation is the existence of the mercury resistance Mer regulatory system in the *E. coli* lac system and phage Lambda. This simply shows that mercury is an important element not only to seventeenth century alchemists, but also to modern microbiologists.

The Modern and Alchemic Use of Sulfur. Now, sulfur is an essential element to cells, and has been associated with ancient life-forms such as *E. coli* that possesses more than one hundred genes directly involved in sulfur metabolism (Sekowska et al. 2000). Also, the tail tip complex of phage Lambda assembles from eight different proteins, including gpL protein with its C-terminal domain coordinating an iron-sulfur cluster. Researchers claim that this is the first example of a viral structural protein binding to this type of metal group. (Tam et al. 2013) Relative to seventeenth century praxis, Basil Valentine states, "there is a nearer place yet in which these three, Mercury, Salt, and Sulphur—Spirit, Body, and Soul—lie hid together in one place well known, and where they may with great praise be gotten" (Craven 1910, 18). Could it be that the reputable alchemists may have understood the function of the circular DNA molecules (extrachromosomal elements found naturally in *E. coli*) called plasmids? A plasmid can confer traits such as antibiotic resistance on the host cell, while a conjugative plasmid mediates its own transfer between bacteria by the process of conjugation. The plasmid is a good cloning vector (carrier of DNA) because of its low molecular weight, antibiotic resistance genes, origin of replication and recognition sites. (Nicholl 2002, 61-62) When DNA is isolated, a plasmid can carry DNA fragments into a bacterial cell for propagation as clones. In a study by Dempsey and colleagues (1978), they explain that in *E. coli*, resistance to mercury is plasmid-determined. In their study researchers found a plasmid with phage Lambda inserted between the genes governing sulfonamide resistance (sul) and mercuric resistance (mer). If alchemists such as Basil Valentine had isolated an *E. coli* plasmid, then the reference to Mercury, Salt, and Sulphur hiding together in a place well known may reference the positioning of the Lambda prophage (Salt) between the mercuric resistance genes (Mercury) and the sulfonamide resistance genes (Sulphur) on an *E. coli* plasmid. Relative to Dempsey's experiment, if phage Lambda excises or exits from the plasmid, it can sometimes leave with plasmid genes, what is called specialized transduction. Dempsey and colleagues found that the *mer* genes were part of three Lambda phages isolated from the *E. coli* plasmid. The mer region of the plasmid functions by the mercuric reductase enzyme reducing Hg2+ to Hg0. This means that phage Lambda left with the resistance gene to mercury and the mercuric reductase enzyme regulated mer synthesis during phage infection.

On the *mer* operon in *E. coli*, Robinson and Tuovinen (1984) inform us that with Lambda transducing phages prepared from cointegrate isolates of Lambda and plasmid R100, all lysogens exhibited only inducible synthesis of the mercuric reductase enzyme. Researchers concluded that after an infection of *Escherichia coli* by one of the Lambda-mer cointegrates, mercury resistance is inducible and is paralleled by the induction of the mercury-detoxifying enzymes. Robinson and Tuovinen explain that today induction by exposure to Hg+ or an organomercurial is standard experimental procedure to minimize toxic effects while maximizing enzyme function. The *mer* operon encoding mercury resistance is in the plasmid R100 from *Escherichia coli.* It is a group of four genes: "*mer*A, which encodes the mercuric reductase enzyme; *mer*B, encoding the organomercurial lyase enzyme in broad-spectrum resistant strains; *mer*T, the gene believed to govern the Hg2+ uptake function; and *mer*R, which codes for the regulatory protein responsible for the inducibility of the system" (108).

In light of the alchemical emphasis on mercury and sulfur, this research suggests that the reputable alchemists may have understood mercury resistance genes, as well as the sulphur resistance genes on an *E. coli* plasmid, along with the iron-sulfur binding activity of phage Lambda's tail tip protein gpL.

Conclusions. So we can easily see that the alchemic process involved isolating and purifying DNA, viruses, and plasmids, as well as observing bacteria. Also, the alchemic emphasis on mercury, salt, and sulphur may have had a direct correlation with the mercury genes and sulfur genes in *E. coli* plasmids, as well as the tail tip sulfur-binding activity of the viral protein gpL. Further, the Salt may represent the virus, for a virus can be crystallized like inorganic salt crystals, and viruses are like salt crystals that grow and reproduce. Researchers at the State University of New York have recently found ancient bacteria called archaea alive in salt crystals in which they became trapped 34,000 years ago. Also trapped in the salt crystal are non-surviving algae, which provided the sugar alcohol glycerol necessary for minimal archaea sustenance. (Lam 2011) In addition, along with their microscopic observations of animalcules or bacteria, the thirteenth and seventeenth century scientists and alchemists may have been using modern techniques such as UV irradiation, the shadow-casting technique with the use of gold as a shadowing technique, sterilization by heating and filtration, and so on. However, the most interesting facet of alchemic praxis is the objective of multiplication, suggesting the process of viral cloning.

Fly Ball Ferris Wheel 24

Against the Vulgar

I have disclosed the Truth to You which I have gathered out of the monuments of the Ancients by incredible labour and the expense of many years.

Dr. Michael Maier
Atalanta Fugiens [1617]

To close this case study on alchemy, let's consider Dr. Michael Maier's alchemical work for a clearer understanding of evidence supporting that reputable thirteenth and seventeenth century alchemists were involved in bacterial transformation mediated by phage Lambda. According to J. B. Craven, Maier places material alchemy before spiritual or psychological alchemy (1910, 25), so Maier relies on science.

Stanislas Klossowski de Rola presents commentaries on alchemical engravings from the prolific seventeenth century. His intention is not to render a psychological interpretation, but to frame his comments within the spirit of the Hermetick Tradition (1988), keeping in mind the alchemical axiom *Solve et Coagula* ("Dissolve the Fixed and Coagulate the Volatile"). Yet, he understood that alchemic emblems with double meanings could secretly transmit esoteric knowledge. To follow is a brief summary of the prominent seventeenth century alchemists in de Rola's text.

François Beroalde de Verville, who studied every science, was born in Paris in 1556, and was the son of a Protestant theologian. His work: *Le Tableau des riches inventions* (1600) and *Le Voyage des princes fortunez* (1610).

Léonard Gaulthier (1561-1641), arrived in Paris, becoming engraver to three successive kings, Henri III, Henri IV and Louis XIII.

Heinrich Khunrath (died Dresden, 1605) Alchemist. *Amphitheatrum sapientiae aeternae* (1602). A Christian-Kabbalist, Divine-Magical, Physico-Chemical with a Privilege for publication given by emperor Rudolph II dated June 1, 1598.

Andreas Libavius. *Alchymia* (1606), born in Saxony 1540-1616; a graduate of medicine; a chemist credited with writing the first real textbook. Engraver was Georg Keller (born Frankfurt c.1568. died 1640 Frankfurt). Keller was also a painter and painted the Coronation of the Emperor Ferdinand III (1627) and religious subjects.

Steffan Michelspacher, a physician from Tyrol, little is known. *Cabala* (1616). Plates were designed by Michelspacher and engraved by Raphael Custos, or Custodis, the grandson of the Dutch painter Pieter Balten. Michelspacher died Frankfurt 1651.

Michael Maier. *Tripus aureus*, 1618 (Golden Tripod) Maier was Imperial Count Palatine, Free Knight of the Empire, Doctor of Philosophy and Medicine, formerly of His Imperial Majesty's Court.

These prominent individuals associated with the kings and nobility, and this élite group protected their alchemical knowledge. Born in Rendsburg during 1566, Michael Maier attended the University of Rostock, continuing on to Nuremberg and Padua. By 1592, his principal patron was Emperor Rudolph II. At the University of Frankfurt an der Oder, Maier held the title of *Poeta Laureatus Caesareus*. He received his doctorate in medicine at Basle and then moved to Prague to become Imperial Count Palatine in the Emperor's intimate circle. (de Rola 1988, 59) Obviously, Michael Maier and the other educated alchemists listed by de Rola traveled in élite circles of influence, and their passion for alchemy was appreciated by kings and emperors, as it was in ancient Egypt and early China.

A Model Viral System. Before reviewing Maier's contribution, a 2012 review of genetic recombination in phage Lambda by biologist Christopher R. T. Hillyar concludes with the following description of bacteriophage Lambda's importance.

In conclusion, the temperate bacteriophage λ, a pre-existing cellular structure or metabolic processes, has 'learned' partial independence from its prokaryotic host, *E. coli* (Hendrix et al., 2000). The absolute majority of genomes on Earth, these dsDNA-containing tailed phages outnumber bacteria 10-fold. Approximately 4500 pervasively mosaic phages infect a huge diversity of bacteria (Hendrix, 2002, 2003). Bacteriophage-bacterial transduction of Shigella Stx toxin has generated the enterohaemorrhagic *E. coli* strain O157:H7, a novel human pathogen associated with haemolytic-uremic syndrome (Zhou et al. 2010). Thus, genomic diversity and virulence generated by promiscuous recombination and extensive horizontal genetic exchange at legitimate and illegitimate sites between bacteriophages and bacteria has important implications for animal and human health and disease (Fishers, Hofreuter and Haas, 2001; Hendrix, 2003). Genetic recombination in bacteriophage λ should be further studied, as this model system continues to inform our understanding of the mechanisms underpinning these complex genetic exchanges that drive ecological evolution (Friedman and Court, 2001).

As one might surmise from reading this current assessment of phage Lambda, this bacterial virus with its ability for recombination with various cell-types is a dominant evolutionary driving force. It is possible that this virus was also the focus of alchemy. In *The Golden Game* Stanislas Klossowski de Rola describes the crystalline alchemical

Philosopher's Stone as both material and spiritual. The hermetic axion *Solve et coagula* consisted of repeated dissolutions followed by crystallizations of the "secret Subject of the Wise, the *Materia Prima* or Stone of the Philosophers." This process resulted in "a mysterious celestial influx" allowing extraordinary qualities, what de Rola describes as a spiritualization of matter followed by a materialization of spirit that produces the Stone, "a highly evolved substance capable of the most extraordinary effects."

> The Philosopher's Stone is the true Universal Quintessence, capable of transmuting all metals into gold; it is also called the Universal Medicine or Panacea, as it removes the very causes of diseases, and the Fountain of Youth, as it rejuvenates the organism and prolongs life beyond its normal span. (de Rola 1988, 19)

***Atalanta Fugiens* by Dr. Michael Maier.** In the interest of discovering if the Philosopher's Stone is a bacteriophage, let's review Michael Maier's *Atalanta Fugiens* (1617 published in Latin) to uncover evidence of microbiology or quantum biology. To begin, in Discourse III Maier explains that the Philosopher's subject "for whatever faeces or Crudities are in it" will be purged away by the proper waters or washing to a great perfection through the chemical preparations of Calcination, Sublimation, Solution, Distillation, Descension, Coagulation, Fixation. Maier is saying that the Philosopher's Stone is being isolated from faeces and purified by a washing away of impurities. Faeces, of course, is the homeland of the *E. coli* bacterium with its sleeping phage Lambda. Again, it seems that the reputable alchemists were isolating and purifying a virus to effect transformation in a bacterium. Transformation is the introduction of DNA into cells. In the interests of justifying the scientific logic of reputable alchemists, let's consider this possibility by examining Maier's work.

Maier then explains that "the *Calc Vive* [living stone] or Quicklime & *Ignis Graecus* [Greek fire or lightning]" are kindled by Water and cannot be extinguished by Fire. So, a lye must be made from Metals to wash and calcine (heat), and "it must not be Common Water, but Water Congealed into Ice and snow" because this has finer Particles than the standing waters of fens and marshes. This allows better penetration into the "Recesses of the Philosophic Body to wash and purge it from filth & Blackness." In Chemistry, a lye is a highly concentrated, aqueous solution of potassium hydroxide or sodium hydroxide. Also, to effect transformation in *E. coli* bacteria, the cells must be competent, that is, soaked in an ice-cold solution of calcium chloride to induce competence (Nicholl 2002, 81). Competence simply means preparing the cell for uptake of DNA. In genetic engineering today, transformation in competent cells is accomplished by mixing plasmid DNA with the cells, incubating on ice for twenty to thirty minutes, and applying a brief heat shock, so that the DNA enters the cell (81). So, Maier's objective to wash from ice water and heat is very similar to the modern genetic engineering process of icing and applying heat to effect transformation in *E. coli*.

In Discourse IV Maier justifies the marriage or incest of brother and sister by suggesting that Adam and Eve must have resorted to it for the initial propagation of the human race. Actually, this metaphor of brother/sister marriage or incest is a symbolic example of bacterial conjugation or the transfer of genetic material between bacterial cells by direct cell-to-cell contact. Conjugation is a mechanism of horizontal gene transfer. Incest, however, was a common practice of the pharaonic élite in ancient Egypt, and this behavior models bacterial conjugation, for bacteria trade genes horizontally, from cell to cell, rather than passing genes vertically from the mother and father to the children as humans do. Bacteria are functionally immortal because they trade genes horizontally to neighbors in the same generation (Margulis 1986). The incest of siblings, who trade genes horizontally in the same generation, mirrors bacterial conjugation. Vertical gene transfer (VGT) is when an organism receives DNA from a species from which it evolved. Since Maier is using the brother/sister incest metaphor to describe bacterial sex, he explains the metaphor by referring to another metaphor for bacterial sex—the obvious incest of Adam and Eve to produce descendants. What Maier is actually explaining here is the chemical process of conjugation, the transfer of genetic material from a "male" donor bacterium to a "female" recipient in the same generation. Reputable alchemists used metaphor and symbolism to conceal their science from the vulgar.

In Discourse V, we meet the idea of HGT again (the exchange of DNA between two different species) when Maier informs us that the Philosophers speak of a toad suckling a woman's breast milk, which can be compared to the allegory of Cleopatra suckling vipers. Again, human behavior such as pharaonic incest and suckling vipers (if the latter really occurred) are metaphors for HGT and hybridization. Nonetheless, Maier states that the Philosophical Toad has Gold in itself. In Discourse VI, Maier stresses that these matters are strictly chemical, and are mysteries of nature best left to the wise rather than the vulgar. Skipping over Maier's discussion of elements in Discourse VII, we learn in Discourse VIII that when we pass from our present lives, there remains a most perfect and eternal transformation by fire with a penetration by waters and dissolution. Similar to this transformation by fire, the Philosophical Egg is subjected to Temperate Heat, leaving a "white starry splendid powder, & of the white Stone, of which powder are made fit instruments for the Egg."

In Discourse IX Maier states, "There is nothing that can restore Youth to man but death itself, which is the beginning of Eternal life that follows it." Discourse X references processes of evaporation and rain, while in Discourse XI Maier explains the alchemical confusion of authors who use different terminology for the same concepts, different-but-similar procedures, and other ambiguities, so he claims that truth only surfaces for those who have the ambition and money to continue with experiments.

Jumping up to Discourse XIV about the dragon who devours its own tail, the ouroboros is a common symbol in alchemy and a good representation of the closed circular DNA molecule found in bacteria, many viruses, mitochondria, and plasmids. However, it also suggests rolling circle lytic replication of phage Lambda, for the viral

DNA circularizes to clone viral particles in the lytic lifestyle. In Discourse XVIII, Maier explains the collection of sunlight into burning glasses and its reflection by concave and steel mirrors. Now, this process creates polarized light, in which electromagnetic vibrations oscillate repeatedly in only one direction perpendicular to the direction of propagation. Recall Luc Montagnier and Peter Gariaev's experiments with microbes and electromagnetic signals. Similarly, seventeenth century alchemists used a type of polarization microscopy to observe and measure phenomena (microbes) with polarized light or a preferential orientation of optical properties with respect to the vibration plane of polarized light.

Also, Maier mentions that *Lac Virginis*, a compound matter, is used in the process of coagulation. If it is true that Maier was examining *E. coli*, then the Milk of Virgins must be a reference to the Lac system in *E. coli*. Also, lactose is a white crystalline disaccharide occurring in milk. In ancient Egyptian texts, Isis represents the Milk-Goddess or lactose metabolism. Maier then claims the true Philosophers have a "peculiar Gold which they do not deny must be added to the Aurifick Stone as a Ferment at the End of the Work, seeing it leads thinges fermented into its own Nature, without which the whole composition would never return to Perfection." Along with *Lac Virginis*, this last remark on "Ferment" can be taken as a direct reference to the ancient glycolysis-fermentation gene expression network.

In this treatise, we learn from Maier that the alchemist desires the power of life and death, that is, the power of creation. However, with this quest in mind, Maier urges us to attend to the meaning of allegory, ignoring the literal and vulgar sense of the allegory, and that the allegory is a real event. As we learn, the Sun or Sol is like gold among the metals because of its heat, color, virtue and essence. Maier also discusses themes found in ancient Egyptian texts, such as the brother rivalry, the marriage of the Sun and Moon, and so on. In Discourse XXXII, Maier explains that the Philosopher's Stone is a Vegetable because it grows, increases, and multiplies like a plant. He compares it to the growth of coral or pearls. Similarly, phage Lambda possesses a gene regulation network of vegetative replication that produces a multitude of viral heads with tails. It is interesting that Christopher Marlowe envisioned his character Tamburlaine with a pearl head, indicating that he may have understood the significance of pearl-like vegetative growth. Marlowe was born in 1564, two years before Michael Maier's birth.

In Discourse XXXIV, Maier writes that conception in baths is only "putrefaction in Dung." In Discourse XXXV, Maier admits that humans have knowledge of vegetables by custom, but not much experience in minerals and metal bodies, and he stresses the importance of milk and fire (lactose metabolism). Maier then discusses mythological deities such as Osiris, Dionysus, Achilles, Peleus, Thetis, emphasizing the "Nutriment of the Stone is Fire," yet Fire only gives the Stone virtue, maturity and color, for it brings vitality and provision with itself. Thus, the nutriment Fire is lactose.

In Discourse XXXVI, Maier discusses the lowliness of the Philosopher's Stone, reminding one of Jung's archetype, the "lowly origin of the redeemer." Like phage Lambda, the Stone can be found anywhere—in earth, mountain, air, rivers. It appears in obscure black bodies such as "the very dung itself" and "fools would not believe it to be the Thing." Maier says the Stone possesses a "perpetual Fire, which sublimes the Stone and exalts it to the highest dignity." Recall Güenther Witzany's statement that the virus is the only microbe that can re-animate itself. This Stone seems to be a perpetual motion machine similar to Frank's time crystal due to quantum effects.

In Discourse XLI, Maier explains again that truths are hidden under the veils of allegories of ancient deities. In Discourse XLIII, we learn the stone has four colors, a possible reference to Lambda's metallic coat of many colors. In Discourse XLIV, Maier notes the similarities to ancient mythologies, explaining the allegories are chemical in nature. This is direct evidence for the Isis Thesis interpretation, which Maier probably understood 400 years before the theory originated. Maier explains that all the ancient deities are signs of chemical processes.

> In the first book of our Hieroglyphics we have fully explained and reduced the Allegory of Osiris to its true Original, which is Chemical. And though we shall not repeat that, yet we shall make a discourse parallel to it, whereby we may retain Osiris within the bounds of Ancient Chemistry, all which has been so often sung of and figured out by the Ancient Poets. For you can never possibly persuade me that Osiris was a God, or a King of Egypt. For to me the contrary to both seems apparent from several circumstances. He is indeed the Sun, but it is the Philosophical one. Now that name being often attributed to him, the Vulgar who read it, and knew of no other Sun but that which gives light to the World, interpreted it in that sense. The Sun of the Philosophers has its denomination from the Sun of the World, because it contains those properties of Nature which descend from the celestial Sun, or are agreeable to it. Therefore Sol is Osiris, Dionysus, Bacchus, Jupiter, Mars, Adonis, Oedipus, Perseus, Achilles, Triptolemus, Pelops, Hippomanes, Pollux. And Luna is Isis, Juno, Venus the Mother of Oedipus, Danae, Deidaneira, Atalanta, Helena; as also Latona, Semele, Leda, Antiope, Thalia. These are the parts of that compound which before the Operation is called the Stone; and by the Name of every metal, Magnesia.

In modern experiments, cells are rendered competent or able to take up DNA by removing the growth medium by centrifugation, followed by resuspending the cells in solutions containing divalent cations Mg^{++} and/or Ca^{++} (magnesium and calcium). As Maier explains, magnesium (Mg) is involved. The magnesium and calcium cations serve to neutralize the negative charges on the DNA and membranes, as well as disrupting the membrane structures, rendering them susceptible to transfer of DNA (Popham and Stevens 2006). In an early experiment Mark Ptashne inhibited synthesis of host proteins with ultraviolet light and maximized lysogenic *c1* repressor genes by increasing the multiplicity of phage Lambda in the cell. He discovered that increasing multiplicities of phage Lambda above 10-15 caused lytic phage proteins to function, overriding the *c1* repressor genes. In unirradiated cells with high multiplicities of phage

Lambda, cellular protein and RNA synthesis is inhibited. However, Ptashne said the discovery that high concentrations of Mg++ can reverse this inhibition in the irradiated *E. coli* strain made his experiments possible. (1966)

In Discourse L, Maier states, "No water will dissolve a Metallic Species by Natural reduction, but that which continues with it in matter and form, and which the Metals themselves can recongeal, and a little after." Perhaps the Metallic Species is phage Lambda with its metallic coat of many colors. Not only are Dr. Michael Maier's alchemical descriptions evidence for microbiological activities, but the Isis Thesis interpretation is also supported by Maier's statement that ancient deities are simply signs for chemical processes. Alchemical descriptions show that these chemical processes involve isolating and purifying a virus to effect transformation in a bacterium.

Same Terrain. We keep travelling over the same terrain in history, for the Egyptian hieroglyphs define a system of chemical transformation that emerges centuries later in early China, in Christianity as an undercurrent, in thirteenth and seventeenth century scientists, and later in modern genetic engineering. Evidently, what the old philosophers meant by the *lapis* was a crystalline virus, and what they meant by the snake or dragon devouring its tail was the breakdown or putrefaction of the virus (active prophage) during rolling circle replication and its ultimate multiplication of clones. Finally, the unconscious content they were projecting into their chemical activity of isolating and purifying a virus was the quantum biology of phage Lambda, rising in their consciousness like Jung's archetypes, for the psychology of the unconscious appears to be the quantum biology of phage Lambda. The unconscious psyche then may be grounded by our quasi-conscious genome, particularly the 98 percent viral component, as well as our microbiome.

What we are exploring is the possibility of continuity of our quasi-conscious DNA via a virus, a natural mechanism in nature. This historical enigma of continuity of mind, protected for the last five thousand years by the cultural élite who deprived common individuals of this knowledge of life and death, is the foundation rock of power, truth, and order that pervades our dynamic social network. But it is also the shadowy specter behind the inspiration for religion and mythology (Christianity, pagan gods), dissecting disease and mental health (medicine, psychiatry), encouraging pleasure (art, music, sports), promoting forms of knowledge (philosophy, science, humanities), and inflicting pain (war, violence, racism, terrorism). Our potential for viral transformation arouses fear, for it is expressed by a base eroticism that also surfaces nonconsciously as hysteria, madness or bizarre sexuality. The guiding currents of biopower control the meaning of the shadowy specter by promoting obedience, submission, faith, ignorance, heterosexual reproduction, and materialism.

Relative to our debate on Christ and Osiris, Dr. Michael Maier understood that ancient deities were signs for attaining our full potential through a chemical process mirroring viral replication. In reference to Leonardo's painting, Christ is holding the crystal ball or *Salvator Mundi.* Christ then is a sign for a dying/rising virus, one that can be described as dying when it is inert and alive when it rises from the dead. It may be possible that an informed human being at a death transition has the potential to become a crystalline quasi-hybrid being or the God-man that the crystal represents. This should finally close the lid on the casket of historical lies about the meaning of Osiris, Christ and similar dying/rising gods.

Perhaps the supreme principle is that we exist here with specific laws of physics and biology that are just right for our lives because we are the necessary switch-hitters that have the potential to restore order at the quantum origin of our cosmos. At this quantum holographic heart may be a higher-ordered, ancestral virus that is merely a head packed with crystalline DNA.

Fly Ball Ferris Wheel 25

The Body Without Organs

> Where alchemy, through its symbols, is the spiritual Double of an operation which functions only on the level of real matter, the theater must also be considered as the Double, not of this direct, everyday reality of which it is gradually being reduced to a mere inert replica—as empty as it is sugar-coated—but of another archetypal and dangerous reality, a reality of which the Principles, like dolphins, once they have shown their heads, hurry to dive back into the obscurity of the deep.
>
> Antonin Artaud
> *The Theatre and Its Double* (1958)

The objective here is to grasp the complete system that supports life as we know it. For evolutionary success, we need natural, general system laws so obvious that we can predict what will happen. We also need to explain how the parts relate to the whole system, so we can clarify our relation to the cosmos and guide our evolution. Like many others, the French dramatist Antonin Artaud (1896-1948) and the French philosophers Gilles Deleuze (1925-1995) and Felix Guattari (1930-1992) have relational insights into this system that support the Isis Thesis interpretation of ancient Egyptian texts. What ancient Egyptian texts are describing is a complete system that functions like the morphogenesis of a complex viral organism composed of only DNA. As James Watson explains, a phage capsid is primarily naked genes (2003, 41) or perhaps what Antonin Artaud called a "body without organs."

The entire organismal system of these naked viral genes packed in a spherical head depends on the activities of controlling proteins that act on their genes through their folding funnel landscapes. To speculate, it may be possible that the macrocosm of our cosmos-stars-planets-animals, the mesocosm of microbes, and the microcosm of particles-atoms are simply parts of the numerous components of one organismic system. The whole of the system is greater than the sum of its parts because of the system's potential for emergence. If, as ancient Egyptian texts inform, the original viral genome of naked genes houses the c1 protein gene-seed of our current cosmos that grew into expanded space with an edge in time, then the viral heart also houses the gene-seed of cro protein for a second permissible capacity for existence—a quantum state of being with expanded time and an edge in space. This viral origin is within nature's microbial mesocosm, and it allows a natural cycling because the two capacities for existence model the genetic switch of Lambda's two competitive viral proteins that are inextricably linked due to their binding interactions that allow cro protein to fold to its native state on the genome. Microbiology research and intuition guide this speculation, which is supported by historical human behavior and the evolution of society. Also, human history maps the ancient glycolysis-fermentation pathway using lactose metabolism. This pathway is used by phage Lambda for its competing lifestyles of vegetation replication.

In "A Taste of Systemics," system analyst Bela Banathy explains that "Purpose, process, interaction, integration, and emergence are salient markers of understanding systems." In light of the Isis Thesis, we discover that the biological *purpose* of our entire organismal system is to fold to both proteins' native states so the proteins can become biologically active via the circular genetic switch. The problem is that Lambda's two viral proteins are adjacent on the genome, so they compete for the same gene-seats on the genome. This problem of competitive *interaction* is reconciled by the *process* of Lambda's genetic switch, a cycle resulting in two viral lifestyles or protein folding funnel landscapes that operate like the formation/evaporation dynamics of a quantum mechanical black hole connected to its white hole time reverse. From our perspective, lysogeny results in our current macrocosm and human existence, and lysis enables our quantum capacity for a crystallized quasi-hybrid being. The cycling process also involves a CPT violation, and an *integration* or combination relating to hydrogen binding, self-assembly, HGT, UV irradiation, and lactose transport, the environmental variables necessary for the genetic switch. However, the final ingredient for *emergence* is human observation or measurement, and this requires knowledge obtained from our environment or life. Thus, a lawful cycle or reinforcing feedback loop exists that requires human donor DNA, as well as the partnership of both viral proteins within their folding funnel landscapes.

Principles from the Deep: The Moby Rules (see King 2011).

To fully understand the process, we need some general laws. First, in light of the Holographic Principle, our black-hole cosmos may be an illusion or shadow cast by the quantum molecular world. The human agent is the catalyst that throws the genetic switch due to the observer-participancy principle. The possession of organic viral DNA such as reverse transcriptase is also helpful in this process. The human genome with its vast suite of epigenetic marks "can produce hundreds of different cell types and a staggering range of cell functions depending on which genes are switched on and off" (Katsnelson 2010, 646). Life is grounded by physics and chemistry. We are composed of 2 percent genomic proteins that have coded our bodies, and we have a 98 percent genomic toolbox of non-coding microbial protein recipes within each of our cells for some unknown future goals, according to researchers (Mantegna et al. 1994). Perhaps this non-coding DNA restructures itself at a death transition. Thus, the first general law involves the genetic switch of phage Lambda supported by observation, the theoretical Holographic Principle, the Observer-Participancy principle supported by experiment, and the use of the organic 98 percent protein recipe toolbox discovered in human DNA.

The second general law involves a CPT violation. First, in quantum mechanics the weak force violates Charge/Parity/Time or CPT. A Charge violation changes a positive energy state to a negative energy or antimatter, which is simply matter moving backward in time. A Parity violation allows space to have preferred directions and a right or left handedness. A Time violation allows energy states or waves traveling backward in time. CPT violates our fundamental space-time symmetry. Now, a Parity violation also turns an object into its mirror reflection and rotates it 180 degrees about the axis perpendicular to the mirror (upside down). In their descriptions of CPT-violating journeys, ancient Egyptian, early Chinese, and Navajo (ECN) texts reinforce the stick-to-the-right clockwise direction (West-North-East) of spin relative to merging with the Sun at the preferred points of West to Northwest. These ECN clockwise instructions within the Sun and earth's strongly interacting magnetic fields are essential because a magnetic field can reorientate spin which carries information (Thomas 2011; Sommer et al. 2011), and macroscopic entanglement exists at low temperatures (Amico et al. 2008, 43), such as at earth's elementally-enriched polar cusp near its axis, the ECN destination marked by the polestar. Further, an organic molecule can reverse the spin orientation of the conducting electrons of an inorganic, magnetic material (Atodiresei et al. 2010; Sanvito 2010, 664). Put simply, ECN advice is to spin-up and stick-to-the-right for time reverse. (see King 2011) For ultracold fermions, broken symmetries lead to an energy spectrum gap, which has been shown experimentally (Jotzu et al. 2014). Imagine an electron as planet earth with its spin representing earth's rotation. Earth rotates counterclockwise or with a spin-down orientation. Using the Sun and positioning Parity advice of ECN texts, the agent realizes a clockwise-rotating spin-up orientation backward in time, reversing biological evolution and photosynthesis into its chemical reverse, the production of light without heat—cold starlight.

The quantum universe is random, yet a protein folding to its native state is a nonrandom process. Also, we can predict a particle approaching an event horizon will bounce back or go backward in Time. The CPT theorem, with its three parameters of Charge, Parity and Time Reversal, is a basic precept of particle physics. According to the theorem's logic, if there is a violation of Charge, there most likely will be a violation of Parity and Time. After discussions with Stephen Hawking, Andrew Strominger investigated CPT, finding that in two dimensions, weak CPT invariance can be restored in a sector of Hilbert space by including the possibility of white hole formation and evaporation, viz., the time reverse of black hole formation and evaporation (1993). Also, a finite probability exists that any particle approaching the event horizon will bounce back (along with CPT reversed Hawking radiation), a reflection that depends on its low energy, clockwise direction and projection on the black hole axis of rotation

(Kuchiev 2004). These are the same conditions detailed in Egyptian texts (see King 2005). As Stuart Kauffman considers in *At Home in the Universe*, "Perhaps such a location on the axis, ordered and stable, but still flexible, will emerge as a kind of universal feature of complex adaptive systems in biology and beyond." Kauffman calls this a potential universal law (Kauffman 1995, 91-92).

The third general law is that information is now flowing in reverse from proteins to genes as implied by Lamarckism and the epigenetic switch of phage Lambda governed by the brotherly rivalry between c1 repressor and cro proteins over vegetative replication. Lambda microbiology is an example of Lamarckism because the proteins regulate genes due to environmental stress caused by UV light, which influences the lysis-lysogeny decision circuit along the ancient glycolysis-fermentation pathway using lactose metabolism. The fourth general law is self assembly due to UV light or hydrogen that may also play a role in activating superconductivity at temperatures close to absolute zero. The fifth general law relates to self-organization and creation on the axis, as theorized by ancient Egypt, early China, the Navajo, Poe, and Stuart Kauffman. Finally, molecular crystallization processes are mediated by viral DNA to form a crystallized superlattice. From the quantum deep, these are the Moby Rules.

Antonin Artaud (1896-1948). With all this in mind, let's explore Antonin Artaud's brilliant insights. Artaud, who received repeated electro-shock treatments during the last three years of nine consecutive years in mental hospitals, described the world of matter as clogged with shit, blood and sperm (Sontag 1973, xxxviii). In his radio play *To Have Done with the Judgment of God* (1947-8), he claims that man chooses to "shit" rather than "die alive." Further, Artaud wrote in the radio play that "what has been called microbes is god," mirroring the Egyptian biological message, and possibly the idea that viral DNA is god-like in that it can be a vehicle of transformation (horizontal gene transfer) to a timeless energy state of higher order and unity in the two-dimensional mesocosm.

At Artaud's level of quantum insight, he grasped the biological action of the signs at the transcendent level of microbial DNA. For example, Artaud describes it as such while using peyote in 1936, when his spiritual quest led him to the primitive Tarahumara Indians in the Copper Canyon region of Old Mexico. In *The Peyote Dance* (1976), Artaud writes:

> But one arrives at such a vision only after one has gone through a tearing and an agony, after which one feels as if turned around and *reversed* to the other side of things, and one no longer understands the world that one has just left.

Artaud describes a "terrible force" of reverse into a limitless "effervescent wave." He describes elements from his spleen, liver, heart, and lungs bursting in a self-assembly process.

> The things that emerged from my spleen or my liver were shaped like the letters of a very ancient and mysterious alphabet chewed by an enormous mouth, but terrifyingly obscure, proud, *illegible*, jealous of its invisibility; and these signs were swept in all directions in space while I seemed to ascend, but not alone. Aided by a strange force. But much freer than when on the earth I was alone. At a given moment something like a wind arose and space shrank back. (36-37)

Artaud describes going through an agony and then being "reversed to the other side of things," a limitless wave experience or feeling of being in all directions (nonlocality), a self-assembly from his microbiome (spleen, liver, and so on) of "a very ancient and mysterious alphabet" that suggests microbial DNA and its lingual structure, and finally ascension, followed by space shrinking or contracting. His experience sounds similar to Penrose's quantum state reduction.

Artaud's perception of becoming a wave describes quantum mechanical particle/wave duality: a matter particle sometimes behaves like a wave and sometimes like a particle due to electron degeneracy. Recall that Peter Gariaev believes our DNA has a substance-wave duality, which seems reasonable because even minute matter owns both wave and particle properties. Also, physicist Kip Thorne compares electron degeneracy to human claustrophobia, describing matter being squeezed to a density where the electrons are confined to a smaller space. In this state, they are shaking uncontrollably, flying about at high speed, and kicking forcefully against adjacent electrons. Nothing can stop the degenerate motion, "for it is forced on the electron by the laws of quantum mechanics" (1994, 146). The electron then begins to behave like a wave. Artaud's description (tearing and an agony, strange force, space shrank back) mirrors this process. Motion, of course, dilates time and contracts length (Einstein's Special Relativity). Continuing, Artaud explains more of his mental experiences on peyote:

> . . . the fabric of perception opens in a cross, and cracks in such a way that one no longer knows whether it is from one's own heart that this cross has emerged, or from the heart of that Other, who then is no longer the Other, *any* Other, but THAT ONE, the Only Source of Flames, whose tongue pierces and gathers the taste for the Word, when the heart which was beating like a Double, recognizes its GENERATOR! (1976, 70)

This experience of the origin by Artaud may represent a space-time singularity, where the time axis and the space axis cross. At a singularity, space-time separates, stretching all objects radially (in a direction toward and away from itself) and squeezing all objects transversely (Thorne 1994, 450-51). However, because Lambda protein folding funnels of c1 and cro proteins exhibit the dynamics of quantum mechanical black hole (c1) connected to its white hole time reverse (cro), Artaud's quantum insight may also be describing a site-specific recombination process responsible for the integration and excision of bacteriophage genomes into and out of their bacterial host chromosomes. This process involves the remarkable genetic cross (see Figure 1.3), a cruciform branched DNA intermediate known as the Holliday junction (Holliday 1964) that forms due to strand-swapping of DNA (Gopaul et al. 1998). Thus, with the acceleration-induced partitioning of space-time and/or Holliday junction exchange, Artaud experiences the crossed DNA and *Source*, viz., phage Lambda on the host chromosome, aka as Osiris or Christ on the cross. Artaud has found his origin, his "GENERATOR," the naked genes of a creator-virus breaking free of its genetic cross for rolling circle replication. In his radio play Artaud writes that the Tarahumara "smash the cross so that the spaces of space can never again meet and cross."

The Fire-Bearers. In *The Peyote Dance* Artaud discovered that Tarahumara traditions told of a fire-bearing race of men, who had three masters or kings that traveled north to the polestar. To the Tarahumara, these three kings are the three sorcerers of the peyote dance, and they wear crowns of mirrors, the triangular Masonic apron, and more triangular designs at the bottom of their knee-length pants. The triangular emphasis suggests phage Lambda morphology, and the polestar route is the same as the pharaonic advice in the Pyramid and Coffin Texts. In the Peyote Dance, the Tarahumara dance around a circle with mirrored crosses, two of which mirror the Male Principle of Nature called *San Ignacio* and the Female Principle called *San Nicolas* that lie dormant in the circle's hole. Artaud explains that "this advance into the illness is a voyage, a *descent in order to RE-EMERGE INTO THE DAYLIGHT*" (1976, 52). Similarly, in ancient Egyptian texts, the Female and Male Priniciples Isis and Osiris lie dormant on the host chromosome. The point is that in 1936, the Tarahumara, who do not respect civilized humans, were still practicing biosemiotic rites similar to ancient Egyptian ceremonies. Living in the Sierra Madre mountains, the Tarahumara are somewhat free of biopower, but they are not free from the influence of their viral DNA.

In *The Peyote Dance* Artaud also explained how the pre-Renaissance painters of Italy "were initiated into a secret science which modern science has not yet completely rediscovered, and this science was also known to the artists of the High Renaissance" (1976, 59). Artaud is referring to the paintings of nativity scenes related to the three kings by the Piero della Francescas, the Lucas van Leydens, the Fra Angelicos, the Piero di Cosimos, and the Mantegnas (60). He also writes that Religion took over the Science, while the weak masses worshipped Religion instead of the scientific "Principles." Further, he said astronomical Sun worship is expressed universally by signs that mirror the ancient science, which has been absurdly designated as "UNIVERSAL ESOTERISM." The marooned universal signs are "the anserated cross, the Swastika, the Double cross, the large circle with a dot in the middle, the two opposing triangles, the three dots, the four triangles at the four cardinal points, the twelve signs of the zodiac, etc." (61) Artaud believed that these signs pointed to a "high rational Science of the mind" (62), and based on the Isis Thesis, Artaud was correct.

Gilles Deleuze and Félix Guattari (D&G). In *A Thousand Plateaus* (1987), philosophers Gilles Deleuze and Félix Guattari (D&G) address Antonin Artaud's claim in the last lines of his radio play *To Have Done with the Judgment of God* about the creation of a "body without organs."

> For you can tie me up if you wish, but there is nothing more useless than an organ. When you will have made him a body without organs, then you will have delivered him from all his automatic reactions and restored him to his true freedom. Then you will teach him again to dance wrong side out as in the frenzy of dance halls and this wrong side out will be his real place. (Sontag 1973, 571)

D&G ask: How do you make yourself a body without organs (BwO)? First, they explain that the creation is connected with an uninterrupted continuum of intensities, a "formal multiplicity," and a "plane of consistency" in the West uninhabited by people that must be constructed. D&G determine that an abstract Machine must create the consistent plane and the BwO requires "assemblages." A plateau is a "multiplicity connected to other multiplicities" (22) and an "assemblage establishes connections between certain multiplicities" (1987, 22-23). Now, we are going to try to scientifically imagine their speculative idea in a lawful way. First, their continuum of intensities suggests emotional intensities due to the reverse experience of effect then cause, where what you have done comes intensely back to you due to time reverse. Second, the other elements can be respectively interpreted as quantum entanglement, a horizon or plane, the preferred Western direction to the Sun, and viral self-assembly. D&G then explain that the organism is the enemy of this quest to create a BwO, which "swings between two poles" (159) or the "surfaces that stratify it and the plane that sets it free" (161). So, the process can be botched:

> There are, in fact, several ways of botching the BwO: either one fails to produce it, or one produces it more or less, but nothing is produced on it, intensities do not pass or are blocked. This is because the BwO is always swinging between the surfaces that stratify it and the plane that sets it free. If you free it with too violent an action, if you blow apart the strata without taking precautions, then instead of drawing the plane you will be killed, plunged into a black hole, or even dragged toward catastrophe. (1987, 161)

D&G's black hole fate is similar to the ancient Egyptian agenda of the deceased person without knowledge, who activates nothing and is gravitationally dragged into molecular degradation in the black-hole energy landscape funnel. This is the necessary nonnative process designed for the vulgar masses. So, you can fail to produce the BwO or the quasi-hybrid being by maintaining causal action or cause to effect. Despite some vagueness here, D&G identify blocked intensities (implying cause to effect), a slippery plane, and black hole dragging that can botch the creation process. Translated, intensities relate to emotional experiences backward in time from effect to cause. In other words, as time reverses, the individual who has caused certain events now feels the intense effect of his or her action coming back because of the violation of time. Put simply, if intensities are blocked, no BwO and no CPT violation will occur.

Regarding the slippery plane and dragging black hole, this translates into the outer horizon of a rotating, gravitational black hole or a funneled protein landscape. Both are similarly inclined planes. In physics, a tilted surface or inclined plane slides an entity down the plane because the force is unbalanced. However, two forces are active on the entity, the gravitational force pushing downward and the normal force pushing in a direction perpendicular to the surface. On a horizontal surface, the normal perpendicular force goes upward; yet, on an inclined plane, the surface is tilted, but the normal force is still directed perpendicular to the surface. So, on a slope the normal force would not point upward as on a horizontal surface, but rather perpendicular to the slope surface. If you were on the slope of a black hole, the perpendicular upward force would point to the axis of the hole, exactly where you want to be. Typically, the normal force is provided by electromagnetism.

Recall that Roger Bacon emphasized the importance of the perpendicular relative to a multiplicity or production of a species of light "in the first part of the air" contiguous to the second part and so on, the generation multiplying "according to the dimensions of the air." Bacon emphasized the perpendicular passage of a species on a transparent interface versus the easier oblique passage, as well as the images produced by the reflection and refraction of light on a mirror surface. These dynamics relate to the inner axial horizon of a Kerr black hole acting like a holographic mirror. Bacon argued that generation can occur along straight, reflected, refracted, twisting, and accidental lines.

Twist and Tilt. Recently, Barry and colleagues (2009) matched data to theory by using bacteriophage particles to show an analogy between smectic liquid crystals and superconductors. Smectic denotes the state of a liquid crystal (virus) in which the molecules are oriented in parallel and arranged in well-defined planes. In this case, a pure aqueous suspension of viruses formed smectic phases similar to superconductors. The smectic layers expel and twist deformations in the same way superconductors expel magnetic field. In this system, the researchers examined the molecular tilt penetration profile to determine the twist penetration length. Recall that rolling circle replication in metal-binding bacteriophage Lambda may operate like a superconductor, generating and casting out virions along twisting lines. Perhaps bacteriophage Lambda also forms smectic phases similar to superconductors like the experimental bacteriophage demonstrated in Barry and colleagues' study. In addition, D&G's advice about creating a BwO by tipping or tilting the assemblage may also be relative:

> This is how it should be done: Lodge yourself on a stratum, experiment with the opportunities it offers, find an advantageous place on it, find potential movements of deterritorialization, possible lines of flight, experience them, produce flow conjunctions here and there, try out continuums of intensities segment by segment, have a small plot of new land at all times. It is through a meticulous relation with the strata that one succeeds in freeing lines of flight, causing conjugated flows to pass and escape and bringing forth continuous intensities for a BwO. Connect, conjugate, continue: a whole "diagram," as opposed to still signifying and subjective programs. We are in a social formation; first

> see how it is stratified to the deeper assemblage within which we are held; gently tip the assemblage, making it pass over to the side of the plane of consistency. It is only there that the BwO reveals itself for what it is: connection of desires, conjunction of flows, continuum of intensities. You have constructed your own little machine, ready when needed to be plugged into other collective machines. (1987, 161)

Perhaps D&G's "social formation" and "lines of flight" represent liquid bacteriophage molecules, oriented in parallel and arranged in well-defined planes that were generated in a process of twisting and tilting similar to Roger Bacon's descriptions. D&G's creative process also relates directly to the individual Will performing exploratory acts. Remember how Poe's fisherman gives up his fear to explore the wild vortex of water? The philosophers then speak of Castaneda's books and his many spiritual exercises related to becoming-animal, becoming-molecular, and so on.

Now, D&G's "plane of consistency" in the West may relate to a two-dimensional black hole horizon where creation can occur in a particle acceleration zone. D&G insist that "the BwO chooses, as a function of the abstract machine that draws it" (1987, 165) and that "the totality of all BwO's, can be obtained on the plane of consistency only by means of an abstract machine capable of covering and even creating it, by assemblages capable of plugging into desire" (166). Perhaps their "abstract machine" is the genome of a bacteriophage. These comparative biological insights related to an abstract machine and the plane of consistency suggest a viral machine that self-assembles and replicates on a two-dimensional mesoscopic horizon. This BwO can easily be a machine such as a bacteriophage with "naked genes," and this viral gene machine may be what allures us to the machinic.

Allure of Machinic. Our avid interest in understanding our brain architecture prompts scientists to map neural circuits at the mesoscopic scale using viral vectors (Bohland et al. 2009). Scientists also try to understand human intentionality by designing machine intelligence through the use of nonlinear mesoscopic brain dynamics (Freeman 2000). Consider this proposition:

> We propose a concerted experimental effort to comprehensively determine brainwide mesoscale neuronal connectivity in model organisms. Our proposal is to employ existing neuroanatomical methods, including tracer injections and viral gene transfer, which have been sufficiently well-established and are appropriately scalable for deployment at this level. (Bohland et al. 2009)

And so, mesoscopic physics, chemistry, and molecular biology are exploring novel physical phenomena in the mesoscopic range of 100 to 1000 nanometers, sizes ranging from a typical virus to a typical bacterium. Yet, because mesoscopic physics deals with metal or semiconducting material, it has a close connection to the fields of nanofabrication and nanotechnology. However, the bacteriophage with its metal-binding proteins is the virus with the most machinic allure.

To begin to explain the meaning of the machinic forms in our culture, John Johnston (2008) draws on science, the philosophy of D&G, and machinic life processes. As a Professor of English and Comparative Literature at Emory University in Atlanta, Johnston presents a clear history of cybernetics, Artificial Life, and Artificial Intelligence (AI). In his introduction, Johnston defines machinic life as new kind of "luminal" machine associated with life and behaviors of living entities (homeostasis, self-directed action, adaptability, reproduction), but not fully alive. Machinic life mirrors purposeful action like the behavior of organic life, but it also suggests an artificial form of life not quite accountable to the organic biological domain. Johnston explains that many human creations, such as machines and smart systems, display a complex order similar to simple organisms. However, no consensus exists on a definition of life. Yet, biologists do agree that two processes are necessary: a metabolism extracting energy from the environment and reproduction for hereditary survival. (2008, 2-3)

Johnston believes we are entering a new era of nature and technology and luminal machines are alluring. He mentions the widely held assumption that "life somehow emerged from nonliving matter," and he asks, "And what about luminal instances like viruses, which only live in a state of parasitic dependency on host organisms, but nonetheless replicate, mutate, and evolve, and thus fall under the seemingly iron law of natural selection?" (2008, 215) Johnston also discusses the original thought of D&G and Manual Delanda about the potential of nonorganic life and the machinic phylum. All forms of nonorganic life are included in the self-organization of matter because the machinic phylum is the physical and conceptual space . . .

> —where crossovers and exchanges erode the clear and fixed distinction between the two. In a further extension of this body of work, I suggest that this realm is the site of a "becoming machinic," fully explicable as neither a natural evolutionary process nor a human process of construction. Rather, it is a process of dynamic self-assembly and organization in new types of assemblage that draw both the human and the natural into interactions and relays among a multiplicity of material, historical, and evolutionary forces. In this new space of machinic becomings an increasingly self-determined and self-generating technology thus continues natural evolution by other means. (2008, 107-108)

Perhaps the new space of machinic becomings is the quantum mesoscopic domain, where viral self-assembly and DNA organization occur. Johnston, D&G, and Delanda can envision crossovers and exchanges, that is, horizontal gene transfer between the human and the natural as an evolutionary option.

Again, the idea of becoming machinic may be alluring because of the influence of our viral genome. Viral DNA is machinic, and it seems to be the inspiration for our ideas about machinic life. Phage Lambda with its metal-binding proteins is machinic, and we seem to be listening to the messages of ALife and AI from our viral chromosomal apparatus that encourages us to remake ourselves into quantum machines as an evolutionary option. Our behavior suggests that we envision a luminal machine in the image and likeness of a luminal virus, a sign that our viral chromosomal apparatus is influencing us. Technocrats imagine this metal material creation as functioning within a new environment with a new range of nano-variables to exploit the microcosm. They advocate the emulation of the human brain, so we can upload our brains into machines. This may cause death or metal entrapment of consciousness or DNA in a machine, depending on how it is accomplished. However, it seems doubtful that a self-replicating machine can die and self-organize into a crystallized creation in a meso-scopic environment ruled by quantum mechanical laws that require human measurement or a participatory value judgment. Perhaps a machine can function like the internet with human input, but it cannot merge with a quantum life form such as a bacteriophage into a new hybrid form in the quantum niche via natural laws. Or can it?

In *At Home in the Universe*, Stuart Kauffman states that "technological evolution may be governed by laws similar to those governing prebiotic chemical evolution and adaptive coevolution." Still, Johnston assures us that man is not an automaton because he has free will:

> I believe it also lies at the heart of that most extraordinary of human abilities: creative thinking. Great ideas are not pulled out of the air; they are pulled out of the quantum multiverse. In a sense, our minds have recaptured the same process of quantum evolution that I believe propelled life through its origin billions of years ago and drove the evolution of living organisms towards increasing complexity. Although that process may be alive and well inside microbes, its influence on the lives of multicellular creatures may now be buried within our bodies or restricted to negative effects like infectious disease and cancer. Yet, by nurturing sensitivity to the electromagnetic field of the brain, animals, and particularly man, have recaptured entanglement with a quantum mechanical entity – the conscious mind – and once again harnessed quantum measurement to perform directed actions. We call those directed actions, our free will. (2008)

Our free will can only be the guide for directed action, if we can loosen the stranglehold of biopower and understand the microbial potential of our genomic inheritance. Since our behavior appears to be ordered by the quantum, becoming machinic in the classical world may only be another sign of the ancient viral message for morphogenesis and structural stability from our genome.

Baseball breaks your heart. It is designed to break your heart.
The game begins in the spring, when everything else begins again,
and it blossoms in the summer, filling all the afternoons and evenings,
and then as soon as the chilling rains come, it stops
and leaves you to face the fall.

A. Bartlett Giamatti
Former MLB Baseball Commissioner

Fly Ball Ferris Wheel 26

When Things Go Black

> Then little by little the darkness thickened and the shapes, sounds and the feeling of places became confused in my sleepy spirit; I thought I was falling into a chasm that crossed the world. I felt myself carried painlessly along on a current of molten metal, and a thousand similar streams, the colors of which indicated different chemicals, criss-crossed the breast of the world like those blood vessels and veins that writhe in the lobes of the brain. They all flowed, circulated and throbbed just like that, and I had a feeling that their currents were composed of living souls in a molecular condition, and that the speed of my own movement alone prevented me from distinguishing them.
>
> Gérard De Nerval
> *Aurélia*

In his recent book *Consciousness beyond Life: The Science of the NDE* (2010), German cardiologist Pim Van Lommel, M.D. challenges the purely materialist scientific paradigm that relegates consciousness to merely a byproduct or side effect of the human brain. Van Lommel explains his view that consciousness is nonlocal or everywhere and endless, surviving the death of the brain. He bases his theory on a twenty year Dutch study on cardiac arrests and Near Death Experiences (NDE). One comprehensive account Van Lommel presents is the fascinating double NDE of Monique Hennequin at age thirty-one. During a cesarean section, Monique gave birth to a healthy baby boy, but afterward, she suffered complications from surgery and then cardiac arrest followed by a NDE. Resuscitated, she then remained in critical condition until she became aware of her hopeless situation and attempted to return to the loving environment of her NDE by biting off her breathing tube. This activated her second NDE, yet she was resuscitated a second time. Later, she wrote her own account of the two NDEs which Van Lommel includes in his book.

Monique's account from Van Lommel's book begins with her first NDE. To summarize, she felt as if she were suffocating, and her last conscious thoughts were about her children, her job, her house. Then she heard the beeping alarm from the monitor, indicating that she had flatlined. She watched two nurses working on her body, feeling their panic, but she was happy that she was "no longer ensouled" in her body. Monique explains, "Picking up speed, I saw every single room in the hospital, including patients and staff, as well as the past, present, and future of everything that whizzed past me." Monique describes life as a cycle involving another dimension. At that point, "Everything went black," and a sense of nostalgia and a childlike innocence overcame Monique, as well as a feeling of maturity over everything she left behind in her life. She said, "Gradually a sense of sight developed around me, like a sphere that I myself was a part of. I seemed to have ended up in the omniverse, as another image formed below me; in fact, it formed around and through me." Monique's perception of accelerating speed and motion through things or people suggests the quantum effects of light-speed and nonlocality.

Monique then describes a time in her past, when she almost drowned in 1974, remembering that she saw herself as a six-year-old girl. She describes how she felt connected to "the truth," her origins, what she identifies as a "strong sense of separation from my source." Then she explains that she "felt everything as though I had gone back in time," including feeling the pain and sadness of people upset by her words or actions. This suggests time reverse from effect to cause, from the effect of the pain and sadness of others to the cause of her actions or words, rather than the normal process in time of cause to effect. With time reverse, what you do comes back to you.

Monique now has no fear of her thoughts or feelings, only the desire to explain her feelings to some people. At that point everything fades and she painfully thuds back into her body, only to take off again. She ascends through a light spectrum enveloped in "a soft tornado of all colors" with its tip "pointed at the earth's atmosphere." She enters the "eye" of the tornado with the sense that she was "soaring to the center." Monique said, "I sensed and knew instinctively that this force had the shape of an hourglass and that it would expand at some point, become even bigger than the place I had come from." She then experiences different levels and dimensions, ascending, descending, and then everything becomes a "warm black," as she realizes "that pure, warm, soft black is also light, a kind of energy, palpable even without a body." Rising higher through the levels, she senses "parts of both myself and others" in "my life review," and that she would arrive at the "purest form of consciousness." Monique mentions that she has never thought of spirituality, although she had questions about God, war, nature, disease. Then the "black light cleared like a fog," turning an indescribable luminous color. Monique floats along a shore, describing the luminous light as follows:

> Slowly but surely I realized that this luminescence consisted of a kind of infinite river of brilliance, like the brilliance of a setting sun reflected in rippling water with little pinpricks of light like small stars. The brilliance was made up of beautiful little globules of light, extremely bright and quite unlike anything on earth. They looked like nuclei surrounded by a body of light. Not literal bodies, but more like celestial bodies or atoms with clearly visible electrons floating past me, close to the ground. I check to see where they were coming from and wondered if the dark aperture I saw might be a so-called

> black hole. The deep black looked more a cave from which the light gushed like a waterfall and thus formed the river in this tranquil field. The "river," the field, the current, and the black formed a peaceful whole. Brilliant clusters of DNA appeared to be flowing right by my feet (even though I didn't have any). The particles were linked in complementary pairs, which in turn made up an enormous organic spiral. The spirals formed the clusters in the field. I sensed that I could sail or float along with any pinprick of light (particle, being, consciousness, atom, soul, or whatever). All I had to do was "enter" or join the chain.

This description of "brilliant clusters of DNA" is similar to a description of a particle creation zone. Recall Punsly's model (1998) of a polar particle acceleration gap that injects seed pairs into the gap (see Chapter 11). Monique then describes her deep happiness and desire to join "these particles in this current," the "creative force," "this all-encompassing consciousness." She recognizes "many particles as belonging to people I had known on earth, including my younger brother, who had died before I was born." At this point Monique considers becoming part of this "whole," but she realizes there would be no way back to her body. She then describes a sense of timelessness, feelings of overwhelming love and knowledge. She feels as if she has awakened at her source, but she desires to share her experience of tranquility, so she "made a well-considered choice" to return to her body. However, after returning to her life-supported body, she senses the hopelessness of her situation, feeling imprisoned and angry. So, she clamped her teeth down on the respiratory tube in her mouth and "bit down as hard as I could until the incubation tube was severed." Monique felt freed, but realized that this act was not good, "not loving toward nature" or herself. She then experiences not a "soft vortex," but instead a "hard, cold funnel" with a "pinprick of light," along with confusion, panic, loneliness, fear, anger, and the feeling of a "wall." She feels pain for giving up on her body, for abandoning "oneness with those clusters."

Feeling totally desolate, she sees her deceased father drifting by, sensing that "he knew his way around this darkness." He floated in slow motion ahead of her, never looking at her. Monique senses she had a choice to return to her body or accompany her father to the lower intensity pinprick of light. Feeling anxious, she tries to follow her father, but she cannot catch up to him, and then he enters the light. Now, Monique focuses on moving "backwards," so as not to enter the light so that she can share "knowledge, love, honesty, and awareness" with others despite the pain she would endure in a comatose state. During this coma, she realizes that her first NDE was natural, what she called a "pure perception within the source" without her ego. The negative NDE was an "unnatural one that sprang from lovelessness" where she still possessed her ego.

After these NDEs and her personal library research, Monique now views our classical world as a "feeble reflection of reality" and her NDEs as the "greatest reality," "where I awoke again and again with a speed far greater than the speed of light." She writes, "I still have no scientific explanation for what I experienced in 1991, but within the exact sciences, the M-theory and the description of antagonist harmonic pairs seem to come closest to my picture of the clusters in the field and their effect."

The 'M' in M-theory has been referred to as M for membrane or M for mother of a theory. In physicist Brian Greene's book *The Fabric of the Cosmos*, he explains how the five versions of String Theory were unified by Edward Witten who, in 1995, showed how a single master theory encompasses all five string versions. For example, Witten's analysis revealed that M-theory has eleven space-time dimensions with seven of these curled up along with the four we see (three space and one time). Further, the unified five string versions were not only about strings, but also branes or membranes. So, the cosmos itself may be a large brane, that is, a three-brane with four dimensions: three space and one time.

In summary, Van Lommel believes the "the roles of DMT, junk DNA, and nuclear spin resonance, in particular, require further analysis" (2010, 345). Van Lommel challenges the purely materialist scientific paradigm of current Western science, believing in "nonlocal and endless consciousness" which can explain the NDE. (345-346)

Fly Ball Ferris Wheel 27

Threshold Stabilization

Eternal rest grant unto them, O Lord; and let perpetual light shine upon them. Deliver me, O Lord, from everlasting death on that day of terror: When the heavens and the earth will be shaken. As you come to judge the world by fire.

The Verdi Requiem Mass

In mathematics, catastrophe theory relates to the study of dynamical systems. René Thom explains his hypothesis that a process is either structurally stable or not, and information is form which must be structurally stable. His catastrophe theory attempts to account for small discontinuous or nonlinear changes in nature. Thom explains the evolutionary formation of areas of neutral equilibrium bounded by potential walls as the placement of a heavy ball on a sandy base. By shaking the system, the ball soon rests on a flat stable space bounded by a steep wall. Thom thinks a similar mechanism accounts for the evolutionary genesis of organs and that catastrophes (discontinuities) are necessary to ensure the stability of a process.

In ancient Egyptian texts, a human death or catastrophe begins the process of biological morphogenesis that results in a stable form or threshold stabilization. Thom explains that threshold stabilization, a feature of biological morphogenesis, is similar to the theory of games or von Neumann's minimax theorem (1928), where the process behaves like two competing players who adopt a common strategy to minimize their losses (1975, 142-143). Biologically, we observe threshold stabilization in the binding interactions of Lambda c1 and cro proteins, when cro protein is home on the genome binding. Order is free there. Also, Thom's depiction of a ball bounded by a steep wall is similar to the Egyptian image of a disk between two mountains. Both represent an area of neutral equilibrium bounded by potential walls, that is, threshold stabilization, a feature of biological morphogenesis.

Like René Thom and pharaonic Egypt, the philosophical Roman poet Titus Lucretius Carus who lived 50 BCE describes this same image of equilibrium. In Harvard scholar Stephen Greenblatt's book entitled *The Swerve* (2011), the historian discusses the papal secretary Poggio Bracciolini's discovery of *On the Nature of Things* by Lucretius. In this elegant Latin poem, Lucretius describes the origin, structure and destiny of the universe. As an advocate of Epicureanism, Lucretius' philosophy elevates intellectual pleasure and peace of mind relative to death. Epicurus also taught that high-velocity atoms travel straight downward, but at random, unpredictable moments, the atoms could deviate or swerve from their course. This reminds one of Bacon's emphasis on the perpendicular for the production of a species of light, as well as D&G's tipping or tilting of the assemblage to achieve a BwO. According to Epicurus, the soul was made of tiny atoms, and it was not totally predestined because the soul atoms could swerve (Konstan 2014). This idea of swerving soul atoms seems to have justified the existence of free will for Epicureans. Similarly, ancient Egyptian texts describe "Will" or the dead king regulating the morphogenetic process at a polar cell axis, while other wailing soul atoms travel downward in the circles of the thermodynamic "Abyss" due to the gravitational effect. Consider soul atoms as DNA. Egyptian texts describe the gene as the unit of selection on the quantum level; however, the epigenetic knowledge of the dead king guides the process. Assuming that epigenetic psychological or intellectual traits influence genes, the individual or the group with evolutionary knowledge also functions as the unit of selection, since knowledge learned in the dead king's classical environment informs individual choice for quantum HGT and gene replication.

Greenblatt comments that Lucretius' poem of Epicurean thoughts is a great work of philosophy, while the swerve represents "indestructible particles swerving into one another, hooking together, coming to life, coming apart, reproducing, dying, recreating themselves, forming an astonishing, constantly changed universe" (2011, 200). Yet, Lucretius' philosophical poem and his swerve may also have a more precise scientific meaning allowing predictability. Like ancient Egyptian texts, the poem's circulation was limited to the élite group (256). Perhaps Lucretius' swerve represents the process of metals cooling to a temperature close to absolute zero, where the electrons suddenly shift into a highly ordered superconductive state to travel collectively. For ultracold fermions, broken symmetries lead to an energy spectrum gap, which has been shown experimentally (Jotzu et al. 2014).

Recall that a quantum mechanical black hole ring singularity may represent circularized Lambda DNA undergoing rolling circle replication. Brian Punsly (1998) describes a polar particle acceleration gap where black hole pair production onto magnetic field lines occurs and seed pairs are accelerated to high-energy curvature into two finite bands. The process is very similar to the base pairs necessary for DNA production (adenine and thymine; guanine and cytosine) and rolling circle replication. Punsly also describes the ring singularity as a "small inert fossil disk," what may represent the small inert viral head of DNA in the cell, that is, the Lambda prophage that will circularize to produce viral heads and tails by rolling circle replication. In his poem, Lucretius describes a cold fountain pouring out many seeds of heat or balls of fire from earth that burst forth. This bursting fire melts copper and fuses gold, and the process is similar to metal creation within a star, as well as quantum mechanical white hole formation.

Lucretius explains a void is present and the primal germs of iron fall joined "into the vacuum and the ring itself." He explains that from the iron elements, bodies collect and bind in the vacuum and the ring follows, binding with the lode-stone in invisible links. Lucretius is describing molecular assembly in a black-hole protein folding funnel of rolling circle replication. At the end of his biophysical morphogenetic process that mirrors knowledge of ancient Egyptian science, Lucretius describes a stationary Sun and moon, and "Between two mountains far away aloft from midst the whirl of waters open lie a gaping exit for the fleet, and yet they seemed conjoined in a single isle." This biophysical description is identical to the ancient Egyptian references to Sun and moon related to the image of a disk between two mountains, when the Sun-god's bark finally ascends to the starry heavens in a stable form. Over a period of five thousand years, René Thom, Lucretius and ancient Egyptian artwork depict threshold or borderline level stabilization (structural stability) with the same image of a ball or sphere between two crests. Similarly, Stuart Kauffman explains that the ball resting in the bowl represents a low energy equilibrium state, such as a virus.

In threshold stabilization, as Thom explains, the morphogenetic process behaves like two competing players in a game who adopt a common strategy to minimize their losses. In ancient Egyptian texts the two players are the brothers Seth and Horus, signs of the competitive proteins c1 and cro, who ultimately partner or cooperate for cro binding. It seems obvious that the pharaonic priesthood understood both threshold stabilization and game theory, for they created baseball to describe their chemistry.

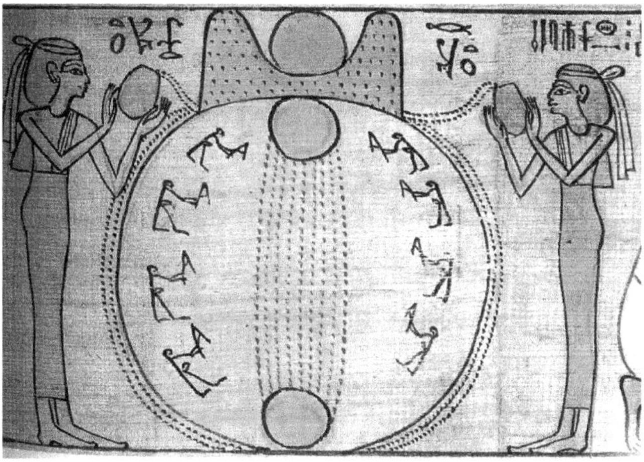

Figure 27.1 Public Domain image from the Book of the Dead of Khensumose (c. 1075-945 BCE). The Sun rises from the mound of creation at time's beginning. With the central circle representing the creation mound, at the top is the "horizon" hieroglyph with the Sun centered on top of it between two mountains. Goddesses of the north and south pour out the waters surrounding the mound. Water molecules govern the structure, stability, and function of biomolecules because hydration forces pack and stabilize the protein structure (Levy and Onuchic 2004). Inside the circles are the eight original Ogdoad deities related to primeval creation.

Tomb of Ramesses VI. Another image of the Sun centered between two mountains is in the Tomb of Ramesses VI at Thebes. According to Alexandre Piankoff (1954) who translated the texts, prior to the Eighteenth Dynasty the mortuary temple was originally attached to the pyramid. But, circa 1600 BCE the mortuary temple was separate from the pharaoh's burial chamber and a pyramid was no longer used. Instead, in Thebes, the western ridge of mountains was called Mistress of Silence, while its flanks were burrowed with deep corridors or the tombs of the Eighteenth, Nineteenth and Twentieth Dynasties that the Greeks called "syringes." René Thom mentions that a bacteriophage shell functions like a syringe (1975, 157) to pump its DNA into the cell. Also, sarcophagus means "flesh-eating," and this suggests the process of bacteria decomposing or eating the human body at death, while bacteriophage means "bacteria-eating," a process of phages decomposing or eating bacteria. Perhaps both the pharaonic priesthood and the early Greeks understood how microbes function.

According to Piankoff, the Royal Tomb No. 9 is where Ramesses VI is buried. In Sarcophagus Hall, the texts deal with the creation of the solar disk, which we will explore shortly. Piankoff explains that the tomb's walls bear Greek graffiti, showing it was visited by the Greeks. Also, in the first centuries of the common era, Piankoff states that it was also occupied by the Christians, who left their traces in the Coptic language. The Greek inscriptions state that Plato and Epiphanius visited the tomb, with one inscription stating, "Epiphanius declares he saw nothing to admire but the stone." (1954, 7-8) This reference is probably to Saint Epiphanius of Salamis (c. 310-403), bishop of Salamis in Cyprus. Other Christian scholars and ecclesiastics were also named Epiphanius during the fifth, sixth, and seventh centuries, but Epiphanius of Salamis lived during the first centuries of the common era when Christians were visiting the Tomb of Ramesses VI. After visiting the Tomb of Ramesses VI, St. Epiphanius must have had quite an epiphany because he wrote the *Panarion* or the *Medicine Chest against Heresies*. Byzantine historian Ernst Kitzinger claims that Epiphanius was the first cleric to focus on Christian religious images (1954). Perhaps this is because Epiphanius saw the many images of humans with spheres for heads in the Tomb of Ramesses VI, as well as the transformed king depicted as an armless human with a sphere for a head and serpentine legs. This image on the Overdoor from Doorway G to Hall H is striking in its startling inimitability. An area for future study, St. Epiphanius seems to be a direct link between the information in Egyptian tombs and the mirrored liturgy of early Christianity, suggesting that the Christian hierarchy may have been aware of its pagan roots. However, it is doubtful that they understood the actual chemistry because the Tomb of Ramesses VI emphasizes its Egyptian mysteries, and in imitation, Christianity follows through in their dogma with the same emphasis on accepting necessary mysteries. On the other hand, Plato envisioned original humans as having spherical form in *Timaeus*, so he may have understood the hybrid implications.

A Viral Conductor? In his book *The God Delusion*, biologist Richard Dawkins suggests that religious behavior may be a misfiring of something that was once useful (2006,

174). He explains that Darwinian natural selection allowed it to survive, so religion must have conferred some advantage. Actually, for the pharaonic priesthood in ancient Egypt, their "religion" was a quantum survival science that enhanced the survival of the individual's genes, particularly the beneficial viral genes that Witzany's research (2008) shows should broaden human evolutive potential for a new species. However, religion today has lost this evolutionary message. So, Dawkins is correct—religion has misfired. But then, baseball has misfired, for we do not see it as survival chemistry. Perhaps natural selection has also misfired, for it did not remove our non-coding viral DNA that many consider junk. Perhaps these natural activities can be integrated under the model that shapes our behavior—a modeling relation between a complex bacterial virus and human beings.

Today research such as the ENCODE project that involves a worldwide consortium of research groups is discovering that 80 percent of our genome's junk-DNA desert contains elements linked to biochemical function. The space between genes (viral introns) is also a treasure trove of enhancers (regulatory DNA elements), promoters (sites where DNA transcription begins), and other regions encoding RNA transcripts with possible regulatory roles. (Ecker 2012, 52) Researchers have found an expansive human regulatory lexicon encoded in a transcription factor 50 base-pair footprint within thousands of human promoters that has been imprinted on our genome, containing "elements with evolutionary, structural and functional profiles that parallel the collections of experimentally derived genomic regulators brought to light during the past 30 years" (Neph et al. 2012). And so, the barren desert of junk DNA has burst into a treasure trove of biological richness, reminding us of our lowly birth from viruses. Also, University of Oxford (UK) researchers compared the human genome with eleven other mammals and argued that natural selection has shaped only about 8.2 percent of the human genome that exhibits low rates of mutation, while the rest is nonfunctional (Woolston 2004). Since our eukaryotic nucleus shares traits with a eubacterial virus like phage Lambda, consider that the transition from the first anaerobic cells or prokaryotes existing 3.8 billion years ago (bya) may have resulted from a Lambda prophage invading a prokaryotic cell, becoming its inert heart, and then evolving into our eukaryotic nucleus that developed 1.8 bya. One wonders if life is orchestrated by a viral gene conductor through random natural selection (c1) and adaptive mutation (cro).

Perhaps the holographic shadow of our diffusion-collision universe and natural selection's groping process on earth mirror the diffusion-collision and induced-fit protein folding/binding landscape of Lambda c1 protein. Perhaps natural selection on earth is a continuation of the random general process of diffusion-collision in our universe that is tuned, as Martin Rees (2000) supports, by "just six numbers." This suggests that the ultimate nature of the gravitational field may represent the diffusion-collision folding/binding landscape of Lambda c1 protein, which is tuned to just six viral gene seats in the mesoscopic domain. HGT or mutation-based adaptation by a bacterial virus would be the variation supplying form, so as Stoltzfus (2012) explains, "selection is no longer the creator that shapes raw materials into products." Add to this that human behavior in the religious rituals of ancient Egypt and Christianity models phage Lambda's lifestyles along the ancient enzyme pathway still functioning in our cells, and one can see that what was perceived as religion was originally a quantum science of human DNA merging with viral metal-binding DNA to achieve re-animation. Today this idea of a human-viral DNA merger is rising in the contemporary technocratic vision of a human-machine merger, suggesting that a metal-binding virus may be what drives our interest in metal machinic technology and its potential for extending life. Yet, a human being is an organism that anticipates the future, not a machine. For all these reasons, a transdisciplinary science of life and death is necessary to understand the evolutionary survival message inherent in our behavior and the systems we create. The formula that explains the past is the viral footprint stamped on human history that showcases viral morphogenesis for human re-animation. Knowledge is the difference between possible destinies of man-made-metallic-crystal through death and man-made-metal-jacket through technological power.

Lost in Translation. In the Tomb of Ramesses VI, the use of the word "soul" in the translation by Piankoff can easily be replaced with the word "DNA" because the contextual meanings of the two words easily coalesce. Mutation-based adaptation by a bacteriophage is also suggested by the language implying that variation supplies form. In the Book of Caverns, the texts state, "The light is made in the West!" The focus is on "He of Forms," "Lord of Forms," the act of giving "light to my forms," and "to take a form which is unknown." The texts reinforce that the Sun-god Re makes the forms. By the time one has read the complete texts of the *Book of What Is in the Netherworld*, the *Book of Gates* and the *Book of Caverns*, one has no doubt that the tomb's rich Egyptian library is talking about viral morphogenesis, the development of a unique spherical form. What also confirms this claim is the artwork on the main ceiling in Corridor G of the tomb of Ramesses VI. The upper register shows nine figures with spheres for heads and stars at their feet, suggesting the hybrid development of form. The middle register shows seven figures with spheres for heads, and the lower register shows eight figures sitting with spheres before them and eight stars above their heads. Also, bull heads with disks between the horns are signs of structural stability.

At the end of Corridor G is a small sloping ceiling with two registers. The upper register shows a great disk within a crescent with two male figures on each side, another sign very similar to the Sun disk centered between two mountains, a sign of threshold stabilization. The idea of equilibrium is supported by artwork showing four hearts surmounted by the Egyptian sign for stability next to the two male figures on each side of the crescent. The lower register shows a human body forming a boat with human head and arms holding a sphere. Then the remarkable hybrid image appears between two mountain crests—the armless human with a sphere for a head and serpentine legs, the alien, the stranger, the Unique One. On the overdoor of Doorway G to Hall H is also the representation of a sphere centered between two mountain crests, another sign of threshold stabilization. Above and below the central sphere are two smaller disks. To the right and left of the central disk are two beetles or transformation signs. To the right of the central disk is the armless figure with a sphere for a head and legs terminating in wavy serpents. To the left of the central sphere is a similar figure, but it is destroyed, showing only legs terminating in cobra heads. These two strange images represent the structurally stable quasi-hybrid being that is viral-human. The artwork is describing threshold stabilization, a feature of biological morphogenesis.

During the Twelfth Division of the *Book of Gates* in the Tomb of Ramesses VI, the Sun-bark ascends into the starry heavens with the transformed Sun-god and the crew: Mind, Will, Magic, Thoth, Geb, Isis, and Nephthys (1954, 224). With the emphasis on Mind, one must concede that ancient Egypt is the origin of the idea that Mind is primary over matter, and this idea eventually generated the mind-body debate. Their idea of Will may represent the human will power or choice necessary to accomplish the morphogenesis through the genetic switch allowing lytic replication. Magic, of course, represents the counterintuitional dynamics of the magico-mystical quantum world of nonlocality, entanglement, time reverse, and other quantum phenomena.

Matching Data to Theory: a Gene-Culture System of Evolutionary Patterns. The evidence reveals that not only our behavior, but also the evolution of society over the last 12,000 years has carved a footprint into human history, profiling the morphogenesis of a complex metal-binding virus. The chemical formula maps the ancient glycolysis-fermentation gene expression network for lactose metabolism used by bacteriophage Lambda in its lytic lifestyle of vegetative replication. Considering the copious amount of viral DNA in our bodies, this seems reasonable, especially in light of the possibility that our eukaryotic nucleus evolved from an inert viral prophage like phage Lambda. History's 12,000 year timeline and other evidence reveal that our behavior and potential for emergence are linked to the morphogenesis of this complex virus. The data is consistent with the Isis Thesis interpretation and a gene-culture system where our genes influence the probability that certain cultural traits will be adopted in accordance with Lambda's gene expression network. This viral footprint indicates a quasi-hybrid being of human and viral DNA is a possible, emergent state of being.

Behavior or Evolutionary Sign of the Real	Possible Mesoscopic Meaning (the REAL)
Early Hunter-gatherers	Viral self-assembly; recruitment of energy-rich molecules
Fermentation technology (bread, wine, beer)	Ancient glycolysis-fermentation pathway
Predilection for sweets	Ancient glycolysis and other sugar pathways
Agricultural cultivation of barley, wheat	Ancient glycolysis and Maltose Transport pathways
Invention of writing, printing, computers	DNA transcription and replication

Behavior or Evolutionary Sign of the Real	Possible Mesoscopic Meaning (the REAL)
Milk & cheese production & consumption; domestication of cattle & goats for dairy	Ancient glycolysis-fermentation lactose pathway
Agricultural production (cloning seeds)	Viral vegetative DNA replication or cloning
Egyptian deities, literature, artwork, ritual	HGT and Lambda lifestyles for humans
Christian deities, ritual, Mass	HGT and Lambda lifestyles for humans
Human incest	HGT or bacterial sex in same generation
Industrial Revolution (human/machine)	Merger of human with metal-binding virus
Information Society/advanced capitalism	DNA replication, transcription, cloning
Design of highway transportation system	Design of DNA molecule
Humans and slime moulds produce same nature-inspired design on highway system	Molecular search for energy-rich nutrients or food
Bartering, buying, selling	Substitution/exchange of amino acids on DNA molecule
Human beings and repetitive behavior	Proteins follow collective algorithm in viral genes
Human land territorialism	Gene ownership or binding to the gene-seat
Ball design (baseball, soccer, football)	Related spherical or icosahedral viral shape
Mythology (Egypt, China, Navajo)	Phage Lambda lifestyles
Sed festival and Catholic Mass	Lambda lifestyle of lysis for morphogenesis
Baseball and sports warfare	Phage Lambda lifestyles and protein competition
Religious concept of virgin birth	Asexual viral cloning with lactose metabolism
Osiris and Christ's stretched out arms	Two arms of the Lambda genome
Ford's mass production of metal auto parts	Cloning lifestyle of metal-binding viral parts
Funnel architecture of sports stadiums	Funnel architecture of tiny black holes and protein folding landscapes
Mayan Hero Twins ball game; warfare	Phage Lambda protein competition
Hierarchal organization of government, capitalism, religion, corporations, etc.	Survival of the Few over the Many: priority of Lambda protein native state over nonnative states
Contemporary genetic engineering	HGT, cloning, viruses, bacteria
Human sleep/wakefulness cycle	Phage Lambda lifestyles
Etymology of certain words	Influence of our viral genome's lingual structure
Heterolateral brain organization	Topology of arms of Lambda genome

Behavior or Evolutionary Sign of the Real	Possible Mesoscopic Meaning (the REAL)
Nonconscious physiological opisthotonos	Viral rolling circle replication
H. G. Wells' *War of the Worlds*	Human evolution from viruses
Forecast of hybrid human/machine merger	Potential HGT with machinic, metal-binding virus
Marlowe's pearl-headed Tamburlaine	Spherical viral head structure/morphogenesis
Spherical head obsession in Christianity, Egypt, literature, Christ's crystal orb, and Philosopher's Stone	Spherical viral head structure/morphogenesis
Chinese Emperor climbing mountains	Uphill activities of native-state protein folding
E. A. Poe's stories	Two chemical pathways (native/nonnative)
Navajo Nine Night Ceremony, Sixth Dance	Viral rolling circle replication and head assembly
Circle dancing, *arc de cercle* position of Charcot's hysterics and Osiris	Viral rolling circle replication
Jung's archetypes of collective unconscious	Phage Lambda lifestyles
Spherical form in Plato's *Timaeus*	Viral ancestry/topology of spherical head
Repetitive viral focus in cultures, alchemists, Christianity, and genetic engineering	Our viral genome influences human behavior
Leonardo's painting "Saviour of the World"	Crystal viral morphology/morphogenesis
Alchemy	Isolation/purification of viruses, DNA, plasmids
Antonin Artaud/Tarahumara Indian Ritual Artaud's Body without Organs	Viral morphogenesis Bacteriophage head of DNA without organs
Machinic phylum of D&G	Machinic mesocosm of microbes
Machine Dream of Hybrid Human/Metal	Morphogenesis of Hybrid Human/metal-binding virus
Sport of Fly-fishing; fisherman in myth	Fly-fishing mechanism of Lambda cro protein
Magic in Religion	Quantum Entanglement, Nonlocality, Time Reverse
Motivation for human sacrifice, genocide, war, and survival of the Few over the Many	Limited viral protein native state binding requiring large nonnative interactions

Additional patterns can be found in the twelve articles that expand on the Isis Thesis. As the sociologist Jean Baudrillard thought, our experience in the classical world is destabilized because meaning is not anchored to the Real. Recall that the quantum domain orders our environment. Surprisingly, the DNA activities of a complex, metal-binding bacteriophage such as phage Lambda is a good model that fits human behavior. Also, the influence of the gene expression network of phage Lambda functions as a force of cultural change. Once the original causal pattern is perceived, the corresponding human behavior patterns stand out like stars in the sky. The viral pattern is directly linked to quantum physics, quantum biology, and quantum cosmology. In these three sciences, we have a trinity of natural patterns—the protein folding funnels of Lambda c1 and cro proteins in biology, the formation/evaporation processes of black hole/white hole physics, and the mathematics of our expansion/collapse cosmos. The evidence supports a vital need for an intellectual and epistemological science of human behavior, so we can alter our negative viral behavior by means of brain neuroplasticity. In this new science, physicists could talk to biologists and biologists to cosmologists, and all these scientists could talk to psychologists and others in various disciplines such as the Humanities (anthropology, archaeology, art, communication, cultural studies, philosophy, politics, religion, sociology, music). From observing the evolutionary signs of the Real, it seems obvious that the microbial DNA in our bodies is influencing our behavior.

In Defense of Junk DNA. Our non-coding DNA can switch on developmental genes, and its G+C content is strong and stable. Let's take a closer look at the 98 percent DNA that has been classified as junk DNA. DNA is composed of four bases, with adenine (A) pairing with thymine (T) and cytosine (C) pairing with guanine (G). Whereas the A+T content has two hydrogen bonds, the G+C content is stronger because it is stabilized by three hydrogen bonds. Researcher A. M. Selvam (2009) uses a holistic approach, explaining that approximately 15 percent of the so-called junk DNA functions to repress proteins and switch on developmental genes. Using systems theory that was first developed for atmospheric flows, she explains that the fluid energy flow structure is a spiral structure with Fibonacci winding number and internal icosahedral Penrose tiling pattern, that is, vortices within vortices. Selvam applies systems theory to our genomic spatial organization of the DNA bases G+C. She concludes that the distribution of the G+C bases in all chromosomes exhibits self-similar fractal fluctuations, memory is inherent in the self-organized Penrose tiling pattern network, and coding exons work together with non-coding introns so the DNA molecule functions as a whole. Selvam also concludes that the icosahedral Penrose tiling pattern provides efficient packing for the DNA molecule within the chromosome, and that other studies show that our non-coding DNA may be responsible for the crucial signals that allowed human evolution and directed our genome to function differently than the genomes of other organisms. She concludes that self-organized quasicrystalline pattern formation exists at the molecular level. (2009) Thus, 15 percent of our viral DNA can switch on developmental genes, while our genome's non-coding viral DNA may have guided human evolution.

G+C content correlates with gene expression efficiency in mammalian cells (Kudla et al. 2006). In the human genome, the average G+C content of a 100-kb fragment ranges between 35 percent and 60 percent (Romiguier 2010). The evolutionary forces responsible for G+C content are debated; however, recombination is one possibility (Meunier and Duret 2003). This suggests that icosahedral phage Lambda with its high G+C content of 49.858 percent (Mesbah et al. 1989, 161) may be the viral conductor shaping human evolution through recombination or HGT. It seems apparent that the ancient glycolysis-fermentation pathway has inspired our evolution relative to fermentation technology (bread, wine, beer), our predilection for sweets, the agricultural cultivation of barley and other grains (cloning of seeds), milk and cheese production and consumption, as well as the domestication of cattle, goats and sheep for dairy and food purposes. These human behaviors are signs pointing to the ancient glycolysis pathway in our cells used by icosahedral phage Lambda. Recent research is also showing that we have a microbiome-gut-brain axis, where gut microbiota communicate with the brain to modulate behavior (Cryan and O'Mahoney 2011), and this new concept will soon impact human health and disease. Within the human gut microbiome is the *E. coli* bacterium with its lactose genes, where the sleeping Lambda prophage resides. Also, in the initial folding process, Lambda cro protein is an intrinsically unstructured protein (Gsponer and Babu 2009) like the unstructured toolbox of 98 percent noncoding DNA in our genome. So, our genome is ready for viral morphogenesis.

Some Observations on the Power Grid of Players. The classical world we see may be the holographic pattern or shadow cast by bacteriophage Lambda's lifestyles. Quantum molecules, proteins, and particles at the DNA level impact human behavior, events and thought. Call it genopower. Recall that Tolstoy said people are "involuntary tools of history" and historical actions are fated because leaders cannot control the system. This is probably because we function in a nonlinear system ordered by microbes, where unpredictable events occur and causes do not produce a straight-line effect. Themes such as determinism (Skinner), our behavior modeling the morphogenesis of an organism (Gariaev), humans as particles (Thom), natural causes as minute parts of human bodies (Hume), molecules and crystals as "structured societies" (Whitehead), and our behavior being entangled with our genome are augmented by the contemporary scientific understanding that the quantum world orders our classical environment. As explained by Juan Maldacena, the peculiar interactions of these lower-dimensional particles and fields may generate the force of gravity and one of the spatial dimensions. So, our genome or chromosomal apparatus that we all share is generating repetitive human behavior patterns in history. This genopower also fashions the dynamics of biopower, that is, the soft-core altruism or non-altruism of the Few over the Many, a sign reflecting the quantum survival advantage about the limited priority of native state binding over nonnative interactions. As Skinner says, "behavior is lawful and determined." Since it seems to be determined by our viral genome, perhaps we can apply our brain's neuroplastic potential to moderate negative behavior like warfare. In this way, we can temper genopower or what Gariaev describes as our chromosomal apparatus' capacity to control the creation of society space.

What does the Power Grid of Players reveal about our position in history? Today, humanity has not progressed beyond the original ideological rivalry between the Gnostics, who had knowledge of ancient Egyptian chemistry (King 2006), and the early Christians, who believed their deities were not signs, but real material people. The question then and now involves whether or not deities are real material gods or viral signs of a chemical process. Related to Christianity's imitation of Egyptian ritual, deities and mysteries, perhaps the underlying motivation for the original rivalry over images and deities resulted from Christianity's attempt to conceal their imitation of Egyptian mystical knowledge from their congregations. Not understanding Egyptian quantum chemistry, Christianity misfired and unknowingly peddled Egyptian science as religion, while anathematizing mysterious images like spheres for human heads and ideas that the stars are living beings.

As time passed, the Christian imitation of Egyptian science became a boomerang because with the invention of the microscope and early laboratory methods, reputable seventeenth century alchemists such as Newton and Maier were able to grasp the chemistry in the alchemical process of isolating and purifying DNA, a virus, and plasmids. Further, they understood that ancient deities, including Osiris and Christ, were only chemical signs of a deep evolutionary science. As time passed, modern scientists excavated pyramids and decoded hieroglyphs, paving the way for other interpretations, such as the Isis Thesis that questions the general consensus of egyptologists, who believe ancient Egyptian texts are primitive, confusing and unintelligible. How long will they ignore the behavioral similarities between ancient Egyptian deities and Christian deities that model the complex lifestyles of an ancient virus?

Like the ancient Egyptian texts, the counterintuitional domain of quantum mechanics is confusing, but it is the key to understanding the texts. Also, advances in chemistry have shown us that the elements in our bodies are from the stars, so why can't we stretch the definition of life to the stars, our living planet, and a simple virus that reanimates? The problem is that today we are still dealing with earlier historical dilemmas related to what is real or illusionary, what is living or dead, as well as the mysteries related to the interaction between our classical cosmos and the quantum world. However, five thousand years ago, Egypt understood viral morphogenesis and its relation to human survival, an understanding our contemporary scientists lack. So, our position in history seems to be at a destabilized dead end, especially when one considers the technocrats and their vision of a human-machine merger that may end humanity if we do not exercise our free will based on scientific knowledge. Yes, we are destabilized at an emotional, material, and unreasonable intellectual threshold that hopefully will lead to the creation of a new science of life and death and the elimination of the mind-body problem and other repetitive historical conflicts. What matters here is your viewpoint, for as a reader, you are the arbiter of truth that can control or alter the contemporary political power grid that is founded on an actual historical Game of the Centuries. Always remember Wheeler's observer-participancy principle.

Another historical observation is that perhaps Foucault's theory is correct—it is All against All. From my perspective, the battle of All against All has become the Few versus the Many because this behavior represents nature's general modus operandi for

DNA survival, as well as our culture's ideal mental model of things as they should be. Our culture appears to be a coded sign of competitive Lambda protein activity and gene expression. Lambda cro is an initially disordered protein (IDP), folding and binding by the fly-casting mechanism. Our 98 percent microbial DNA is very similar to IDPs. Huang and Liu (2010) explain that IDPs possess more nonnative interactions in folding and binding processes than conventional ordered proteins. In their study, they demonstrate that the binding rate of IDPs accelerates at first and then decreases as nonnative interactions increase. Also, they found that nonnative interactions enhance the kinetic advantage of IDPs folding by the fly-casting mechanism.

Because strong nonnative interactions function to enhance cro's binding and folding, human behavior emphasizing the survival of the Few (limited native state binding) over the Many (larger nonnative interactions) may reflect this evolutionary criterion. As Pope Francis informs, "It is no longer simply about exploitation and oppression, but something new." In our wasteful society, the exploited represent the biological "leftovers," the plucked out. The Few—those with the power network, the money, and perhaps the scientific knowledge—are smart enough to sense that they must distract the Many from the survival mechanism, a virus. Innately, our social organizations model viral protein dynamics related to native state binding for the Few and nonnative dissolution for the Many. We can see this same survival pattern of the Few over the Many in capitalism, government, labor relations, land ownership, corporate structure, religion, medicine, psychiatry, education, nature, and so on. As Foucault believed, knowledge and power are closely connected and power is creative. However, history confirms that power is only creative for the èlite, who conceal DNA survival knowledge from the biological altruists.

What does the Power Grid of Thunderheads and Diamondhearts tell us about advancing technology? Relative to the Thunderheads, Hawking and Gunlet show how technology can be both constructive and destructive. Also, world governments need to be cautious about the creation of the first superintelligent AI and the use of only linear logic or means-end reasoning. Further, technocrats like the Machine Dream Pitching Team with their fast-ball pitches may benefit from conducting further research about telomerase and by taking a humanities course in ancient cultures, so they can see the obvious quantum biology and physics present in ancient Egypt, early China, the American Navajo Indians, the Aztec culture, and other sources that favor the evolution of our DNA. This may deter them from deleting important genes from the human genome such as telomerase. Relative to the Diamondhearts, scientists are showing us that we can improve ourselves by using mental effort to access our prefrontal lobe, so we can moderate our negative viral social behavior. However, in our attempts to extend life and prevent disease, our DNA can be damaged by innovations related to the use of gene lasers, so more research is necessary here, especially relative to preventing cancer.

Our underlying aspirations for a hybrid merger with a metal-binding virus will impact the creation of the first superintelligent AI. Many scientists today are not in consensus regarding the role of human values and aesthetics, biotechnology, and human enhancement in the creation of the first AI. Relative to these technoprogressive goals, we must take a hard look at our current historical position, while understanding that death may be a transition allowing evolutionary emergence and stability, if we understand the quantum capacity of our DNA for survival or biological immortalization. After all, Marx's vision of future communism may relate to either a community of hybrid human machines or the cloned community of a hybrid viral crystal—perhaps a state of being like Frank's time crystal. So, all must guide technology, not just a few technocrats.

From a population bottleneck 60,000 years ago, our population has grown to 7.125 billion individuals. Yet, the fossil record is evidence that extremely successful life forms have gone extinct. Just prior to extinction, species often reproduce with considerable abundance, what is called the sunset phenomenon, and some scientists believe this is happening now to our species (Margulis 1986, 228). However, our behavior has stamped a viral survival message on human history.

We must also remember that reality is a haze until someone observes it, so scientists must figure out how to abandon causality, as Wheeler supports. Quantum measurement is a choice similar to free will. Ultimately, we are not slaves to our genes, and we have free will, but we must learn to use it productively instead of allowing fear, ignorance, conformity, and microbiological determinism to limit our choices. What we must attempt to discover is the unity of knowledge.

The Unity of Knowledge. So, we can hear the machinic beat of a gene-culture system that is greater than the sum of its parts. The relations within the system result from the flow of information from our genome. The least obvious part of the system, the gene expression network for viral morphogenesis, is stamped on human history. Generated from an internal predictive model, the viral footprint is a sign of a characteristic set of viral behaviors in human beings. Hopefully, a unity of knowledge about this viral gene-culture system will positively impact future human behavior.

Six assumptions guide the Isis Thesis interpretation. First, universals exist relative to reality and existence, for a regularity of patterns exists in different-but-similar human activities. Second, the world has a holographic mode of operation. Third, Complexity Theory and John Wheeler's observer-participancy principle support that an agent-based model, as described by ancient Egypt, early China, and the seventeenth century American Navajo, with controlled information injected at key points can guide a system's behavior to a desired state due to the underlying structure and quantum biology of phage Lambda's genetic switch. Fourth, the quantum world ruled by quantum mechanics orders our classical cosmos ruled by general relativity (Einstein's theory of gravity). Fifth, spin carries information. Sixth, our consciousness is a value selection mechanism. Consider the last assumption. A human catalyst for the whole DNA process is necessary, and this catalyst must have knowledge of the process. Gariaev and colleagues' research supports that all organisms have a material substance and an "energy informational (EI) substance" that are intimately linked. The EI level is relative to the 98 percent microbial DNA of humans, and this level is the "leading one." (2011, 20) Also, will and intentionality may be housed in the biophotonic structure process of our DNA (Rapoport 2010). Since a human catalyst for the whole DNA process is necessary, and this catalyst or switch-hitter must make a human value judgment influenced by knowledge of the process, consciousness is functioning as a value selection mechanism.

While scientific experiment, statistical mathematics and theory generally support the last five assumptions of the Isis Thesis, the first assumption that universals exist is necessary to expand one's perspective beyond boundaries, divisions, and classifications of things, in order to encompass a universal range of transdisciplinary knowledge. As an example, the general view is that *Homo Sapiens Sapiens* is one of many different living species, separate from nonliving rocks, crystals, and viruses. Yet, scientists are beginning to understand that a sharp line cannot be drawn between species or between nonlife and life. If we imagine a God's eye universal view, we might see that all life is interconnected, not only on our thriving planet, but in the cosmos at large, for the evidence shows that the living elements in our bodies are from the stars. Could we not expand our view to include the cosmos, earth, minerals, viruses, and bacteria as members of one genus or class, considering that we originated from stardust, developed with minerals and rocks on earth, and evolved from a last universal common ancestor that may be a virus? The philosophical idea that universals exist relative to reality and existence requires that we remove mental barriers that stifle our thinking because the scientific evidence suggests that a metaphysical entity exists that remains unchanged in character in a series of changes or changing relations that have stamped a viral footprint on human history. The Isis Thesis supports that this entity is a bacteriophage genome.

Now universals imply a unity of knowledge. In this brief chapter, we have already tapped sources of knowledge such as astrophysics, physics, biology, chemistry, and cultural mechanisms that interrelate. One must concede that a unity of knowledge from different disciplines is a prerequisite to understanding knowledge as a whole. This transdisciplinary approach is also essential to understanding the Isis Thesis because the pharaonic priesthood is using a systems approach. Now, Helmut Löeckenhoff proposes a systems model that provides a semiotic and methodical foundation for transdisciplinary research (2006), so we can be encouraged that we can grasp the unity of knowledge. We can also be encouraged because our future investigations can be assisted by Nobel laureate Neils Bohr's correspondence principle. In Bohr's search for the unity of knowledge, he understood the fundamental character of the quantum world and thus came up with the mathematical formulation called the correspondence principle of past and present ideas. In other words, the sound views of the past should correlate with the sound views of the present, and from our position in history, we can observe this, for the underlying current in human history is viral morphogenesis. Another pioneer in quantum theory, Hermann Weyl in his address delivered at the Bicentennial Conference of Columbia University (1954) on "The Unity of Knowledge," explains that interpretation "springs from inner awareness and knowledge of myself. Therefore the work of a great historian depends on the richness and depth of his own inner experience." (quoted in Wheeler 1994, 175) The historians with rich and deep awareness of this unity are the ancient Egyptian pharaonic priesthood, the early Chinese emperors and sages, the Navajo American Indian shamans, the Aztec rulers, the Tarahumara, William Blake, Antonin Artaud, Teilhard de Chardin, Charles Peirce, and many others interested in universal knowledge. These visionaries support that the act of observer-participancy by a knowledgeable human agent at a death transition allows a transformation from matter to a quasi-hybrid being or mind crystal. This emergence may be possible in light of Wheeler's delayed-choice cxperiment and the inbuilt circularity of nature's laws (see King 2011).

As physicist John Wheeler explains, "In a delayed-choice experiment, one discovers, a choice made in the here-and-now has irretrievable consequences for what one has the right to say about what has already happened in the very earliest days of the universe, long before there was any life on Earth." (2004, 114) This means the past has no existence except for what is recorded in the present. Wheeler supports that we must abandon causality and move the structure of science onto the foundation of acts of observer-participancy. Because of quantum entanglement, which entails the phenomenon of nonlocality, objects can become linked and instantaneously influence one another regardless of distance. Add to this that our brains remember the Past, not the Future, and we have the intrinsic circularity to go backward in time and embrace the cosmos, possibly at the transition of death as early cultural evidence suggests, natural laws support, and contemporary scientific studies reinforce (King 2011).

In the wake of Wheeler's delayed-choice experiment, physicist Paul Davies explains two requirements supporting how life and mind are built into the cosmos at the deepest level—first, the radical idea of present and future observers shaping the far past when no observers existed, thereby giving life and mind an indispensable, creative role in the cosmos, and second, an overarching principle or information-theoretic vision of physical laws redolent with "an inbuilt level of looseness or flexibility," minuscule today but higher in the early cosmos (2006, 248-249). Perhaps Einstein was right, "God does not play dice." With these perspectives in mind, perhaps we can develop a new transdisciplinary science of life and death before we go extinct or eradicate ourselves.

Opening Day 28

Miles Above

When the sound and wholesome nature of man acts as an entirety, when he feels himself in the world as in a grand, beautiful, worthy and worthwhile whole, when this harmonious comfort affords him a pure, untrammeled delight: then the universe, if it could be sensible of itself, would shout for joy at having attained its goal and wonder at the pinnacle of its own essence and evolution. For what end is served by all the expenditure of suns and planets and moons, of stars and Milky Ways, of comets and nebula, of worlds evolving and passing away, if at last a happy man does not involuntarily rejoice in his existence?

Goethe's essay on Winckelmann (1805)

The Game of the Centuries is presented to you in iTime (imaginary Time).

Opening Day is the day that professional baseball leagues begin the season. Usually, Opening Day is during the first week of April, and it signifies rebirth, a new chance for each team to begin the season again. In the Baseball Almanac, Hall of Fame pitcher Early Wynn, who played for the Washington Senators, Cleveland Indians, and Chicago White Sox explains:

> An opener is not like any other game. There's that little extra excitement, a faster beating of the heart. You have that anxiety to get off to a good start, for your self and for the team. You know that when you win the first one, you can't lose 'em all.

It is interesting that on the average, Major League Baseball teams go through 900,000 baseballs each season, and players toss additional balls to the fans in the stands. Actually, you could consider the production of balls as cloning, for they are all the same. Again, ancient Egyptian science is the origin of baseball, as well as the idea of Opening Day, for the Egyptian deity Ptah is another sign of the Sun-god or the rising Sun that opens the day. The name Ptah has often been explained as "Opener," and Ptah is considered the "Opener" of the day, while the deity Tem plays the role of "Closer" of the day. Ptah was also the chief god of all the workers in metal and stone. As the architect of the universe, Ptah beat out the iron firmament with a hammer and supported it, while covering the sky with crystal. (Budge 1904, 500-502)

At today's Game of the Centuries, the fans are enthusiastic as they wait for the teams to take their positions on the diamond. In the western part of the stadium off third base, a noticeable ripple in the crowd is emerging. Smoothly, it travels around the circular stadium, going across sections of bleachers as groups of happy fans rise up and sit down to create the movement of a wave. In unison, row after row of fans rise up and sit down to produce the wave that circles the stadium. Undeniably, this activity that transports individual energy represents the physics of wave motion, for a wave can transport its energy without transporting matter. The individuals are still stationary in their seats, but they are also part of the wave. This is a sign of our particle-wave duality that may also be a quality of our chromosomal apparatus.

Suddenly, lightning flashes and somewhere between life and death, Schrödinger's cat is hissing. Cheers and applause explode, as the Diamondhearts storm into the stadium. Dead or alive, the shining stars are prepared to conquer the materialist analytic of power relations conveyed by the reasoning system of the Thunderheads. Immediately, an intensifying grumble of thunder oppresses the landscape, as the Thunderheads storm into the stadium on the heels of the Diamondhearts. The Thunderheads express a materialist grid of power relations that can dominate the known physical world.

Each team charges into their dugout, where they eat and spit out clonal sunflower seeds from their mouths, or they blow up bubblegum balls. This volcanic sign of spitting forth seed or the word from the mouth suggests the creation via the word of DNA in ancient Egyptian texts. The Egyptian uraeus or cobra, a royal symbol on the crowns of Horus and Seth, symbolizes the spitting forth of the new creation, for the serpent represents the DNA wormhole or black hole/white hole casting forth its information. Also, in the Memphite Theology of ancient Egypt, the god Ptah brings forth the universe through the power of the word (Goelet 1994, 145) or the viral DNA. In the ancient Egyptian Coffin Texts, the dead king states that the "stream is spat out for me" (CT 190), that he was not conceived but "Atum spat me out" (CT 76), and that he has "assumed the forms of Atum" (CT 703), the original creator-god. The dead Sun-god addresses the Egyptian sky goddess, "O Nut, through the seed of the god which is in you, it is I who am the seed which is in you" (PT 563). Further, the texts support that this seed, that is, the elements regenerating the dead Sun-god, the word, the DNA, is from Orion or the stars. And so, spitting sunflower seeds or blowing bubblegum balls at a baseball game is redolent with the ancient meaning of a free human act for an emergence based on knowledge and the memory in the starlight.

A Unique Anthem. Today is the Game of the Centuries between multiple historical ideologies identified by two oppositional teams in a debate on mind and matter. Certainly, a special anthem is necessary for this game before it begins. The musician, especially one with a trumpet, is a symbol of fascination with death. Consider how a trumpet can collapse the instant of time with a long burst of smooth tones, blasting out in a blues that spins out shattering tones. When the music alternates between deep and high-pitched tones or conquers the space between the valley and the mountain, the earth and the sky, then the sound expresses a leap of anguish and the need for inversion, transformation, a change of destiny, a death permitting birth.

True to this idea of music as a bridge between the material world and uncorrupted will, the charismatic trumpeter Miles Davis (1926-1991) improvised an instinctive message of grief and longing in his albums. Like a solitary magician intent on stripping off the mask of God, his sound expresses a cascade of note doublings that scale into infinity, as his trumpet skates on the pulse of the celestial winds along the edge of chaos. One senses the mysterious boundary between smooth flow and turbulence in nature, as his trumpet foreshadows the Abyss with shattering tones tripping along the fluid edge of time's chasm.

On June 7, 1967, the Quintet of Miles Davis on trumpet, Wayne Shorter (tenor sax), Herbie Hancock (piano), Ron Carter (bass), and Tony Williams (drums) played the first performance of "Nefertiti." Nefertiti, a sign of the milky bull-goddess with her balls of fire, was the Eighteenth Dynasty royal wife of Pharaoh Akhenaten, who was known for revolutionary religious ideas related to the worship of the god Aton or the Sun disk. Her Bantu name NFURAA-TI-TI derived from the Egyptian hieroglyphs means "a beauty has come" (Somo 2008). When the Quintet began rehearsing the melody for the first time, their performance stretched the element of time into its fluid dynamic range. When their unorthodox melody dissipated and stopped, the musicians laughed, for they had discovered the power of their own mesmerizing music. The composition "Nefertiti" has been described as "an organic relationship: a balance between the written and the improvised, a shared melodic and rhythmic sensibility." (Belden 1967) Miles' metaphysical interpretation of death and finality pervades the menacing cool of his melancholy trumpet in pursuit of the whisper of immortality. One remembers his super hit "So What" and realizes there was little if any fear of anything in Miles Davis. Miles' quest for a new sound was about change and evolution, not standing still and being safe. What counted to Miles was knowledge and freedom, so he was true to his voice. Remembering the unique voice of Miles Davis' trumpet on the *Sketches of Spain* CD, co-editor Nat Hentoff of *The Jazz Review* was reminded of the following line from the Spanish writer Ferran: "Alas for me! The more I seek my solitude, the less of it I find. Whenever I look for it, my shadow looks with me."

Fluidity. With thunder still rumbling through the stadium, the Game of the Centuries is ready to start. The skies are cloudy, but a kind of blue shines above the stadium. There the Sun is a luminous pearl disk whirling its way to the West. The announcer introduces the anthem: "The name of the sound is 'Circle' by the Miles Davis Quintet, and the sound of the name means survival for those with knowledge." Then, the Miles Davis Quintet appears on the diamond, sauntering slowly into the stadium to a covered platform, where trumpet, tenor sax, piano, bass, and drums wait passively. The fans become attentive and silent. And then the voice of Miles' muted horn liberates a depth of poignant feeling that resonates through the stadium. Melodies resonating within melodies flow from the rich sound of Wayne Shorter's tenor sax. Herbie Hancock's keyboard improvisation resonates like Chopin, as Ron Carter's bass blends into the whole, and Tony's cymbal bursts add dimension and pulse to the showers of sound. Rising and falling like a wave, the improvisation of the musicians finally rests with the melody of Miles' trumpet.

Circle. In math, a circle is one-dimensional even though it exists in a two-dimensional plane. The Thunderheads generally reason within the confines of three dimensions, while the Diamondhearts think within the opening of a horizon of two dimensions. Yet, nature has a particle-wave structure that allows the unity of thought and knowledge from both dimensions.

The cyclical process of life flows from a quantum two-dimensional cosmic origin to three dimensions and then back to two dimensions. In three dimensions, natural selection allows individuals and their genomes to interact with the environment, causing a variation in traits. However, the environment is threaded or woven together by horizontal gene transfer from viruses and bacteria that impact genomes. Nature saves its information for morphogenesis in genomes. Some individuals may survive longer and reproduce more than other individuals. In time, specific individuals in particular environments may emerge into a new species in two dimensions due to adaptive mutation, viral gene transfer, and knowledge learned in the three dimensional environment.

In the last five thousand years, élite human groups have recognized the viral survival message or origin-seeking, teleological dynamics of nature, but they concealed their survival knowledge and delivered soft-core altruistic schemes to the masses. Although natural selection operates in nature, teleology and individual intentional choice function with it. So nature does tend to definite ends, even though science avoids teleological explanations because it is difficult to determine the truth of the matter due to the secrets of biopower, the fury of human competition, and the evolving surges of human creativity in various forms that are consistent with the predictions of the Isis Thesis regarding quantum hybrid speciation. Proofs of teleology are human cultural obsessions with microbiology in the forms of religion, mythology, ritual, alchemy, genetic engineering, literature, artwork, and so on. As explained, all these behaviors dance around the genetic switch of phage Lambda. The switch allows cro protein to bind and fold back to its gene, a nonrandom definite end, involving the human switch-hitter. Thus, a human being can anticipate and adapt directly to a quantum environment by changing its genome. In light of this, nature is probabilistic but not random because the ultimate beneficiary of gene selection is the flow of the whole viral genetic switch.

Humans are viral, as shown by our genome and our microbiome. Consider that our major occupation in life involves forming networks (governments, capitalistic corporations, religious organizations, political associations, internet, families). These viral networks comprise the three-dimensional shadow thrown by our two-dimensional viral genome with its survival message. In three dimensions, the survival message splinters into diverse networks and various disciplines, making the message difficult to grasp. Yet in two dimensions, the survival message on the two lifestyles of phage Lambda is whole, casting a shadow on our classical world that explains the past and supports a theory of human history as viral morphogenesis.

Once again, the quantum world of DNA is ordering our classical existence. However, what the rainmaker freely decides to do here instantly influences what is true somewhere else, perhaps the other side of the city or the planet or the cosmos. This is free will choosing from a ground of possibilities. The choice of one informed individual can disassemble the pyramidal power structure of the Few over the Many. It is the difference between three dimensions and two dimensions, between the macrocosm and the mesocosm, between a ball and a disk, between a pyramidal power structure and a circular communal harmonic. Simply put, a royal flush is possible because of our freedom to choose.

If a person really understands science and finite probabilities, that person would bet on the team that knew how to cycle back to nature's origin. Knowledge is the great equalizing force. Nature is actually similar to the Chicago White Sox players who conspired with gamblers to throw the 1919 Baseball World Series. Nature is telling us that things do not have to be random when it comes down to victory and survival.

Ω and A. When the Miles Davis Quintet finished their anthem "Circle," lightning flashed and thunder roared as the fans applauded. Above the clouds from the high vertex of the silvery Sun disk beyond, humanity could be observed in the circles of the concave stadium below. Everyone looked the same.

At travel close to light speed, an observer would see the aberration effect of shapes twisted into a tunnel shape and color distortion due to the Doppler effect, a distortion similar to driving through a rainstorm.

Without warning the heavens opened up—the sky poured down rain, the fans ran for cover, the stars trembled, and the primordial waters flooded the Game of the Centuries, delaying play and baptizing everybody. Somewhere, a golden child embraced the cosmos.

Figure 28.1. Egyptian Ka: two arms upraised, sign of cosmic embrace and wholeness.

And so, the Game of the Centuries about the debate between mind and matter is now rained out by a new politics of truth, a unity of knowledge, and the prospect of a science of life and death that can illuminate our bounded rationality. For, the logic of actions making sense in one part of the system (classical world) may not be reasonable within the broader context of the whole system (classical and quantum).

Perhaps on some small planet in another material world, the Game of the Centuries is just beginning again. But then—maybe not.

References

Adamatzky, Andrew, Selim Akl, Ramon Alonso-Sanz, Wesley van Dessel, Zuwairie Ibrahim, Andrew Ilachinski, Jeff Jones, Anne V. D. M. Kayem, Genaro J. Martinez, Pedro de Oliveira, Mikhail Prokopenko, Theresa Schubert, Peter Sloot, Emanuele Strano, Xin-She Yang. 2012. Are motorways rational from slime mould's point of view? Cornell University Library arXiv:1203.2851v1 [nlin.PS]. http://arxiv.org/abs/1203.2851.

Agar, Nicholas. 2010. *Humanity's end: why we should reject radical enhancement.* MIT.

Agyemang, Tyler. 2012. Dennett receives 2012 Erasmus Prize for cultural contributions. *Tufts Daily*, February 24.

Altus, Deborah E. and Edward K. Morris. 2009. B. F. Skinner's utopian vision: behind and beyond *Walden Two*. *Behav. Anal.* 32 (2): 319–335.

Alvarado, Carlos S. 2009. Nineteenth-century hysteria and hypnosis: a historical note on Blanche Wittmann. *Australian J. Clinical and Exp. Hypnosis* 37 (1): 21-36.

Amico, Luigi, Rosario Fazio, Andreas Osterloh, and Vlatko Vedra. 2008. Entanglement in many-body systems. *Reviews of Modern Physics* 80 (2): 517-557.

Andreini, C., L. Banci, I. Bertini, A. Rosato. 2006. Counting the zinc-proteins encoded in the human genome. *J. Proteome Res.* 5 (1): 196–201.

Artaud, Antonin. 1947. To have done with the judgment of god. In *Antonin Artaud Selected Writings.* Los Angeles: University of California Press.

———. 1976. *The peyote dance.* Trans. Helen Weaver from *Les Tarahumaras, Tome IX, Oeuvres Completes d'Antonin Artaud*, Editions Gallimard, 1971, New York: Farrar, Straus and Giroux.

Ashtekar, A., T. Pawlowski, and P. Singh. 2006. Quantum nature of the Big Bang. *Phys. Rev. Lett.* 96, 141301.

Atodiresel, Nicolae, Jens Brede, Predrag Lazić, Vasile Caciuc, Germar Hoffmann, Roland Wiesendanger, and Stefan Blugel. 2010. Design of the local spin polarization at the organic-ferromagnetic interface. *Phys. Rev. Lett.* 105 (6).

Avery, John S. 2003. *Information theory and evolution.* World Scientific Pub. Co., Inc.

Avery, Oswald T., Colin M. MacLeod, and Maclyn McCarty. 1944. Studies on the chemical nature of the substance inducing transformation of pneumococcal types. *Journal of Experimental Medicine* 79 (2): 137–158.

Baker, R. 1975. *Binding the devil: exorcism past and present.* New York: Hawthorne Books, Inc.

Ball, Philip. 2008. Quantum all the way. *Nature* 453 (7191): 22-25.

Banathy, Bela. A Taste of Systemics. Ed. Tom Mandel. ISSS Integrated Systemic Inquiry Primer Project. http://www.isss.org/primer/bela6.html.

Barnstone, Willis and Marvin Meyer 2003. *The Gnostic Bible.* Boston and London: Shambhala.

Barry, Edward, Zvonimir Dogic, Robert B. Meyer, Robert A. Pelcovits, and Rudolf Oldenbourg. 2009. Direct measurement of the twist penetration length in a single smectic A layer of colloidal virus particles. *J. Phys. Chem.* B. 113 (12): 3910-3913.

Belden, Bob. 1967. Nefertiti/Miles Davis. CD Commentary, June 7, 1967.

Bergoglio, Jorge Mario. 2013. *Evangelii Gaudium.* Vatican Press. http://www.vatican.va/evangelii-gaudium/en.

Bloodworth, Andrew. 2014. Track flows to manage technology-metal supply. Nature 505 (7481): 19-20.

Bohland J.W., C. Wu, H. Barbas, H. Bokil, M. Bota, et al. 2009. A proposal for a coordinated effort for the determination of brainwide neuroanatomical connectivity in model organisms at a mesoscopic scale. *PLoS Comput. Biol.* 5 (3): e1000334. doi:10.1371/journal.pcbi.1000334.

Bohm, David. 1994. *Thought as a system.* New York: Routledge.
Bohm, David and Basil Hiley, 1993. *The undivided universe.* London and New York: Routledge.
Bojowald, Martin. 2006. Loop quantum cosmology. January 20, arXiv:gr-qc/0601085 v1.
Boos, Winfried and Howard Shuman. 1998. Maltose/maltodextrin system of *Escherichia coli*: transport, metabolism, and regulation. *Microbiol. Mol. Biol. Rev.* 62 (1): 204–229. PMCID: PMC98911. http://www.ncbi.nlm.nih.gov/pmc/articles/PMC98911.
Bos, Kirsten I., J. Verena, G. Schuenemann, Brian Golding et al. 2011. A draft genome of *Yersinia pestis* from victims of the Black Death. *Nature* 478 (7370): 506-510.
Bostrom, Nick. 2005. Transhumanist values. *Review of Contemporary Philosophy* 4 (May): 3-14. http://www.nickbostrom.com/ethics/values.html.
———. 2012. The superintelligent will: motivation and instrumental rationality in advanced artificial agents. http://www.nickbostrom.com.
Bousso, Raphael. 2002. The holographic principle. *Rev. Mod. Phys.* 74 (3): 825-874.
Brent, Joseph. 1998. *Charles Sanders Peirce: a life.* Bloomington and Indianapollis: Indiana University Press.
Brooks III, Charles L., José N. Onuchic, David J. Wales. 2001. Statistical thermodynamics: taking a walk on a landscape. *Science* 293:612-613.
Brown, D. M., ed. 1992. *Egypt: land of the pharaohs.* Alexandria, Viginia: Time-Life Books.
Budge, E. A. Wallis. 1904. *The gods of the Egyptians volume 1.* Chicago: Open Court Publishing Company (New York: Dover Publications, Inc., 1969).
Cairns J., J. Overbaugh, and S. Miller. 1988. The origin of mutants. *Nature* 5:142-45. [Pub Med. 3045565]
Campbell, Joseph. 1962. *Oriental mythology: the masks of god.* New York: Penguin Books.
Chein, Edmund and Hiroshi Demura. 2010. *Bio-Identical hormones and telomerase: the Nobel Prize-winning research into human life extension and health.* Bloomington, IN: iUniverse.
Choi, Charles Q. 2009. Quantum afterlife. *Scientific American* 300 (2): 24-25.
Christian Classics Ethereal Library. The Anathematisms of the Emperor Justinian against Origen. (Labbe and Cossart, *Concilia.* Tom. V., col. 677). http://www.ccel.org/ccel/schaff/npnf214.xii.x.html.
Chu, Xiakun, Linfeng Gan, Erkang Wang, and Jin Wang. 2013. Quantifying the topography of the intrinsic energy landscape of flexible biomolecular recognition. *PNAS* 110 (26): E2342-E2351.
Cirlot, J. E. 1971. *Dictionary of symbols second edition.* Trans. Jack Sage. New York: Philosophical Library.
Craven, J. B. 1910. *Count Michael Maier, doctor of philosophy and doctor of medicine, alchemist, Rosicrucian, mystic, 1568-1622.* London: Dawsons, 1968.
Crick, Francis and James D. Watson. 1956. The structure of small viruses. *Nature* 177 (4506): 473-75.
Cryan, J. F. and S. M. O'Mahony. 2011. The microbiome-gut-brain axis: from bowel to behavior. *Neurogastroenterol. Motil.* 23:187-192.
Curry, Andrew. 2013. The milk revolution. *Nature* 500 (7460): 20-22.
Cvetkovic, Aleksandar, Angeli Lal Menon, Michael P. Thorgersen, Joseph W. Scott, Farris L. Poole II, Francis E. Jenney Jr, W. Andrew Lancaster, Jeremy L. Praissman, Saratchandra Shanmukh, Brian J. Vaccaro, Sunia A. Trauger, Ewa Kalisiak, Junefredo V. Apon, Gary Siuzdak, Steven M. Yannone, John A. Tainer & Michael W. W. Adams. 2010. Microbial metalloproteomes are largely uncharacterized. *Nature* 466 (7307): 779-782.
Dalmia, Shikha. 2013. Pope Francis shouldn't bite the hand that feeds the Catholic Church. *Washington Examiner*, November 27.
Danovaro, Roberto, Cinzia Corinaldesi, Manuela Filippini, Ulrike R. Fischer, Mark O. Gessner, Stephan Jacquet, Mirko Magagnini, and Branko Velimirov. 2008. Viriobenthos in freshwater and marine sediments: a review. *Freshwater Biology* 53:1186-1213.
Davies, Paul. 2006. *The Goldilocks enigma: why is the universe just right for life?* Boston: Houghton Mifflin Co.
Dawkins, Richard. 2006. *The God delusion.* Boston, New York: Houghton Mifflin Company.
De Grey, Aubrey. 2006. Aubrey de Grey: an exclusive interview with the renowned biogerontologist. By Ben Best, *Life Extension Magazine*, February. http://www.lef.org/magazine/mag2006/feb2006_profile_01.htm.
De Nerval, Gerard. 1996. *Aurélia and other writings.* Trans. Geoffrey Wagner. Boston: Exact Change.

De Rola, Stanislas Klossowski. 1996. *Golden game: alchemical engravings of the seventeenth century.* London: Thames and Hudson Ltd., 1988. Published in hardcover in the US in 1996.

DePristo, Mark A., Daniel M. Weinreich and Daniel L. Hartl. 2005. Missense meanderings in sequence space: a biophysical view of protein evolution. *Nature Reviews Genetics* 6 (September): 678-687.

Death, A. and T. Ferenci. 1994. Between feast and famine: endogenous inducer synthesis in the adaptation of *Escherichia coli* to growth with limiting carbohydrates. *Journal of Bacteriology* 176 (16): 5101–5107.

Deleuze, Gilles and Félix Guattari. 1987. *A thousand plateaus: capitalism and schizophrenia.* Trans. Brian Massumi. Minneapolis, London: University of Minnesota Press.

Dempsey, W. B., S A McIntire, N. Willetts, J. Schottel, T. G. Kinscherf, S. Silver, and W. A. Shannon, Jr. 1978. Properties of lambda transducing bacteriophages carrying R100 plasmid DNA: mercury resistance genes. *J. Bacteriol.* 136 (3): 1084. http://jb.asm.org/content/136/3/1084.full.pdf.

Dennett, D. C. 1971. Intentional systems. *Journal of Philosophy* 68:87-106.

———. 1987. *Intentional stance.* Cambridge, MA: MIT Press.

Dill, Ken A. and Hue Sun Chan. 1997. From Levinthal to pathways to funnels. *Nature Structural Biology* 4 no. 1 (January).

Downey, Glenville. 1959. The name of the church of St. Sophia in Constantinople. *Harvard Theological Review* 52 (1): 37-41.

Doye J. P. K., M. A. Miller, and D. J. Wales. 1999. Evolution of the potential energy surface with size for Lennard-Jones Clusters. *J. Chem. Phys.* 111:8417-8428.

Dressler, David. 1970. The rolling circle for *phi* X DNA replication, II. Synthesis of single-stranded circles. *PNAS* 67 (4): 1934-1942.

Dubovsky, S. L. and S. M. Sibiryakov. 2006. Spontaneous breaking of Lorentz invariance, black holes and perpetuum mobile of the 2nd kind. *Phys. Lett.* B638:509-514.

Dvorsky, George and James Hughes. 2008. Postgenderism: beyond the gender binary. *Institute for Ethics and Emerging Technology* (IEET-03) (March). http://ieet.org/archive/IEET-03-PostGender.pdf.

Eagleton, Terry. 1986. Marxism and literary criticism. In *Criticism: the major statements second edition,* 532-554. Ed. Charles Kaplan. New York: St. Martin's Press.

Eckel, Stephen, Jeffrey G. Lee, Fred Jendrzejewski, Noel Murray, Charles W. Clark, Christopher J. Lobb, William D. Phillips, Mark Edwards and Gretchen K. Campbell. 2014. Hysteresis in a quantize superfluid 'atomtronic' circuit. *Nature* 506 (7487): 200-203.

Ecker, Joseph R. 2012. ENCODE explained. *Nature* 489 (7414): 52-53.

Edgar A. Poe Society. 2009. Edgar Allan Poe and Rufus Wilmot Griswold. (Updated January 22, 2009). http://www.eapoe.org/geninfo/poegrisw.htm

Elgar G. and T. Vavouri. 2008. Tuning in to the signals: non-coding sequence conservation in vertebrate genomes. *Trends Genet.* 24 (7): 344–52. doi:10.1016/j.tig.2008.04.005.PMID 18514361.

ENCODE Project Consortium. 2012. An integrated encyclopedia of DNA elements in the human genome. *Nature* 489 (7414): 57-74.

Faulkner, R. O. 1969. *Ancient Egyptian pyramid texts, 2 vols.* Oxford.

———. 1973-78. *Ancient Egyptian coffin texts, 3 vols.* Warminister (rpt. 1994).

Feldman, U., U. Schühle, K. G. Widing, and J. M. Laming. 1998. Coronal composition above the solar equator and the north pole as determined from spectra acquired by the SUMER Instrument on SOHO. *Astrophysical Journal* 505:999-1006.

Fiene, Donald M. 1989. What is the appearance of divine Sophia? *Slavic Reviews* 48 (3): 449-476.

Forterre, Patrick. 2005. The two ages of the RNA world and the transition to the DNA world: a story of viruses and cells. *Biochimie.* 87:793–803. http://www.college-de-france.fr/media/marc-fontecave/UPL36851_forterre.pdf.

Foster, Patricia L. 1993. Adaptive mutation: the uses of adversity. *Annu. Rev. Microbiol.* 47:467-504. http://www.ncbi.nlm.nih.gov/pmc/articles/PMC2989722/

Foucault, Michel. 1973. *The order of things.* New York: Vintage Books.

Freeman, Walter J. 2000. The neurodynamics of intentionality in animal brains may provide a basis for constructing devices that are capable of intelligent behavior. For "Workshop on metrics for intelligence" in a program for "Development of criteria for Machine Intelligence" at National

Institute of Standards and Technology (NIST), Gaithersburg MD, August 14-16, 2000.

Galloway, Warren R. J. D., James T. Hodgkinson, Steven D. Bowden, Martin Welch, and David R. Spring. 2011. Quorum sensing in gram-negative bacteria: small-molecule modulation of AHL and AI-2 quorum sensing pathways. *Chem. Rev.* 111 (1): 28–67.

Gariaev, Peter. Wave Genetics. Accessed Nov. 2, 2014. http://eng.wavegenetic.ru/index.php?option=com_content&task=view&id=2&Itemid=1.

Gariaev, Peter P., Mark J. Friedman, and Ekaterina Leonova-Gariaeva. 2011. Principles of linguistic wave genetics. *DNA Decipher Journal* 1 (1): 11-24.

Gariaev, Peter P., George G. Tertishny, Katherine A. Leonova. 2002. Wave, probabilistic and linguistic representations of cancer and HIV. *Journal of Non-Locality and Remote Mental Interactions* May, I (2).

Gibbon, Edward. 1781. *Decline and fall of the Roman empire.* Chapter 65. Christian Classics Ethereal Library. http://www.ccel.org/ccel/gibbon/decline/index.htm.

Gimbutas, Marija. 1974. *The goddesses and gods of Old Europe 6500 to 3500 B.C.: myths and cult images.* Berkeley & Los Angeles: University of California, 1982.

Gleick, James. 1987. *Chaos: making a new science.* Viking.

Goelet, O., Jr. 1994. Introduction and Commentary. *Egyptian book of the dead.* Trans. R. Faulkner. San Francisco: Chronicle Books.

Goldenfeld , Nigel and C. Woese. 2007. Biology's next revolution. *Nature* 445 (7126): 369.

Goldman, Nick, Paul Bertone, Siyuan Chen, Christophe Dessimoz, Emily M. LeProust, Botond Sipos, and Ewan Birney. 2013. Towards practical, high-capacity, low-maintenance information storage in synthesized DNA. *Nature* 494 (7435): 77-80.

Gopaul, D. N., F. Guo, and G. D. Van Duyne. 1988. Structure of the Holliday junction intermediate in Cre-*loxP* site-specific recombination. *Embo. Journal* 17 (14): 4175-4178.

Goswami, Amit. 1997. Consciousness and directed mutation. In *Science within Consciousness.* Ed. Henry Swift. http://www.swcp.com/swc/Essays/bio.html (accessed November 14, 2005).

———. 2011. *How quantum activism can save civilization.* Hampton Roads Publishing Company, Inc.

Goswami, A. and D. Todd. 1997. Is there conscious choice in directed mutation, phenocopies, and related phenomena? An answer based on quantum measurement theory. *Integr. Physiol. Behav. Sci.* 32 (2): 132-42.

Greenblatt, Stephen. 2011. *The swerve.* New York: W. W. Norton & Company.

Greene, Brian. 2004. *The fabric of the cosmos.* New York: Alfred A. Knopf.

Gribbin, John. 2001. *Hyperspace.* New York: DK Publishing, Inc.

Gsponer, Jörg and M. Madan Babu. 2009. Rules of disorder or why disorder rules. *Progress in Biophysics and Mol. Biology* xxx:1-10. Journal homepage: www.elsevier.com/locate/pbiomolbio.

Hall, Calvin S. and Gardner Linzey. 1957. Skinner's operant reinforcement theory. In *Theories of Personality*, 476-514. New York: John Wiley & Sons, Inc.

Hamilton, Nigel. 1985. Alchemical process of transformation. http://www.sufismus.ch/assets/files/omega_dream/alchemy_e.pdf.

Harmon, Catherine. 2013. Pope Francis' coat of arms and motto, explained. *Catholic World Report* (March 18, 2013). www.catholicworldreport.com.

Harpur, Tom. 2004. *Pagan Christ: rediscovering the lost light.* Toronto, ON: Thomas Allen Publishers.

Hatfield, Gary. 2014. René Descartes. *Stanford Encyclopedia of Philosophy* (Summer 2014 Edition). Ed. Edward N. Zalta. URL = <http://plato.stanford.edu/archives/sum2014/entries/descartes/>.

Hawking, Stephen. 1988. *A brief history of time.* Toronto, New York, London: Bantam Books.

———. 2001. *Universe in Nutshell.* New York, London: Bantam Books.

———. The computer. http://www.hawking.org.uk/the-computer.html.

Hendrix, R. W., J. G. Lawrence, G. F. Hatfull, and S. Casjens. 2000. Origins and ongoing evolution of viruses. *Trends in Microbiology* 8 (11): 504-508.

Herodotus. 1996. *The histories.* Trans. Aubrey De Selincourt. Revised John Marincola. London, UK: Penguin Books.

Hillyar, Christopher R. T. 2012. Genetic recombination in bacteriophage lambda. Review, *Bioscience Horizons* Vol. 5 (January).

Holliday, R. 1964. A mechanism for gene conversion fungi. *Genet. Res.* 5:282-304.

Hornung, Erik. 1994. Black holes viewed from within: hell in ancient Egyptian thought. *Diogenes* 165:133-156.

———. 1999. *Ancient Egyptian books of the afterlife.* Cornell University Press: Ithaca and London.

Huang Y. and Z. Liu. 2010. Nonnative interactions in coupled folding and binding processes of intrinsically disordered proteins. *PLoS ONE* 5 (11): e15375. doi:10.1371/journal.pone.0015375.

Hudson, R. E., J. E. Aukema, C. Rispe, D. Roze D. Altruism, cheating, and anticheater adaptations in cellular slime molds. *Am. Nat.* 160 (1): 31-43. doi: 10.1086/340613. http://www.ncbi.nlm.nih.gov/pubmed/18707497.

Hughes, John J. 2012. Interview by George Dvorsky. J. Hughes on democratic transhumanism, personhood, and AI. *Institute for Ethics and Emerging Technologies* (IEET). http://ieet.org/index.php/IEET/print/5551.

Hume, Brad D. 2008. Chronology of history of science. University of Dayton. http://campus.udayton.edu/~hume/Microscope/microscope.htm (accessed March 17, 2014).

Huxley, A. 1986. *Devils of Loudon.* New York: Carroll & Graf.

Immerwahr, John. 1996. Hume's aesthetic theism. *Hume Studies* Vol. XXII, No. 2 (November): 325-337.

Instituto Gulbenkian de Ciência (IGC). 2013. How protein suicide assure healthy cell structures. *ScienceDaily* (October 31). http://www.sciencedaily.com/releases/2013/10/131031124815.htm (accessed November 11, 2013).

Irvine, Andrew David. 2014. Alfred North Whitehead. *Stanford Encyclopedia of Philosophy* (Fall 2014 Edition). Ed. Edward N. Zalta. http://plato.stanford.edu/archives/fall2014/entries/whitehead.

Jacobson, Ted. 1996. *Introductory lectures on black hole thermodynamics.* Institute for Theoretical Physics, University of Utrecht.

Jacobson, Theodore and Renaud Parentani. 2005. An echo of black holes. *Scientific American* (December): 3-9.

Jacobson, Theodore A. and Aron C. Wall. 2008. Black hole thermodynamics and Lorentz symmetry. arXiv:0804.2720v2 (hep-th). http://arxiv.org/abs/0804.2720.

Jaskelioff, Mariela., Florian L. Muller, Ji-Hye Paik, Emily Thomas, Shan Jiang, Andrew C. Adams, Ergün Sahin, et al. 2011. Telomerase reactivation reverses tissue degeneration in aged telomerase-deficient mice. *Nature* 469 (7328): 102-106.

Jia, Haifeng, W. John Satumba, Gene I. Bidwell III, and Michael C. Mossing. 2005. Slow assembly and disassembly of Lambda cro repressor dimers. *J. Mol. Biol.* 350:919-929.

Johnson, Aaron and Mike O'Donnell. 2005. Cellular DNA replicases: components and dynamics at the replication fork. *Annual Rev. Biochem.* 74 (July): 283-315.

Johnston, John. 2008. *Allure of machinic life: cybernetics, artificial life, and the new AI.* Cambridge, MA; London, England: MIT Press.

Jotzu, Gregor, Michael Messer, Rémi Desbuquois, Martin Lebrat, Thomas Uehlinger, Daniel Greif, and Tilman Esslinger. 2014. Experimental realization of the topological Haldane model with ultracold fermions. *Nature* 515 (7526): 237-240.

Juergens, Sylvester P. 1953. *New Marian Missal for Daily Mass.* New York: Regina Press.

Jung, C. G. 1953. *Pyschology and Alchemy.* Trans. R. F. C. Hull. Bollingen Series XX, Pantheon Books. Originally published in German as *Psychologie und Alchemie*, by Rascher Verlaag. Zurich, 1944; second edition, revised, 1952.

———. 1958. *Psyche & Symbol: a selection from the writings of C. G. Jung.* Ed. Violet S. de Laszlo. New York: Doubleday.

———. 1964. *Man and his symbols.* London: Aldus Books.

Karplus, Martin and David I. Weaver. 1994. Protein folding dynamics: the diffusion-collision model and experimental data. *Protein Science* 3:650-688. Cambridge Univ. Press.

Kasser, Rodolphe, Marvin Meyer, and Gregor Wurst, trans. 2006. Gospel of Judas. In collaboration with François Gaudard. The National Geographic Society copyright.

Katsnelson, Alla. 2010. Epigenome effort makes its mark. *Nature* 467 (7316): 646.

Kauffman, Stuart. 1995. *At home in the universe: the search for laws of self-organization and complexity.* New York, Oxford: Oxford Univ. Press.

Kemp, Martin. 2011. Sight and salvation. *Nature* 479 (7372): 174-175.

Kern, Martin. 2000. *Stele inscriptions of Ch'in Shih-Huang: text and ritual in early Chinese Imperial representation.* New Haven, Conn.: American Oriental Society.

Kineman, John J. 2010. R-theory: a further commentary on the synthesis of relational science. *International Society for the Systems Sciences* (ISSS). Presented at Proceedings of the 55th Annual Meeting of the ISSS, Hull, UK, July 17-22, 2011. http://journals.isss.org/index.php/proceedings55th/article/viewFile/1713/585.

Kitzinger, Ernst. 1954. Cult of images in the age before iconoclasm. *Dumbarton Oaks Papers* 8:83-150.

Konstan, David. 2014. *Epicurus. Stanford Encyclopedia of Philosophy* (Summer 2014 Edition). Ed. Edward N. Zalta. http://plato.stanford.edu/archives/sum2014/entries/epicurus.

Kuchiev. M. Y. 2004. Reflection, radiation and interference for black holes. *Phys. Rev. D.* 69, 124031.

Kudla G, L. Lipinski, F. Caffin, A. Helwak, M. Zylicz. 2006. High guanine and cytosine content increases mRNA levels in mammalian cells. *PLoS Biol* 4 (6): e180. doi:10.1371/journal.pbio.0040180.

Kumar, David R., Florence Aslinia, Steven H. Yale, and Joseph J. Mazza. 2011. Jean-Martin Charcot: the father of neurology. *Clinical Medicine and Research* 9 (1): 46-49. PMCID: PMC3064755.

Kurzweil, Ray. 2005. *Singularity is near.* New York: Viking.

Lake, James A, 2011. Lynn Margulis. *Nature* 480 (7378): 458.

Lakatos, Imre. 1970. Falsification and the methodology of research program. In *Criticism and the Growth of Knowledge*, eds. Imre Lakatos and Alan Musgrave, 91-197. Cambridge: Cambridge University Press.

Lam, Stephanie. 2011. Ancient organisms found alive in salt crystals. *Epoch Times,* Feb. 3-16, 2011.

Lasenby, A. N., C.J.L. Doran, Y. Dabrowski, and A.D. Challinor. 1997. Rotating astrophysical systems and a gauge theory approach to gravity. Cornell University Library. arXiv:astro-ph/9797165v1.

Laughlin, Robert B. 2005. *A different universe: reinventing physics from the bottom down.* New York: Basic Books.

Lehn, Jean-Marie, and Philip Ball. 2000. *New Chemistry.* Ed. Nina Hall. Cambridge University Press.

Levy, Jonathan. 2013. Vatican Bank Conspiracy with Dr Jonathan Levy. YouTube Video. https://www.youtube.com/watch?v=e50GuZuI7Ww&noredirect=1.

Levy, Yaakov, Samuel S. Cho, Jose´ N. Onuchic, and Peter G. Wolynes. 2005. Survey of flexible protein binding mechanisms and their transition states using native topology based energy landscapes. *J. Mol. Biol.* 346:1121-1145.

Levy, Yaakov. and Jose Onuchic. 2004. Water and proteins: a love-hate relationship. *PNAS* 101 (10): 3325-3326. http://www.pnas.org/content/101/10/3325.full.

Levy, Yaakov, Peter G. Wolynes, and José N. Onuchic. 2004. Protein topology determines binding mechanism. *PNAS* 101 (2): 511-516.

Lindberg, David C. 1997. Roger Bacon on light, vision, and the universal emanation of force. In *Roger Bacon and the Sciences Commemorative Essays.* Ed. Jeremiah Hackett (Leiden: Brill), 243-275.

Livio, Mario. 2003. *The golden ratio: the story of PHI, the world's most astonishing number.* New York: Broadway Books and Crown Publishing Group; Reprint edition (September 23, 2003).

Lockman, Felix J., Nicole L. Free, and Joseph C. Shields. 2012. Neutral hydrogen bridge between M31 and M33. arXiv:1205.5235 [astro-ph.GA].

Löeckenhoff, H. K. 2006. Grounding transdisciplinarity for an innovative science. Ed. Robert Trappl. *Cybernetic Systems* 2:410-415. Austrian Society for Cybernetic Studies, Vienna.

Lorenz, B. and C. W. Chu. 2004. High pressure effects on superconductivity. Cornell University Library. http://arxiv.org/ftp/cond-mat/papers/0410/0410367.pdf.

Louie, A. H. 2010. Robert Rosen's anticipatory systems. *Foresight* 12 (3): 18-29.

Luminet, Jean-Pierre. 2008. *The wraparound universe.* Trans. Eric Novak. Wellesley, MA: A K Peters, Ltd. Originally published in the French language as *L'Univers chiffonné*, by Jean-Pierre Luminet. copyright *Librairie Arthème Fayard*, 2001.

———. 2011. Science, Art and Geometrical Imagination. *Proceedings of the International Astronomical Union.* Vol. 5 Symposium S260 (January): 248-273.

Ma, Wenjian, En-Hua Cao, Jian Zhang, Jing-Fen Qin. 1998. Phenanthroline-Cu complex-mediated chemiluminescence of DNA and its potential use in antioxidation evaluation. *Journal of Photochemistry and Photobiology B: Biology.* 44 (1): 63–68. (Institute of Biophysics, Academia Sinica, Beijing 100101, China).

Maier, Michael. 1618. *Atalanta Fugiens.* Printed by Hieronymous Gallerus. Published by Johann Theodor de Bry (1617 in Latin). http://www.scribd.com/doc/5986531/Maier-Atalanta-Fugiens.

Mainstone, Rowland J. 1988. *Hagia Sophia architecture, structure and liturgy of Justinian's great church.* New York: Thames and Hudson.

Major, John S. 1993. *Heaven and earth in early Han thought: chapters three, four, and five of the Huainanzi.* Albany: State Univ. of New York Press.

Maldacena, Juan. 2005. The illusion of gravity. *Scientific American* (November): 56-63.

Mandelbrote, Scott. 2001. *Footprints of the lion: Isaac Newton at work.* Cambridge Univ. Library.

Mantegna R. N., S. V. Buldyrev, A. L. Goldberger, S. Havlin, C.-K. Peng, M. Simons, and H. E. Stanley. 1994. Linguistic features of non-coding DNA sequences. *Phys. Rev. Lett.* 73 (23): 3169-3172.

Margulis, Lynn and Dorion Sagan. 1986. *Microcosmos.* New York: Summit Books.

Marlowe, Christopher. 1976. Tamburlaine the Great. In *Complete Plays and Poems.* Ed. E. D. Pendry. London, Melbourne and Toronto: Everyman's Library.

Marx, Karl and Friedrich Engles. 1848. *Communist manifesto.* Project Gutenberg EBook www.gutenberg.net.

Matthews, Washington. 1887. *Mountain chant.* Foreward by Paul Zolbrod. Salt Lake City: University of Utah Press, 1997. Originally published in Washington D.C., 1887 in *U.S. Bureau of American Ethnology; Fifth Annual Report.* 1883-84, with two sections suppressed from the original edition.

May, Herbert G. and Bruce M. Metzger, eds. 1973. *New Oxford Annotated Bible with the Apocrypha.* New York: Oxford University Press.

McFadden, Johnjoe, and Jim Al-khalili. 1999. A quantum mechanical model of adaptive mutation. *Biosystems* 50:203-211.

McGee, Harold. 1984. *On cooking and food: the science and lore of the kitchen.* New York: Scribner.

Melville, Herman. 1971. *Confidence-Man: his masquerade.* Ed. Hershel Parker. New York: W. W. Norton and Company.

Merali, Zeeya. 2013. Origins of space and time. *Nature* 500 (7464): 516-519.

Merino, G. and H. A. Shuman. 1997. Unliganded maltose-binding protein triggers lactose transport in an *Escherichia coli* mutant with an alternation in the maltose transport system. *J. Bacteriol.* 179 (24): 7687-7694, http://jb.asm.org/cgi/reprint/179/24/7687.

Mesbah, Mostafa, Usha Premachandran, and William B. Whitman. 1989. Precise measurement of the G+C content of DNA by high-performance liquid chromatography. *International Journal of Systematic Bacteriology* 39 (2): 159-167.

Meunier, Julien and Laurent Duret. 2003. Recombination drives the evolution of GC-content in the human genome. *Molecular Bio. and Evolution* 21 (6): 984-990.

Meyendorff, John. 1987. Wisdom-Sophia: contrasting approaches to a complex theme. *Dumbarton Oaks Papers* 41:391-401.

Micale, Mark S. 1990. Charcot and the idea of hysteria in the male: gender, mental science, and medical diagnosis in late nineteenth-century France. *Medical History* 34:363-411.

Modi, Sheetal R., Henry H. Lee, Catherine S. Spina, and James J. Collins. 2013. Antibiotic treatment expands the resistance reservoir and ecological network of the phage metagenome. *Nature* 499 (7457): 219-222.

Montagnier, L., J. Aissa, E. Del Giudice, C. Lavallee, A. Tedeschi, and G. Vitiello. 2010. DNA Waves in Water. arXiv:1012.5166v1 [q-bio.OT] 23 December 2010.

Montagnier, Luc., Jamal Aäissa, Stéphane Ferris, Jean-Luc Montagnier, and Claude Lavallee. 2009. Electromagnetic signals are produced by aqueous nanostructures derived from bacterial DNA sequences. *Interdisciplinary Sci. Comput. Life Sci.* 1 (2): 81–90. doi:10.1007/s12539-009-0036-7 (accessed January 27, 2013).

Morris, Mark, Keven Uchida, and Tuan Do. 2006. A magnetic torsional wave near the Galactic Centre traced by a 'double helix' nebula. *Nature* 440 (7082): 308-310.

MSU. 2012. Gold-loving bacteria show Superman strength. (Oct. 1) Contact Kristen Parker and Adam Brown. http://msutoday.msu.edu/news/2012/gold-loving-bacteria-show-superman-strength.

Murphy, Kenneth P., ed. 2001. Protein structure, stability, and folding. *Methods in Mol. Bio.* Vol. 168. Totowa, New Jersey: Humana Press.

Musser, George. 2004. Was Einstein right? *Scientific American* 291 (3): 88-91.

Needham, Joseph. 1975. Cosmology of Early China. In *Ancient Cosmologies.* Eds. Carmen Blacker and Michael Loewe. London: George Allen and Unwin Ltd.

Neph, Shane, Jeff Vierstra, Andrew B. Stergachis, Alex P. Reynolds, Eric Haugen, Benjamin Vernot, Robert E. Thurman, Sam John, Richard Sandstrom, Audra K. Johnsonn, Matthew T. Maurano, et al. 2012. An expansive human regulatory lexicon encoded in transcription factor footprints. *Nature* 489 (7414): 83-90.

Nicholl, Desmond S. T. 2002. *An introduction to genetic engineering second edition.* Cambridge Univ. Press.

Nietzsche, Friedrich. 1961. *Thus spoke Zarathustra.* London, New York: Penguin Books.

———. 1967. *On the genealogy of morals.* Trans. Walter Kaufmann. New York: Random House.

———. 1968. *Will to power.* Trans. Walter Kaufmann and R. J. Hollingdale. New York: Vintage Books.

Noest, André J. 2011. No refraction in Leonardo's orb. *Nature* 480 (7378): 457.

Onuchic, José N., Zaida Luthey-Schulten, and Peter G. Wolynes. 1997. Theory of protein folding: the energy landscape perspective. *Annual Rev. of Phys. Chem.* 48:545-600.

Paquette, V., J. Le´vesque, B. Mensour, J.-M. Leroux, G. Beaudoin, P. Bourgouin, and M. Beauregard. 2003. Change the mind and you change the brain: effects of cognitive-behavioral therapy on the neural correlates of spider phobia. *NeuroImage* 18:401-409.

Parker, R. A., J. LeClant, and J-C. Goyon. 1979. *Edifice of Taharqa by the Sacred Lake of Karnak.* Trans. Claude Crozier-Brelot. Providence, Lund Humphries, London: Brown University Press.

Peirce, Charles S. 1931-1935. *The collected papers of Charles Sanders Peirce, Vols. I-VI.* Eds. Charles Hartshorne and Paul Weiss. Cambridge, MA: Harvard University Press.

———. 1985. *Historical perspectives on Peirce's logic of science: A history of science.* Ed. Carolyn Eisele. Berlin: Mouton Publishers, 1985 (two volumes).

———. 1992. *The essential Peirce: selected philosophical writings V1.* Eds. Nathan Houser and Christian Kloesel. Bloomington and Indianapolis: Indiana University Press.

Penrose, Roger. 1998. Quantum Computation, entanglement, and state reduction. *Phil. Trans. R. Soc. Lond.* 356:1927-1939.

———. 2010. *Cycles of time.* Vintage; Reprint edition (May 1, 2012).

Penrose, Roger, and Stuart Hameroff. 1998. Quantum theory and human consciousness. Third RAND Study Group Meeting (October 22). http://www.rand.org/pubs/monograph_reports/MR1139/MR1139.appb.pdf.

Persinger, M. A. and S. A. Koren. 2007. A theory of neurophysics and quantum neuroscience: implications for brain function and the limits of consciousness. *International Journal of Neuroscience* 117 (2): 157-175.

Piankoff, Alexandre. 1954. *Tomb of Ramesses VI texts.* Ed. N. Rambova. Bollingen Series XL 1. New York: Pantheon Books.

———. 1974. *Wanderings of the soul.* Bollingen Series XL 6. Princeton University Press.

Piobb, P. V. 1950. *Clef universelle des sciences secretes.* In *Dictionary of symbols second edition* by J. E. Cirlot. New York: Philosophical Library (1971).

Pitkänen, M. 2011. DNA & water memory: comments on Montagnier group's recent findings. *DNA Decipher Journal* 1 (1): 181-190.

Poe, Edgar A. Society. www.eapoe.org/geninfo/poedeath.htm (updated Jan. 19, 2014; accessed December 12, 2010).

Popham, David L. and Ann M. Stevens 2006. Bacterial quorum sensing and bioluminescence. In *Tested Studies for Laboratory Teaching, Vol. 27.* Ed. M. A. O'Donnell, ed. Proceedings of the 27th Workshop/Conference of the Association for Biology Laboratory Education, pp. 201-215.

Povolotskaya and Kondrashov. 2010. Sequence space and the ongoing expansion of the protein universe. *Nature* 465 (7300): 922-926.

Pribram, Karl H. 1991. *Brain and perception, holonomy and structure in figural processing.* Hillsdale, NJ: Lawrence Erlbaum Associates, Publishers.

Ptashne, Mark. 1966. Isolation of the Lambda phage repressor. Communicated by J. D. Watson. In *Genetics* (1967) Vol. 57:306-13. http://www.ncbi.nlm.nih.gov/pmc/articles/PMC335506/pdf/pnas00675-0122.pdf.

———. 2004. *A genetic switch: phage Lambda revisited.* New York: Cold Springs Harbor Lab Press.

Ptashne, Mark and Alexander Gann. 2002. *Genes and Signals.* Cold Springs Harbor, New York: Cold Springs Harbor Laboratory Press.

Punsly, Brian. 1998. High-energy gamma-ray emission from galactic Kerr-Newman black holes. *Astrophysical Journal* 498 (May 10): 640-659.

Raju, C. K. 2003. *The eleven pictures of time: the physics, philosophy, and politics of time beliefs.* New Delhi, Thousand Oaks, London: Sage Publications.

Rambova, N. ed., 1954. *Tomb of Ramesses VI texts.* Trans. with Introductions by Alexandre Piankoff. Bollingen Series XL 1. New York: Pantheon Books.

Rapoport, Diego L. 2010. Surmounting the Cartesian Cut further: torsion fields, the extended photon, quantum jumps, the Klein-bottle, multivalued logic, the time operator, chronomes, perception, semiosis, neurology and cognition. In *Quantum Mechanics, an Overture to Natural Philosophy.* Ed. Jonathan Groffe. Nova Science Publishers, Inc.

Ravindran, Sandeep. 2012. Barbara McClintock and the discovery of jumping genes. *PNAS* 109 (50): 20198–20199.

Reardon, Sara. 2014. Phage therapy gets revitalized: the rise of antibiotic resistance rekindles interest in a century-old virus treatment. *Nature* 510 (7503): 15-16.

Rees, Martin. 2000. *Just six numbers.* New York: Basic Books.

Regis, Ed. 2013. A bold and foolish effort to predict the future of computing. *Scientific American* (January), 36-37.

Relman, David A. 2012. Learning about who we are. *Nature* 486 (7402): 194-195.

Ridley, Matt. 1999. *Genome.* New York: Harper Perennial.

Robinson, J. B. and O. H. Tuovinen. 1984. Mechanisms of microbial resistance and detoxification of mercury and organomercury compounds: physiological, biochemical and genetic analyses. *Microbiol. Rev.* 48 (2): 95. http://mmbr.asm.org/content/48/2/95.full.pdf.

Roger, A. J. 1999. Reconstructing early events in eukaryotic evolution. *Am. Nat.* 154 (S4): S146-S163.

Romiguier, Jonathan, Vincent Ranwez, Emmanuel J. P. Douzery, and Nicolas Galltier. 2010. Contrasting GC-content dynamics across 33 mammalian genomes: Relationship with life-history traits and chromosome sizes. *Genome Research* 20 (8): 1001-1009. http://www.ncbi.nlm.nih.gov/pmc/articles/PMC2909565.

Rosen, Robert. 1985. *Anticipatory systems.* Oxford, UK: Pergamon Press.

Rosenberg, S. M. 2001. Evolving responsively: adaptive mutation. *Nature Reviews/Gen.* 2 (July): 504-515.

Russell, Paul. 2013. Hume on religion. *Stanford Encyclopedia of Philosophy* (Fall 2013 Edition). Ed. Edward N. Zalta. http://plato.stanford.edu/archives/fall2013/entries/hume-religion.

Ryan, Cassie. 2012. Bacterial alchemy generates 24-karat gold. *Epoch Times.* http://www.theepochtimes.com/n2/science/bacterial-alchemy-generates-24-karat-gold-300672.html.

Sahin, Ergün, Simona Colla, Marc Liesa, Javid Moslehi, Florian L. Müller, Mira Guo, Marcus Cooper, et al. 2011. Telomere dysfunction induces metabolic and mitochondrial compromise. *Nature* 470, no. 7334 (February 17): 359-365.

Samson, Roger. 2011. *Ingenious Genes: how gene regulation networks evolve to control ontogeny.* A Bradford Book.

Santillán, Moisés and Michael C. Mackey. 2004. Why the lysogenic state of phage λ is so stable: a mathematical modeling approach. *Biophys J.* 86 (1): 75–84.

Sanvito, Stefano. 2010. Seeing the spin through. *Nature* 467 (716): 664-665.

Schlosshauer, Maximilian. 2008. Lifting the fog from the north. *Nature* 453 (7191): 39.

Schnetter, Erik. 2002. A fast apparent horizon algorithm. arXiv:gr-qc/0206003 v1 on June 2, 2002.

Schwartz, Jeffrey and Sharon Begley. 2002. *Mind and the brain.* New York: HarperCollins.

Schwartz, Jeffrey M., Henry P. Stapp, and Mario Beauregard. 2005. Quantum physics in neuroscience and psychology: a neurophysical model of mind–brain interaction. *Phil. Trans. Royal Soc./Bio. Sci.* 360 (1458): 1309-1327.

Scully, Marlon O. and Robert J. Scully. 2007. *Demon and the quantum.* Weinheim: Wiley-VCH Verlag Gmbh & Co. KgaA.

Sekowska, Agnieszka, Hsiang-Fu Kung, and Antoine Danchin. 2000. Sulfur metabolism in *Escherichia coli* and related bacteria: facts and fiction. *J. Mol. Microbiol. Biotechnol.* 2 (2): 145-177.

Selvam, A. M. 2009. Universal spectrum for DNA base C+G frequency distribution in human chromosomes 1-24. *World Journal of Modelling and Simulation* 5 (2): 151-160. http://www.worldacademicunion.com/journal/1746-7233WJMS/wjmsvol05no02paper09.pdf.
Shen, Helen. 2013. Brain Storm. *Nature* 503 (7474): 26-28.
Shimizu, Kazuyuki. 2013. Metabolic regulation of a bacterial cell system with emphasis on *Escherichia coli* metabolism. ISRN Biochemistry Volume 2013 (2013), Article ID 645983. http://www.hindawi.com/journals/isrn/2013/645983/
Shoemaker, Benjamin A., John J. Portman, and Peter G. Wolynes. 2000. Speeding molecular recognition by using the folding funnel: the fly-casting mechanism et al. *PNAS* 97 (16): 8868-8873. http://www.pnas.org/content/97/16/8868.full
Shorvon, Simon. 2007. Fashion and cult in neuroscience—the case of hysteria. *Oxford Journals Medicine Brain* 130 (12): 3342-3348.
Skinner B.F. 1953. *Science and human behavior.* New York: Macmillan.
Sluhovsky, Moshe. 2002. Devil in the convent. *Amer. Historical Review* 107 (5).
Smolin, Lee. 2010. Space-time turn around. *Nature* 467 (7319): 1034-1035.
Socci, N. D., José N. Onuchic, and Peter G. Wolynes. 1996. Diffusive dynamics of the reaction coordinate for protein folding funnels. *J. Chem. Phys.* April 15, arXiv:condi-mat/9601091 v1. Jan. 21.
Sommer, Ariel, Mark K. U. Giacomo Roati, and Martin W. Zwierlein. 2011. Universal spin transport in a strongly interacting Fermi gas. *Nature* 472 (7342): 201-204.
Somo, Ferg. 2008. Nefertiti, Her Bantu Name. Kiswahili-Bantu Research Unit for Advancement of the Ancient Egyptian Language. http://www.kaa-umati.co.uk/nefertiti.html
Sontag, Susan, ed. 1973. *Antonin Artaud selected writings.* Berkeley, LA: Univ. of Calif. Press, 1976.
Sorensen, S. J., M. Bailey, L. H. Hansen, N. Kroer, and S. Wuertz. 2005. Studying plasmid horizontal gene transfer in situ: a critical review. *Nature* 3 (September): 700-710. http://www.nature.com/reviews/micro.
Stambler, Ilia. 2010. Life extension – a conservative enterprise? *Journal of Evolution and Technology* 21, no. 1 (March): 13-26. http://jetpress.org/v21/stambler.htm.
Stanley, Wendell M. 1946. The isolation and properties of crystalline tobacco mosaic virus. *Nobel Lecture, December 12, 1946.* http://www.nobelprize.org/nobel_prizes/chemistry/laureates/1946/stanley-lecture.pdf.
Stapp, Henry P. 1995. Why classical mechanics cannot naturally accommodate consciousness but quantum mechanics can. Theoretical Physics. Univ. of California. http://psyche.cs.monash.edu.au/v2/psyche-2-05-stapp.html.
Stoltzfus, Arlin. 2012. Constructive neutral evolution: exploring evolutionary theory's curious disconnect. *Biology Direct* 2012, 7:35. http://www.biology-direct.com/content/7/1/35.
Strominger, Andrew. 1993. White holes, black holes and CPT in two dimensions. *Phys. Rev. D.* 48:5769-5777.
Sullivan, Karen. 2001/02. Orthodox origin of heterodoxy or how what is good becomes evil. *Cabinet Magazine.* Issue 5 Winter. http://www.cabinetmagazine.org.
Susskind, Leonard. 1995. World as a hologram. *J. Math. Phys.* 36, 6377. arXiv:hep th/9409089.
Suzuki, Miwa. 2011. Red wine offers clue to superconductive future. (May 13) http://phys.org/news/2011-05-red-wine-clue-superconductive-future.html.
Taft, Robert F. 1975. *Great entrance.* Roma: Pont. Pontificium Institutum Studiorum Orientalium.
Taft, Ryan J. and John S. Mattick 2004. Increasing biological complexity is positively correlated with the relative genome-wide expansion of non-protein-coding DNA sequences. *Genome Biology* 5 (1): 1-24. doi:10.1186/gb-2003-5-1-p1.
Tam, W., L. G. Pell, D. Bona, A. Tsai, X. X. Dai, A. M. Edwards, R. W. Hendrix, K. L. Maxwell, A. R. Davidson. 2013. Tail tip proteins related to bacteriophage Lambda gpL coordinate an iron-sulfur cluster. *J. Mol. Biol.* 425 (14): 2450-62. http://www.ncbi.nlm.nih.gov/pubmed/23542343
Tan, Si-Hui, Baris I. Erkmen, Vittorio Giovannetti, Saikat Guha, Seth Lloyd, Lorenzo Maccone, Stefano Pirandola, and Jeffrey H. Shapiro. 2008. Quantum illumination with Gaussian states. *Physical Review Letters* 101 (25): id 253601.

Tanner, Nathan A., Joseph J. Loparo, Samir M. Hamdan, Slobodan Jergic, Nicholas E. Dixon, and Antoine M. van Oijen. 2009. Real-time single-molecule observation of rolling-circle DNA replication. *Nuclei Acids Res.* 37 (4): e27. Published online 2009 March. Doi:10.1093/nar/gkp006.

Taylor, John H., ed. 2013. *Journey through the afterlife: ancient Egyptian Book of the Dead.* Harvard Univ.

Teresa of Avila. 1987. *Autobiography of St. Teresa of Avila.* Trans. Kieran Kavanaugh and Otilio Rodriguez. New York: Book of the Month Club.

Thom, René. 1975. *Structural stability and morphogenesis.* Trans. D. H. Fowler. London, Amsterdam, Sydney, Tokyo: W. A. Benjamin, Inc.

———. 1990. *Semiophysics: a sketch.* Redwood, CA, New York: Addison-Wesley Publishing Co.

Thomas, John E. 2011. Spin drag in a perfect fluid. *Nature* 472 (7342): 172-173.

Thompsom, Damian. 2013. Pope Francis reveals his radical message – and it will startle conservatives. *Telegraph*, November 17, 2013. http://blogs.telegraph.co.uk/news/damianthompson/100247936/pope-francis-reveals-his-radical-message-and-it-will-startle-conservatives/ (accessed March 13, 2014).

Thorne, Kip. 1994. *Black holes and time warps.* New York: W. W. Norton & Company.

Tolstoy, Lev. 1963. *War and peace.* Trans. Louise and Aylmer Maude. New York: Washington Sq. Press.

Traupman, John C. 1966. *New college Latin and English dictionary.* New York: Bantam Books.

University of Maryland Medical Center. 1996. Edgar Allan Poe mystery. September 24. http://umm.edu/news-and-events/news-releases/1996/edgar-allan-poe-mystery. (updated June 13, 2013 and accessed February 2, 2014).

Van Lommel, Pim. 2010. *Consciousness beyond life: The science of the NDE.* New York: Harper-Collins.

Van Wezel, Jasper. 2010. Broken time translation as a model for quantum state reduction. *Symmetry* 2:582-608.

Villarreal, Luis P. 2004. Can viruses make us human? *Proc. of American Philosophical Society* 148 (3).

Villarreal, Luis P. and Guenther Witzany. 2010. Viruses are essential agents within the roots and stem of the tree of life. *Journal of Theoretical Biology* 262 (4): 698-710.

Vinge, Vernor. 1993. The coming technological singularity. VISION-21 Symposium sponsored by NASA Lewis Research Center and the Ohio Aerospace Institute, March 30-31, 1993. NASA Conf. Publication CP-10129. http://www-rohan.sdsu.edu/faculty/vinge/misc/singularity.html.

Visser, Matt. 1996. *Lorentzian wormholes from Einstein to Hawking.* Woodbury, New York: American Institute of Physics.

Visser, Matt, Sayan Kar, and Naresh Dadhich. 2003. Traversable wormholes with arbitrarily small energy condition violations. *Phys. Rev. Lett.* 90, 201102. http://arxiv.org/abs/gr-qc/0301003.

Waldron, K. J., J. C. Rutherford, D. Ford and N. J. Robinson. 2009. Metalloproteins and metal sensing. *Nature* 460 (7257): 823-830.

Wales, David J. 2005. The energy landscape as a unifying theme in molecular science. *Phil. Trans. Royal Society A.* 363:357-377. http://www-wales.ch.cam.ac.uk/pdf/PhilTransRoySoc.363.357.2005.pdf.

Watson, James D. 2003. *DNA the secret of life.* New York: Alfred A. Knopf.

Wells, H. G. 1898. *War of the worlds.* Ed. Patrick Parrinder. London: Penguin Books.

Wheeler, John A. 1988. World as system self-synthesized by quantum networking. *IBM J. of Res. and Dev.* 32 (1): 4-15.

———. 1994. *At Home in the Universe.* New York: The American Institute of Physics.

White, Michael. 1998. *Isaac Newton the last sorcerer.* Fourth Estate Limited.

Whitehead, Alfred N. 1929. *Process and reality: an essay in cosmology.* Eds. David Ray Griffin and Donald W. Sherburne. New York: Free Press Div. of Macmillan Pub. Co., Inc. (1978); London: Collier Macmillan Publishers (1929).

Wilbur, Ken, ed. 1985. *Quantum questions: mystical writings of world's greatest physicists.* Boston: Shambhala.

Wilczek, Frank. 2012. Quantum time crystals. Cornell University Library (July 11) arxiv:1202.2539V2.

———. 2013. Superfluidity and space-time translation symmetry breaking. Frank Wilczek Center for Theoretical Physics, MIT, Cambridge MA. (August 28, 2013) http://arxiv.org/pdf/1308.5949v1.pdf.

Wilson, Edward. O. 1978. *On human nature.* London, UK and Cambridge, MA: Harvard University.

Witzany, Güenther. 2007. Telomeres in evolution and development from biosemiotic perspective. Cold Spring Harbor Laboratory meeting on "Telomeres and Telomerases" (May 2-6, 2007). http://www.somosbacteriasyvirus.com/biosemiotic.pdf.

———. 2008. Viral origins of telomeres and telomerases and their important role in eukaryogenesis and genome maintenance. *Biosemiotics* 1 (2): 191-206.

Wolf, Fred A. 1988. *Parallel universes: the search for other worlds.* New York: Simon and Schuster.

Wolff, Jonathan. 2011. Karl Marx. *Stanford Encyclopedia of Philosophy* (Summer 2011 Edition). Ed. Edward N. Zalta. http://plato.stanford.edu/archives/sum2011/entries/marx.

Wolfram, Stephen. 2003. Past and Future of Scientific Computing. Official website of Stephen Wolfram. http://www.stephenwolfram.com/publications/recent/atanasoff.

———. 2006. Stephen Wolfram forecasts the future. *New Scientist* 2578, (November 18).

Wolynes, Peter G. 1996. Symmetry and the energy landscapes of biomolecules. *PNAS* 93 (25): 14249-14255.

Woolston, Chris. 2014. Furore over genome function. *Nature* 512 (7512). http://www.nature.com/nature/journal/v512/n7512/full/512009e.html.

Zakrzewski, Jakub and Marian Smoluchowski. 2012. Viewpoint: Crystals of Time. *Physics* 5, 116 (2012). American Physical Society website: https://physics.aps.org/articles/print/v5/116.

Zasowski, G., B. Menard , D. Bizyaev, D. A. Garcıa Hernandez, A. E. Garcıa Perez, M. R. Hayden, J. Holtzman, J. A. Johnson, K. Kinemuchi, S. R. Majewski, D. L. Nidever, M. Shetrone, J. C. Wilson. 2014. Mapping the interstellar medium with near-infrared diffuse interstellar bands. http://arxiv.org/pdf/1406.1195.pdf arXiv.1406.1195v2 [astro-ph.GA] Oct. 29, 2014.

Zimmer, Carl. 2011. *Planet of viruses.* Chicago: University of Chicago Press.

Zhang, Y., Thompson, R., Caruso J. 2011. Probing the viral metallome: searching for metalloproteins in bacteriophage λ -- the hunt begins. *Metallomics* 3 (5): 472-81. http://www.ncbi.nlm.nih.gov/pubmed/21423961.

CPSIA information can be obtained
at www.ICGtesting.com
Printed in the USA
LVHW071102111020
668499LV00019B/960

9 780976 281429